Communication Networks

Communication Networks blends control, optimization, and stochastic network theories with features that support student learning to provide graduate students with an accessible, modern approach to the design of communication networks.

- Covers a broad range of performance analysis tools, including important advanced topics that are made accessible to graduate students for the first time.
- Taking a top-down approach to network protocol design, the authors begin with the deterministic model and progress to more sophisticated models.
- Network algorithms and protocols are tied closely to the theory, engaging students and helping them understand the practical engineering implications of what they have learnt.
- The background behind the mathematical analyses is given before the formal proofs and is supported by worked examples, enabling students to understand the big picture before going into the detailed theory.
- End-of-chapter exercises cover a range of difficulties; complex problems are broken down into several parts, many with hints to guide students. Full solutions are available to instructors.

R. Srikant is the Fredric G. and Elizabeth H. Nearing Endowed Professor of Electrical and Computer Engineering, and a Professor in the Coordinated Science Laboratory, at the University of Illinois at Urbana-Champaign, and is frequently named in the university's list of teachers ranked as excellent. His research interests include communication networks, stochastic processes, queueing theory, control theory, and game theory. He has been a Distinguished Lecturer of the IEEE Communications Society, is a Fellow of the IEEE, and is currently the Editor-in-Chief of the *IEEE/ACM Transactions on Networking*.

Lei Ying is an Associate Professor in the School of Electrical, Computer and Energy Engineering at Arizona State University, and former Northrop Grumman Assistant Professor at Iowa State University. He is a winner of the NSF CAREER Award and the DTRA Young Investigator Award. His research interests are broadly in the area of stochastic networks, including wireless networks, P2P networks, cloud computing, and social networks.

Communication Networks

AN OPTIMIZATION, CONTROL, AND STOCHASTIC NETWORKS PERSPECTIVE

R. SRIKANT
University of Illinois at Urbana-Champaign

LEI YING

CAMBRIDGE UNIVERSITY PRESS

Shaftesbury Road, Cambridge CB2 8EA, United Kingdom

One Liberty Plaza, 20th Floor, New York, NY 10006, USA

477 Williamstown Road, Port Melbourne, VIC 3207, Australia

314–321, 3rd Floor, Plot 3, Splendor Forum, Jasola District Centre, New Delhi – 110025, India

103 Penang Road, #05-06/07, Visioncrest Commercial, Singapore 238467

Cambridge University Press is part of Cambridge University Press & Assessment, a department of the University of Cambridge.

We share the University's mission to contribute to society through the pursuit of education, learning and research at the highest international levels of excellence.

www.cambridge.org
Information on this title: www.cambridge.org/9781107036055

© Cambridge University Press & Assessment 2014

This publication is in copyright. Subject to statutory exception and to the provisions of relevant collective licensing agreements, no reproduction of any part may take place without the written permission of Cambridge University Press & Assessment.

First published 2014

A catalogue record for this publication is available from the British Library

Library of Congress Cataloging-in-Publication data
Srikant, R. (Rayadurgam)
Communication networks : an optimization, control, and stochastic networks perspective /
R. Srikant, University of Illinois at Urbana-Champaign, Lei Ying, Arizona State University.
pages cm
Includes bibliographical references and index.
ISBN 978-1-107-03605-5 (Hardback)
1. Telecommunication systems. I. Ying, Lei (Telecommunication engineer) II. Title.
TK5101.S657 2013
384–dc23   2013028843

ISBN   978-1-107-03605-5   Hardback

Additional resources for this publication at www.cambridge.org/srikant

Cambridge University Press & Assessment has no responsibility for the persistence or accuracy of URLs for external or third-party internet websites referred to in this publication and does not guarantee that any content on such websites is, or will remain, accurate or appropriate.

To Amma, Appa, Susie, Katie, and Jenny
RS

To my parents, Lingfang and Ethan
LY

"This book by Srikant and Ying fills a major void – an analytical and authoritative study of communication networks that covers many of the major advances made in this area in an easy-to-understand and self-contained manner. It is a must read for any networking student, researcher, or engineer who wishes to have a fundamental understanding of the key operations of communication networks, from network dimensioning and design to congestion control, routing, and scheduling. Throughout the book, the authors have taken pains to explain highly mathematical material in a manner that is accessible to a beginning graduate student. This has often required providing new examples, results, and proofs that are simple and easy to follow, which makes the book attractive to academics and engineers alike. A must have networking book for one's personal library!"

Ness B. Shroff
The Ohio State University

"*Communication Networks* provides a deep, modern and broad yet accessible coverage of the analysis of networks. The authors, who made many original contributions to this field, guide the readers through the intuition behind the analysis and results. The text is ideal for self-study and as a basis for a graduate course on the mathematics of communication networks. Students in networking will benefit greatly from reading this book."

Jean Walrand
University of California, Berkeley

"*Communication Networks*, by Srikant and Ying, provides a mathematically rigorous treatment of modern communication networks. The book provides the essential mathematical preliminaries in queueing theory, optimization and control, followed by a rigorous treatment of network architectures, protocols and algorithms that are at the heart of modern-day communication networks and the Internet. It is the best textbook on communication networks from a theoretical perspective in over 20 years, filling a much needed void in the field. It can be an excellent textbook for graduate and advanced undergraduate classes, and extremely useful to researchers in this rapidly evolving field."

Eytan Modiano
Massachusetts Institute of Technology

"This book presents a view of communication networks, their architecture and protocols, grounded in the theoretical constructs from optimization and queueing theory that underpin the modern approach to the design and analysis of networks. It is a superb introduction to this approach."

Frank Kelly
University of Cambridge

"This textbook provides a thoughtful treatment of network architecture and network protocol design within a solid mathematical framework. Networks are required to provide good stable behavior in random environments. This textbook provides the tools needed to make this happen. It provides needed foundations in optimization, control, and probabilistic techniques. It then demonstrates their application to the understanding of current networks and the design of future network architectures and protocols. This is a 'must' addition to the library of graduate students performing research in networking, and engineers researching future network architectures and protocols."

Donald F. Towsley
University of Massachusetts Amherst

CONTENTS

Why we wrote this book

Traditionally, analytical techniques for communication networks discussed in textbooks fall into two categories: (i) analysis of network protocols, primarily using queueing theoretic tools, and (ii) algorithms for network provisioning which use tools from optimization theory. Since the mid 1990s, a new viewpoint of the architecture of a communication network has emerged. Network architecture and algorithms are now viewed as slow-time-scale, distributed solutions to a large-scale optimization problem. This approach illustrates that the layered architecture of a communication network is a natural by-product of the desire to design a fair and stable system. On the other hand, queueing theory, stochastic processes, and combinatorics play an important role in designing low-complexity and distributed algorithms that are viewed as operating at fast time scales.

Our goal in writing this book is to present this modern point of view of network protocol design and analysis to a wide audience. The book provides readers with a comprehensive view of the design of communication networks using a combination of tools from optimization theory, control theory, and stochastic networks, and introduces mathematical tools needed to analyze the performance of communication network protocols.

Organization of the book

The book has been organized into two major parts. In the first part of the book, with a few exceptions, we present mathematical techniques only as tools to design algorithms implemented at various layers of a communication network. We start with the transport layer, and then consider algorithms at the link layer and the medium access layer, and finally present a unified view of all these layers along with the network layer. After we cover all the layers, we present a brief introduction to peer-to-peer applications which, by some estimates, form a significant portion of Internet traffic today.

The second part of the book is devoted to advanced mathematical techniques which are used frequently by researchers in the area of communication networks. We often sacrifice generality by making simplifying assumptions, but, as a result, we hope that we have made techniques that are typically found in specialized texts in mathematics more broadly accessible. The collection of mathematical techniques relevant to communication networks is vast, so we have perhaps made a personal choice in the selection of the topics. We have chosen to highlight topics in basic queueing theory, asymptotic analysis of queues, and scaling laws for wireless networks in the second part of the book.

We note that two aspects of the book are perhaps unique compared to other textbooks in the field: (i) the presentation of the mathematical tools in parallel with a top-down view of communication networks, and (ii) the presentation of heavy-traffic analysis of queueing models using Lyapunov techniques.

The background required to read the book

Graduate students who have taken a graduate-level course in probability and who have some basic knowledge of optimization and control theory should find the book accessible. An industrious student willing to put in extra effort may find the book accessible even with just a strong undergraduate course in probability. Researchers working in the area of communication networks should be able to read most chapters in the book individually since we have tried to make each chapter as self contained as possible. However, occasionally we refer to results in earlier chapters when discussing the material in a particular chapter, but this overlap between chapters should be small. We have provided a brief introduction to the mathematical background required to understand the various topics in the book, as and when appropriate, to aid the reader.

How to use the book as an instructor

We have taught various graduate-level courses from the material in the book. Based on our experience, we believe that there are two different ways in which this book can be used: to teach either a single course or two courses on communication networks. Below we provide a list of chapters that can be covered for each of these options.

- A two-course sequence on communication networks.
  - Course 1 (modeling and algorithms): Chapters 1–6 except Section 3.5, and Sections 7.1, 7.2, 7.4.1, and Chapter 8. The mathematical background, Sections 2.1 and 2.3, can be taught as and when necessary when dealing with specific topics.
  - Course 2 (performance analysis): Chapters 9 (cover Section 8.2 before Section 9.14), 10, and 11. We recommend reviewing Chapter 3 (except Section 3.5), which would have been covered in Course 1 above, before teaching Chapter 10.
- A single course on communication networks, covering modeling, algorithms, and performance analysis: Chapters 1–6 except Section 3.5, Sections 9.1–9.10 of Chapter 9, and Chapters 10 and 11.

Acknowledgements

This book grew out of courses offered by us at the University of Illinois at Urbana-Champaign, Iowa State University, and Arizona State University. The comments of the students in these courses over the years have been invaluable in shaping the material in the book. We would like to acknowledge Zhen Chen, Javad Ghaderi, Juan Jose Jaramillo, Xiaohan Kang, Joohwan Kim, Siva Theja Maguluri, Chandramani Singh, Weina Wang, Rui Wu, Zhengyu Zhang, and Kai Zhu in particular, who read various parts of the book carefully and provided valuable comments. We also gratefully acknowledge collaborations and/or discussions with Tamer Başar, Atilla Eryilmaz, Bruce Hajek, Frank Kelly, P. R. Kumar, Sean Meyn, Sanjay Shakkottai, Srinivas Shakkottai, Ness Shroff, and Don Towsley over the years, which helped shape the presentation of the material in this book.

1 Introduction

A communication network is an interconnection of devices designed to carry information from various sources to their respective destinations. To execute this task of carrying information, a number of protocols (algorithms) have to be developed to convert the information to bits and transport these bits reliably over the network. The first part of this book deals with the development of mathematical models which will be used to design the protocols used by communication networks. To understand the scope of the book, it is useful first to understand the architecture of a communication network.

The sources (also called end hosts) that generate information (also called data) first convert the data into bits (0s and 1s) which are then collected into groups called packets. We will not discuss the process of converting data into packets in this book, but simply assume that the data are generated in the form of packets. Let us consider the problem of sending a stream of packets from a source S to destination D, and assume for the moment that there are no other entities (such as other sources or destinations or intermediate nodes) in the network. The source and destination must be connected by some communication medium, such as a coaxial cable, telephone wire, or optical fiber, or they have to communicate in a wireless fashion. In either case, we can imagine that S and D are connected by a communication link, although the link is virtual in the case of wireless communication. The protocols that ensure reliable transfer of data over such a single link are called the *link layer* protocols or simply the link layer. The link layer includes algorithms for converting groups of bits within a packet into waveforms that are appropriate for transmission over the communication medium, adding error correction to the bits to ensure that data are received reliably at the destination, and dividing the bits into groups called frames (which may be smaller or larger than packets) before converting them to waveforms for transmission. The process of converting groups of bits into waveforms is called modulation, and the process of recovering the original bits from the waveform is called demodulation. The protocols used for modulation, demodulation, and error correction are often grouped together and called the *physical layer* set of protocols. In this book, we assume that the physical layer and link layer protocols are given, and that they transfer data over a single link reliably.

Once the link layer has been designed, the next task is one of interconnecting links to form a network. To transfer data over a network, the entities in the network must be given addresses, and protocols must be designed to route packets from each source to their destination via intermediate nodes using the addresses of the destination and the intermediate nodes. This task is performed by a set of protocols called the *network layer*. In the Internet, the network layer is called the Internet Protocol (IP) layer. Note that the network layer protocols can be designed independently of the link layer, once we make the assumption that the link layer protocols have been designed to ensure reliable data transfer over each link. This concept of independence among the design of protocols at each layer is called *layering* and is fundamental to the design of large communication networks. This allows engineers who develop protocols at one layer to abstract the functionalities of the protocols at other layers and concentrate on designing efficient protocols at just one layer.

Next, we assume that the network layer has been well designed and that it somehow generates routes for packets from each possible source to each possible destination in the network. Recall that the network is just an interconnection of links. Each link in the network has a limited capacity, i.e., the rate at which it can transfer data as measured in bits per second (bps). Since the communication network is composed of links, the sources producing data cannot send packets at arbitrarily high rates since the end-to-end data transfer rate between a source and its destination is limited by the capacities of the links on the route between the source and the destination. Further, when multiple source-destination (S-D) pairs transfer data over a network, the network capacity has to be shared by these S-D pairs. Thus, a set of protocols has to be designed to ensure fair sharing of resources between the various S-D pairs. The set of protocols that ensures such fair sharing of resources is called the *transport* layer. Transport layer protocols ensure that, most of the time, the total rate at which packets enter a link is less than or equal to the link capacity. However, occasionally the packet arrival rate at a link may exceed the link capacity since perfectly efficient transport layer protocol design is impossible in a large communication network. During such instances, packets may be dropped by a link and such packet losses will be detected by the destinations. The destinations then inform the sources of these packet losses, and the transport layer protocols may retransmit packets if necessary. Thus, in addition to fair resource sharing and congestion control functionalities, transport layer protocols may also have end-to-end (source-destination) error recovery functionalities as well.

The final set of protocols used to communicate information over a network is called the *application* layer. Application layer protocols are specific to applications that use the network. Examples of applications include file transfer, real-time video transmission, video or voice calls, stored-video transmission, fetching and displaying web pages, etc. The application layer calls upon transport protocols that are appropriate for their respective applications. For example, for interactive communication, occasional packet losses may be tolerated, whereas a file transfer requires that all packets reach the destination. Thus, the former may use a transport protocol that does not use retransmissions to guarantee reliable delivery of every packet to the destination, while the latter will use a transport protocol that ensures end-to-end reliable transmission of every packet.

In addition to the protocol layers mentioned above, in the case of wireless communications, signal propagation over one link may cause interference at another link. Thus, a special set of protocols called Medium Access Control (MAC) protocols are designed to arbitrate the contention between the links for access to the wireless medium. The MAC layer can be viewed as a sublayer of the link layer that further ensures reliable operation of the wireless "links" so that the network layer continues to see the links as reliable carriers of data. A schematic of the layered architecture of a communication network is provided in Figure 1.1. To ensure proper operation of a communication network, a packet generated by an application will not only contain data, but also contain other information called the *header*. The header may contain information such as the transport protocol to be used and the address of the destination for routing purposes.

The above description of the layered architecture of a communication network is an abstraction. In real communication networks, layering may not be as strict as defined

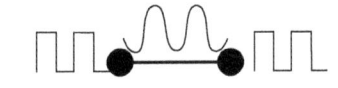

Physical layer: bits over wire/wireless channels.

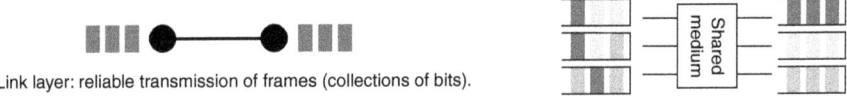

Link layer: reliable transmission of frames (collections of bits).

MAC sublayer: multiple links over a shared medium.

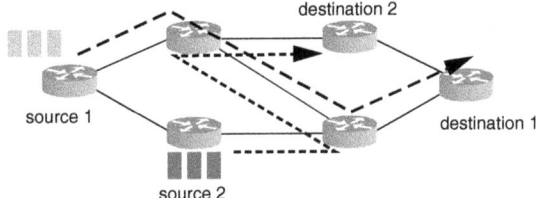

Network layer: data transmitted in the form of packets. Each packet has source and destination addresses, and data. Each node in the network contains routing information to route the packets.

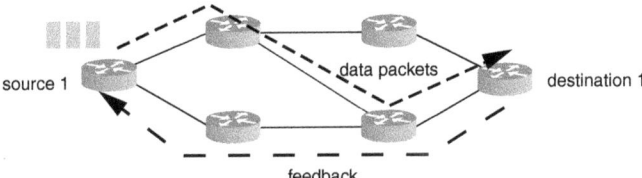

Transport layer: reliable end-to-end data transmission. Sources may use feedback from destinations to retransmit lost packets. Sources may also use the feedback information to adjust data transmission rates.

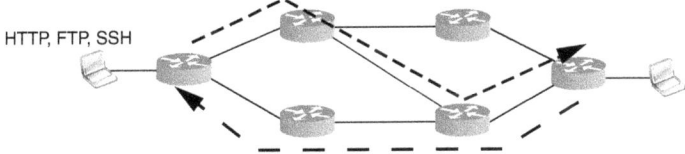

Application layer: applications. Protocols such as HTTP, FTP, and SSH transmit data over the network.

Figure 1.1 Schematic of the layered architecture of a communication network.

above. Some protocols may have functionalities that cut across more than one layer. Such cross-layer protocols may be designed for ease of implementation or to improve the efficiency of the communication network. Nevertheless, the abstraction of a layered architecture is useful conceptually, and in practice, for the design of communication networks.

Having described the layers of a communication network, we now discuss the scope of this book. In Part I, we are interested in the design of protocols for the transport, network, and MAC sublayers. We first develop a mathematical formulation of the problem of resource sharing in a large communication network accessed by many sources. We show how transport layer algorithms can be designed to solve this problem. We then drill deeper into the communication network, and understand the operation of a single link and how temporary overload is handled at a link. Next, we discuss the problem of interconnecting links through a router in the Internet and the problem of contention resolution between multiple links in a wireless network. The algorithms that resolve contention in wireless links form the MAC sublayer. As we will see, the algorithms that are used to interconnect links within a wireline router share a lot of similarities with wireless MAC algorithms. We devote a separate chapter to network protocols, where we discuss the actual protocols used in the Internet and wireless networks, and relate them to the theory and algorithms developed in the earlier chapters. Part I concludes with an introduction to a particular set of application layer protocols called peer-to-peer networks. Traditional applications deliver data from a single source to a destination or a group of destinations. They simply use the lower layer protocols in a straightforward manner to perform their tasks. In Peer-to-Peer (P2P) networks, many users of the network (called peers) are interested in the same data, but do not necessarily download these data from a single destination. Instead, peers download different pieces of the data and share these pieces among themselves. This type of sharing of information make P2P systems interesting to study in their own right. Therefore, we devote a separate chapter to the design of these types of applications in Part I.

Part II is a collection of mathematical tools that can be used for performance analysis once a protocol or a set of protocols have been designed. The chapters in this part are not organized by functionalities within a communication network, but are organized by the commonality of the mathematical tools used. We will introduce the reader to tools from queueing theory, heavy-traffic methods, large deviations, and models of wireless networks where nodes are viewed as random points on a plane. Throughout, we will apply these mathematical tools to analyze the performance of various components of a communication network.

Part I Network architecture and algorithms

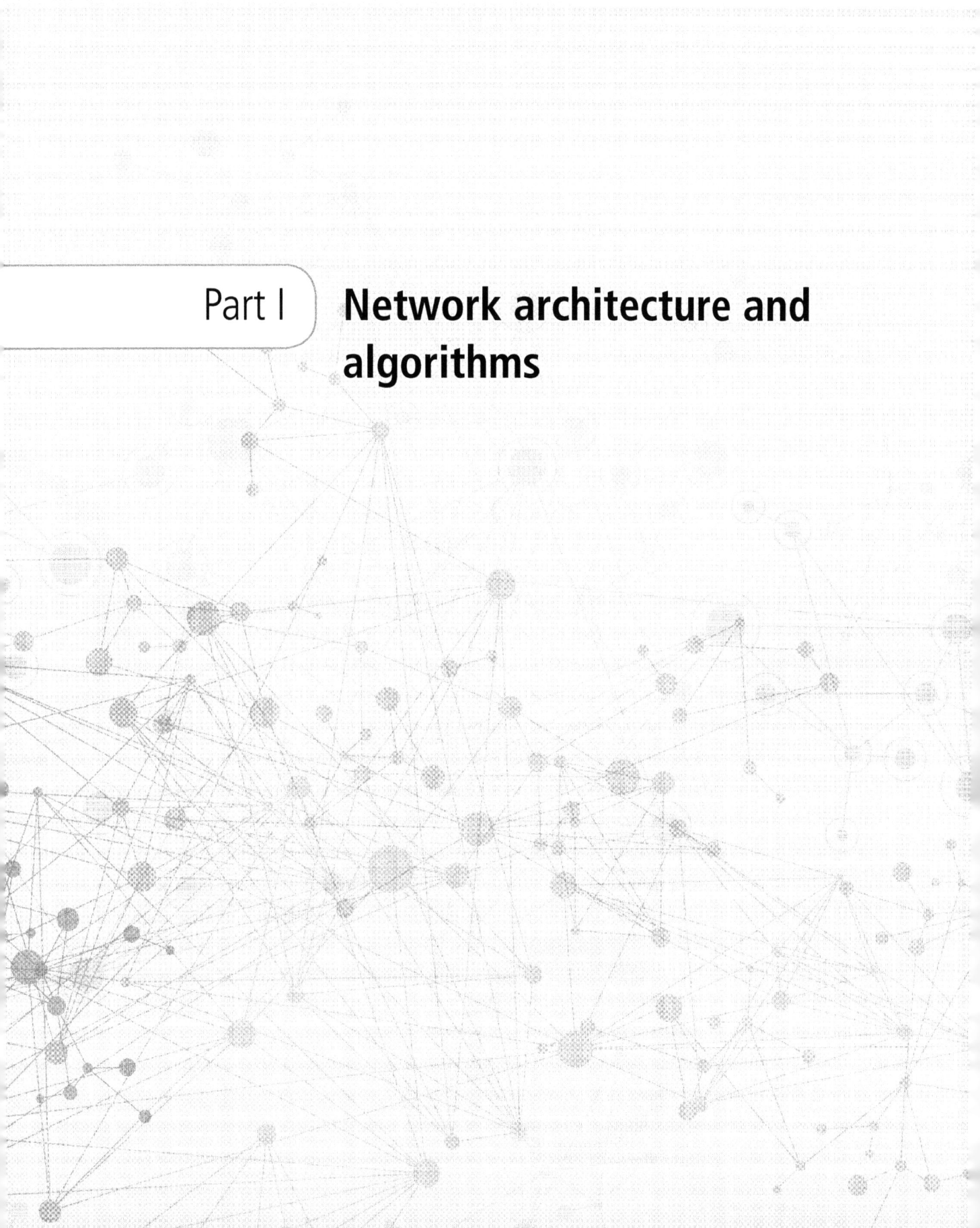

2 Mathematics of Internet architecture

In this chapter, we will develop a mathematical formulation of the problem of resource allocation in the Internet. A large communication network such as the Internet can be viewed as a collection of communication links shared by many sources. Congestion control algorithms are protocols that allocate the available network resources in a fair, distributed, and stable manner among the sources. In this chapter, we will introduce the network utility maximization formulation for resource allocation in the Internet, where each source is associated with a utility function $U_r(x_r)$, and x_r is the transmission rate allocated to source r. The goal of fair resource allocation is to maximize the net utility $\sum_r U_r(x_r)$ subject to resource constraints. We will derive distributed, congestion control algorithms that solve the network utility maximization problem. In a later chapter, we will discuss the relationship between the mathematical models developed in this chapter to transport layer protocols used in the Internet. Optimality and stability of the congestion control algorithms will be established using convex optimization and control theory. We will also introduce a game-theoretical view of network utility maximization and study the impact of strategic users on the efficiency of network utility maximization. Finally, routing and IP addressing will be discussed. The following key questions will be answered in this chapter.

- *What is fair resource allocation?*
- *How do we use convex optimization and duality to design distributed resource allocation algorithms to achieve a fair and stable resource allocation?*
- *What are the game-theoretic implications of fair resource allocation?*

2.1 Mathematical background: convex optimization

In this section, we present some basic results from convex optimization which we will find useful in the rest of the chapter. Often, the results will be presented without proofs, but some concepts will be illustrated with figures to provide an intuitive feel for the results.

2.1.1 Convex sets and convex functions

We first introduce the basic concepts from optimization theory, including the definitions of convex sets and convex functions.

Definition 2.1.1 (Convex set) A set $\mathcal{S} \subseteq \mathcal{R}^n$ is convex if $\alpha x + (1 - \alpha)y \in \mathcal{S}$ whenever $x, y \in \mathcal{S}$ and $\alpha \in [0, 1]$. Since $\alpha x + (1 - \alpha)y$, for $\alpha \in [0, 1]$, describes the line segment between x and y, a convex set can be pictorially depicted as in Figure 2.1: given any two points $x, y \in \mathcal{S}$, the line segment between x and y lies entirely in $\mathcal{S}$.   □

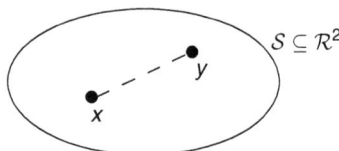

Figure 2.1 A convex set, $\mathcal{S} \subseteq \mathcal{R}^2$.

Definition 2.1.2 (Convex hull) The convex hull of set $\mathcal{S}$, denoted by $Co(\mathcal{S})$, is the smallest convex set that contains $\mathcal{S}$, and contains all convex combinations of points in $\mathcal{S}$, i.e.,

$$Co(\mathcal{S}) = \left\{ \sum_{i=1}^{k} \alpha_i x_i \,\middle|\, x_i \in \mathcal{S}, \alpha_i \geq 0, \sum_{i=1}^{k} \alpha_i = 1 \right\}.$$

See Figure 2.2 for an example.   □

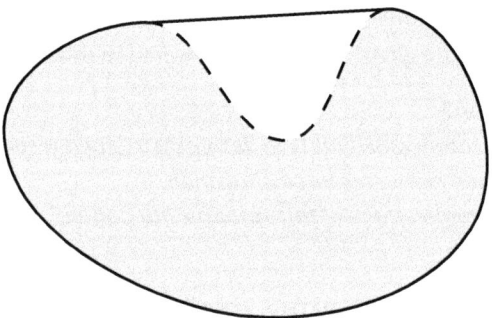

Figure 2.2 The solid line forms the boundary of the convex hull of the shaded set.

Definition 2.1.3 (Convex function) A function $f(x) : \mathcal{S} \subseteq \mathcal{R}^n \to \mathcal{R}$ is a convex function if $\mathcal{S}$ is a convex set and the following inequality holds for any $x, y \in \mathcal{S}$ and $\alpha \in [0, 1]$:

$$f(\alpha x + (1 - \alpha)y) \leq \alpha f(x) + (1 - \alpha)f(y);$$

$f(x)$ is strictly convex if the above inequality is strict for all $\alpha \in (0, 1)$ and $x \neq y$. Pictorially, $f(x)$ looks like a bowl, as shown in Figure 2.3.   □

Definition 2.1.4 (Concave function) A function $f(x) : \mathcal{S} \subseteq \mathcal{R}^n \to \mathcal{R}$ is a concave function (strictly concave) if $-f$ is a convex (strictly convex) function. Pictorially, $f(x)$ looks like an inverted bowl, as shown in Figure 2.4.   □

2.1 Mathematical background: convex optimization

The line segment connecting the two points $(x, f(x))$ and $(y, f(y))$ lies "above" the plot of $f(x)$.

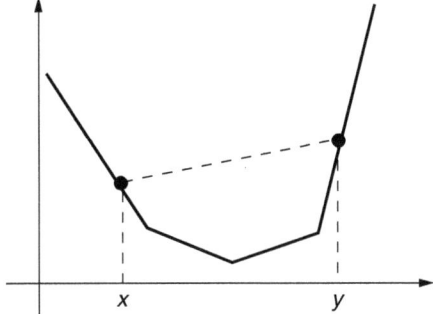

Figure 2.3 Pictorial description of a convex function in $\mathcal{R}^2$.

The line segment connecting the two points $(x, f(x))$ and $(y, f(y))$ lies "below" the plot of $f(x)$.

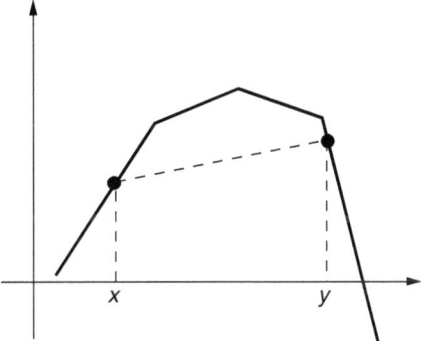

Figure 2.4 Pictorial description of a concave function in $\mathcal{R}^2$.

Definition 2.1.5 (Affine function) A function $f(x) : \mathcal{R}^n \to \mathcal{R}^m$ is an affine function if it is a sum of a linear function and a constant, i.e., there exist $\alpha \in \mathcal{R}^{m \times n}$ and $a \in \mathcal{R}^m$ such that

$$f(x) = \alpha x + a.$$ □

The convexity of a function may be hard to verify from the definition given above. Therefore, next we present several conditions that can be used to verify the convexity of a function. The proofs are omitted here, and can be found in most textbooks on convex analysis or convex optimization.

Result 2.1.1 (First-order condition I) Let $f : \mathcal{S} \subset \mathcal{R} \to \mathcal{R}$ be a function defined over a convex set $\mathcal{S}$. If f is differentiable and the derivative $f'(x)$ is non-decreasing (increasing) in $\mathcal{S}$, $f(x)$ is convex (strictly convex) over $\mathcal{S}$. □

Result 2.1.2 (First-order condition II) Let $f : S \subset \mathcal{R}^n \to \mathcal{R}$ be a differentiable function defined over a convex set S. Then f is a convex function if and only if

$$f(y) \geq f(x) + \nabla f(x)(y - x), \quad \forall x, y \in S, \tag{2.1}$$

where

$$\nabla f(x) = \left(\frac{\partial f}{\partial x_1}(x), \frac{\partial f}{\partial x_2}(x), \dots, \frac{\partial f}{\partial x_n}(x) \right)$$

and x_i is the ith component of vector x. Pictorially, if x is one-dimensional, this condition implies that the tangent of the function at any point lies below the function, as shown in Figure 2.5.

Note that $f(x)$ is strictly convex if the inequality above is strict for any $x \neq y$. ☐

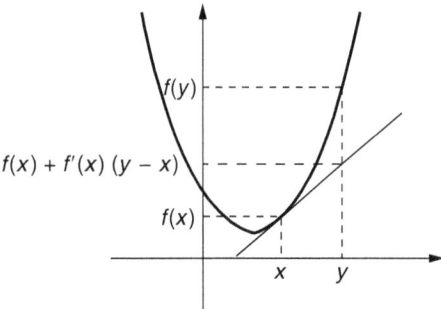

Figure 2.5 Pictorial description of inequality (2.1) in one-dimensional space.

Result 2.1.3 (Second-order condition) Let $f : S \subset \mathcal{R}^n \to \mathcal{R}$ be a twice differentiable function defined over the convex set S. Then, f is a convex (strictly convex) function if the Hessian matrix $\mathbf{H}$ with

$$H_{ij} = \frac{\partial^2 f}{\partial x_i \partial x_j}(x)$$

is positive semidefinite (positive definite) over S. ☐

Result 2.1.4 (Strict separation theorem) Let $S \subset \mathcal{R}^n$ be a convex set and x be a point that is not contained in S. Then there exists a vector $\beta \in \mathcal{R}^n$, $\beta \neq 0$, and constant $\delta > 0$ such that

$$\sum_{i=1}^{n} \beta_i y_i \leq \sum_{i=1}^{n} \beta_i x_i - \delta$$

holds for any $y \in S$. ☐

2.1.2   Convex optimization

We first consider the following unconstrained optimization problem:

$$\max_{x\in S} f(x), \tag{2.2}$$

and present some important results without proofs.

Definition 2.1.6 (Local maximizer and global maximizer) For any function $f(x)$ over $S \subseteq \mathcal{R}^n$, x^* is said to be a *local* maximizer or *local* optimal point if there exists an $\epsilon > 0$ such that

$$f(x^* + \delta x) \le f(x^*)$$

for δx such that $\|\delta x\| \le \epsilon$ and $x + \delta x \in S$, where $\|\cdot\|$ can be any norm; x^* is said to be a *global* maximizer or *global* optimal point if

$$f(x) \le f(x^*)$$

for any $x \in S$. When not specified, maximizer refers to global maximizer in this book. □

Result 2.1.5 If $f(x)$ is a continuous function over a compact set S (i.e., S is closed and bounded if $S \subseteq \mathcal{R}^n$), then $f(x)$ achieves its maximum over this set, i.e., $\max_{x\in S} f(x)$ exists. □

Result 2.1.6 If $f(x)$ is differentiable, then any local maximizer x^* in the interior of $S \subseteq \mathcal{R}^n$ satisfies

$$\nabla f(x^*) = 0. \tag{2.3}$$

If $f(x)$ is a concave function over S, condition (2.3) is also sufficient for x^* to be a local maximizer. □

Result 2.1.7 If $f(x)$ is concave, then a local maximizer is also a global maximizer. In general, multiple global maximizers may exist. If $f(x)$ is strictly concave, the global maximizer x^* is unique. □

Result 2.1.8 Results 2.1.6 and 2.1.7 hold for convex functions if the max in the optimization problem (2.2) is replaced by min, and maximizer is replaced by minimizer in Results 2.1.6 and 2.1.7. □

Result 2.1.9 If $f(x)$ is a differentiable function over set S and x^* is a maximizer of the function, then

$$\nabla f(x^*)dx \le 0$$

for any feasible direction dx, where a non-zero vector dx is called a feasible direction if there exists α such that $x + adx \in S$ for any $0 \le a \le \alpha$.

Further, if $f(x)$ is a concave function, then x^* is a maximizer if and only if

$$\nabla f(x^*)\delta x \leq 0$$

for any δx such that $x^* + \delta x \in \mathcal{S}$.   □

Next, we consider an optimization problem with equality and inequality constraints as follows:

$$\max_{x \in \mathcal{S}} f(x), \tag{2.4}$$

subject to

$$h_i(x) \leq 0, i = 1, 2, ..., I, \tag{2.5}$$

$$g_j(x) = 0, j = 1, 2, ..., J. \tag{2.6}$$

A vector x is said to be *feasible* if $x \in \mathcal{S}$, $h_i(x) \leq 0$ for all i, and $g_j(x) = 0$ for all j. While (2.5) and (2.6) are inequality and equality constraints, respectively, the set $\mathcal{S}$ in the above problem captures any other constraints that are not in equality or inequality form.

A key concept that we will exploit later in the chapter is called Lagrangian duality. Duality refers to the fact that the above maximization problem, also called the *primal* problem, is closely related to an associated problem called the *dual* problem. Given the constrained optimization problem in (2.4)–(2.6), the *Lagrangian* of this optimization problem is defined to be

$$L(x, \lambda, \mu) = f(x) - \sum_{i=1}^{I} \lambda_i h_i(x) + \sum_{j=1}^{J} \mu_j g_j(x), \qquad \lambda_i \geq 0 \ \forall i.$$

The constants $\lambda_i \geq 0$ and μ_j are called Lagrange multipliers. The *Lagrangian dual function* is defined to be

$$D(\lambda, \mu) = \sup_{x \in \mathcal{S}} L(x, \lambda, \mu).$$

Let f^* be the maximum of the optimization problem (2.4), i.e., $f^* = \max_{x \in \mathcal{S}} f(x)$. Then, we have the following theorem.

Theorem 2.1.1 $D(\lambda, \mu)$ is a convex function and $D(\lambda, \mu) \geq f^*$.

Proof The convexity comes from a known fact that the pointwise supremum of affine functions is convex (see Figure 2.6). To prove the bound, note that $h_i(x) \leq 0$ and $g_j(x) = 0$ for any feasible x, so the following inequality holds for any feasible x:

$$L(x, \lambda, \mu) \geq f(x).$$

This inequality further implies that

$$\sup_{\substack{x \in \mathcal{S} \\ h(x) \leq 0 \\ g(x) = 0}} L(x, \lambda, \mu) \geq \sup_{\substack{x \in \mathcal{S} \\ h(x) \leq 0 \\ g(x) = 0}} f(x) = f^*.$$

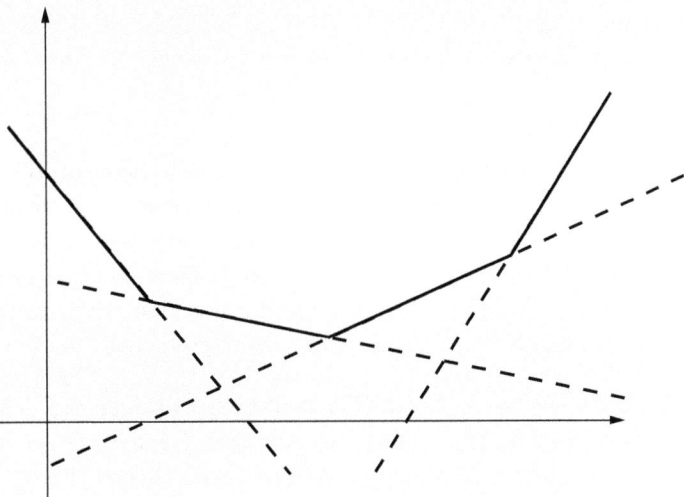

Figure 2.6 The solid line is the pointwise supremum of the four dashed lines, and is convex.

Since removing some constraints of a maximization problem can only result in a larger maximum value, we obtain

$$\sup_{x \in \mathcal{S}} L(x, \lambda, \mu) \geq \sup_{\substack{x \in \mathcal{S} \\ h(x) \leq 0 \\ g(x) = 0}} L(x, \lambda, \mu).$$

Therefore, we conclude that

$$D(\lambda, \mu) = \sup_{x \in \mathcal{S}} L(x, \lambda, \mu) \geq f^*. \qquad \qquad \square$$

Theorem 2.1.1 states that the dual function is an upper bound on the maximum of the optimization problem (2.4)–(2.6). We can optimize over λ and μ to obtain the best upper bound, which yields the following minimization problem, called the *Lagrange dual problem*:

$$\inf_{\lambda \geq 0, \mu} D(\lambda, \mu). \qquad \qquad (2.7)$$

Let d^* be the minimum of the dual problem, i.e., $d^* = \inf_{\lambda \geq 0, \mu} D(\lambda, \mu)$. The difference between d^* and f^* is called the duality gap. For some problems, the duality gap is zero. We say *strong duality* holds if $d^* = f^*$. If strong duality holds, then one can solve either the primal problem or the dual problem to obtain f^*. This is often helpful since sometimes one of the problems is easier to solve than the other. A simple yet frequently used condition to check strong duality is *Slater's condition*, which is given below.

Theorem 2.1.2 (Slater's condition) Consider the constrained optimization problem defined by (2.4)–(2.6). Strong duality holds if the following conditions are true:

- $f(x)$ is a concave function and $h_i(x)$ are convex functions;
- $g_j(x)$ are affine functions;
- there exists an x that belongs to the relative interior[1] of S such that $h_i(x) < 0$ for all i and $g_j(x) = 0$ for all j.   □

As mentioned earlier, when strong duality holds, we have a choice of solving the original optimization in one of two ways: either solve the primal problem directly or solve the dual problem. Later in this chapter, we will see that resource allocation problems in communication networks can be posed as convex optimization problems, and we can use either the primal or the dual formulations to solve the resource allocation problem. We now present a result which can be used to solve convex optimization problems.

Theorem 2.1.3 (Karush–Kuhn–Tucker (KKT) conditions) Consider the constrained optimization problem defined in (2.4)–(2.6). Assume that f and h_i $(i = 1, 2, \ldots, I)$ are differentiable functions and that Slater's conditions are satisfied. Let x^* be a feasible point, i.e., a point that satisfies all the constraints. Such an x^* is a global maximizer for the optimization problem (2.4)–(2.6) if and only if there exist constants $\lambda_i^* \geq 0$ and μ_j^* such that

$$\frac{\partial f}{\partial x_k}(x^*) - \sum_i \lambda_i^* \frac{\partial h_i}{\partial x_k}(x^*) + \sum_j \mu_j^* \frac{\partial g_j}{\partial x_k}(x^*) = 0, \quad \forall k, \tag{2.8}$$

$$\lambda_i^* h_i(x^*) = 0, \quad \forall i. \tag{2.9}$$

Further, (2.8) and (2.9) are also necessary and sufficient conditions for (λ^*, μ^*) to be a global minimizer of the Lagrange dual problem given in (2.7). If f is strictly concave, then x^* is also the unique global maximizer.   □

The KKT conditions (2.8) and (2.9) can be interpreted as follows. Consider the Lagrangian

$$L(x, \lambda, \mu) = f(x) - \sum_i \lambda_i h_i(x) + \sum_j \mu_j g_j(x).$$

Condition (2.8) is the first-order necessary condition for the maximization problem $\max_{x \in S} L(x, \lambda^*, \mu^*)$. When strong duality holds, we have

$$f(x^*) = f(x^*) - \sum_i \lambda_i^* h_i(x^*) + \sum_j \mu_j^* g_j(x^*),$$

which results in condition (2.9) since $g_j(x^*) = 0\ \forall j$, and $\lambda_i^* \geq 0$ and $h_i(x^*) \leq 0\ \forall i$. We remark that condition (2.9) is called complementary slackness.

---

[1] For convex set S, any point in the relative interior is a point x such that for any $y \in S$ there exist $z \subset S$ and $0 < \lambda < 1$ such that $x = \lambda y + (1 - \lambda)z$.

2.2   Resource allocation as utility maximization

The Internet is a shared resource, shared by many millions of users, who are connected by a huge network consisting of many, many routers and links. The capacity of the links must be split in some *fair* manner among the users. To appreciate the difficulty in defining what fairness means, let us consider an everyday example. Suppose that one has a loaf of bread which has to be divided among three people. Almost everyone will agree that the fair allocation is to divide the loaf into three equal parts and give one piece to each person. While this seems obvious, consider a slight variant of the situation, where one of the people is a two-year-old child and the other two are football players. Then, an equal division does not seem appropriate: the child cannot possibly consume the third allocated to her, so a different division based on their needs may appear to be more appropriate. The situation becomes more complicated when there is more than one resource to be divided among the three people. Suppose that there are two loaves of bread, one wheat and one rye, so a fair division has to take into account the preferences of the individuals for the different types of bread. Economists solve such problems by associating a so-called utility function with each individual, and then finding an allocation that maximizes the net utility of the individuals. We now formally describe and model the resource allocation problem in the Internet.

Consider a network consisting of a set of links $\mathcal{L}$ accessed by a set of sources $\mathcal{S}$. We will use the terms source and user interchangeably. Each source is associated with a route, where a route is simply a collection of links. Thus, we assume that the route used by a source to convey packets to their destination is fixed. Since the route is fixed for a source, we use the same index (typically r or s) to denote both a source and its route. We allow multiple sources to share exactly the same route. Thus, two routes can consist of exactly the same set of links. Each user derives a certain utility $U_r(x_r)$ when transmitting at rate x_r. The utility can be interpreted as the level of satisfaction that a user derives when its transmission rate is x_r. We assume that $U_r(x_r)$ is an increasing, continuously differentiable function. It is also usually the case that the rate at which the utility increases is larger at smaller rates than at larger rates. For example, a user's level of satisfaction will increase by a larger amount when the rate allocated to him or her increases from 0 Mbps to 1 Mbps than when the rate increases from 1 Mbps to 2 Mbps. Thus, we also assume that $U_r(x_r)$ is a strictly concave function.

The goal of resource allocation is to solve the following optimization problem, called Network Utility Maximization (NUM):

$$\max_{x_r} \sum_{r \in \mathcal{S}} U_r(x_r) \tag{2.10}$$

subject to the link capacity constraints

$$\sum_{r:l \in r} x_r \leq c_l, \quad \forall l \in \mathcal{L}, \tag{2.11}$$

$$x_r \geq 0, \quad \forall r \in \mathcal{S}. \tag{2.12}$$

Next, we present an example of such a resource allocation problem and its solution.

Example 2.2.1

Consider a network with L links numbered 1 through L, and $L + 1$ sources numbered 0 through L. All link capacities are assume to be equal to 1. Source 0's route includes all links, while source r uses only link r, as shown in Figure 2.7.

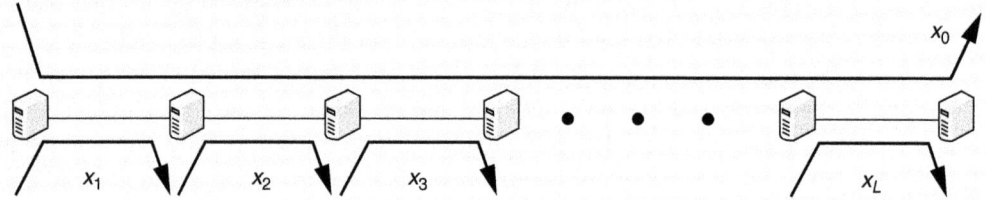

Figure 2.7 An L-link line network with $L + 1$ sources.

Assuming log utility functions, the resource allocation problem is given by

$$\max_{x} \sum_{r=0}^{L} \log x_r \qquad (2.13)$$

with constraints

$$x_0 + x_l \leq 1, \qquad \forall l = 0, 1, \ldots, L$$
$$x \geq 0,$$

where x is the vector consisting of x_0, x_1 through x_L. Now, since $\log x \to -\infty$ as $x \to 0$, the optimal solution will assign a strictly positive rate to each user, and so the last constraint can be ignored.

Let p_l be the *Lagrange multiplier* associated with the capacity constraint at link l and let p denote the vector of Lagrange multipliers. Then, the *Lagrangian* is given by

$$L(x, p) = \sum_{r=0}^{L} \log x_r - \sum_{l=1}^{L} p_l (x_0 + x_l - 1).$$

Setting $\partial L / \partial x_r = 0$ for each r gives

$$x_0 = \frac{1}{\sum_{l=1}^{L} p_l}, \quad x_l = \frac{1}{p_l}, \quad \forall l \geq 1. \qquad (2.14)$$

Further, the KKT conditions require that

$$p_l(x_0 + x_l - 1) = 0 \quad \text{and} \quad p_l \geq 0, \qquad \forall l \geq 1.$$

Substituting for the p_l from (2.14), we obtain

$$p_l = \frac{L + 1}{L}, \quad \forall l \geq 1.$$

Thus, the optimal data rates for the sources are given by

$$x_0 = \frac{1}{L+1}, \qquad x_r = \frac{L}{L+1}, \qquad \forall r \geq 1.$$

We note an important feature of the solution. The optimal rate of each source in (2.14) explicitly depends on the sum of the Lagrange multipliers on its route. Thus, if a simple algorithm exists to compute the Lagrange multipliers on each link and feed back the sum of the Lagrange multipliers on its route to each source, then the source rates can also be computed easily. This feature of the optimal solution will be exploited later to derive a distributed algorithm to solve the resource allocation problem.   □

A closed-form solution, as in Example 2.21, will not be easy to obtain for a network such as the Internet. The number of sources is typically in the millions, the number of links is in the thousands, and there is no central entity that even knows the topology of the network in its entirety. Therefore, we will develop distributed algorithms to solve the resource allocation problem later.

2.2.1   Utility functions and fairness

In our network utility maximization framework, we have associated a utility function with each user. The utility function can be interpreted in one of two different ways: one is that there is an inherent utility function associated with each user, and the other interpretation is that a utility function is assigned to each user by the network. In the latter case, the choice of utility function determines the resource allocation to the users. Thus, the utility function can be viewed as imposing different notions of *fair* resource allocation. Of course, there is no notion of fair allocation that is universally accepted. Here, we will discuss some commonly used notions of fairness.

A popular notion of fairness is *proportional fairness*. Proportionally fair resource allocation is achieved by associating a log utility function with each user, i.e., $U(x_r) = \log x_r$ for all users r. If $f(x)$ is a concave function over a domain $\mathcal{D}$, then it is well known that

$$\nabla f(x^*)(x - x^*) \leq 0, \qquad \forall x \in \mathcal{D}, \tag{2.15}$$

where x^* is the maximizer of $f(x)$. Thus, the optimal rates $\{x_r^*\}$, when $U(x_r) = \log x_r$, satisfy

$$\sum_r \frac{x_r - x_r^*}{x_r^*} \leq 0,$$

where $\{x_r\}$ is any other set of feasible rates. An allocation with such a property is called *proportionally fair*. The reason for this terminology is as follows: if one of the source rates is increased by a certain amount, the sum of the fractions (also called proportions) by which the different users' rates change is non-positive. A consequence of this observation is that, if the proportion by which one user's rate changes is positive, there will be at least one other user whose proportional change will be negative. If the utility functions are of the form $w_r \log x_r$ for some weight $w_r > 0$ for user r, the resulting allocation is called *weighted proportionally fair*.

Another widely used notion of fairness in communication networks is called *max-min fairness*. An allocation $\{x_r^*\}$ is called max-min fair if it satisfies the following property: if there is any other allocation $\{x_r\}$ such that a user s's rate increases, i.e., $x_s > x_s^*$, there has to be another user u with the property

$$x_u < x_u^* \quad \text{and} \quad x_u^* \leq x_s^*.$$

In other words, if we attempt to increase the rate for one user, the rate for a less-fortunate user will suffer. The definition of max-min fairness implies that

$$\min_r x_r^* \geq \min_r x_r,$$

for any other allocation $\{x_r\}$. To see why this is true, suppose that there exists an allocation such that

$$\min_r x_r^* < \min_r x_r. \tag{2.16}$$

This implies that, for any s such that $\min_r x_r^* = x_s^*$, the following holds: $x_s^* < x_s$. Otherwise, our assumption (2.16) cannot hold. However, this implies that if we switch the allocation from $\{x_r^*\}$ to $\{x_r\}$, we have increased the allocation for s without affecting a less-fortunate user (since there is no less-fortunate user than s under $\{x_r^*\}$). Thus, the max-min fair resource allocation attempts first to satisfy the needs of the user who gets the least amount of resources from the network.

Yet another form of fairness that has been discussed in the literature is called *minimum potential delay* fairness. Suppose that user r is associated with the utility function $-1/x_r$. The goal of maximizing the sum of the user utilities is equivalent to minimizing $\sum_r 1/x_r$. The term $1/x_r$ can be interpreted as follows: suppose user r needs to transfer a file of unit size. Then, $1/x_r$ is the delay associated with completing this file transfer since the delay is simply the file size divided by the rate allocated to user r. Hence, the name *minimum potential delay* fairness.

All of the different notions of fairness discussed above can be unified by considering utility functions of the form

$$U_r(x_r) = \frac{x_r^{1-\alpha}}{1-\alpha}, \tag{2.17}$$

for some $\alpha > 0$. Resource allocation using the above utility function is called *α-fair*. Different values of α yield different ideas of fairness. First consider $\alpha = 2$. This immediately yields minimum potential delay fairness. Next, consider the case $\alpha = 1$. The utility function is not well defined at this point, but note that maximizing the sum of $x_r^{1-\alpha}/(1-\alpha)$ yields the same optimum as maximizing the sum of

$$\frac{x_r^{1-\alpha} - 1}{1-\alpha}.$$

Now, by applying l'Hospital's rule, we obtain

$$\lim_{\alpha \to 1} \frac{x_r^{1-\alpha} - 1}{1-\alpha} = \log x_r,$$

thus yielding proportional fairness in the limit as $\alpha \to 1$.

Next, we argue that the limit $\alpha \to \infty$ gives max-min fairness. Let $x_r^*(\alpha)$ be the α-fair allocation. Assume that $x_r^*(\alpha) \to x_r^*$ as $\alpha \to \infty$ and $x_1^* < x_2^* < \cdots < x_n^*$. Let ϵ be the minimum difference of $\{x_r^*\}$, i.e., $\epsilon = \min_r(x_{r+1}^* - x_r^*)$. Then, when α is sufficiently large, we have $|x_r^*(\alpha) - x_r^*| \le \epsilon/4$, which implies that $x_1^*(\alpha) < x_2^*(\alpha) < \cdots < x_n^*(\alpha)$.

Now, by the property of concave functions mentioned earlier (inequality (2.15)),

$$\sum_r \frac{x_r - x_r^*(\alpha)}{x_r^{*\alpha}(\alpha)} \le 0.$$

Considering an arbitrary flow s, the above expression can be rewritten as

$$\sum_{r=1}^{s-1}(x_r - x_r^*(\alpha))\frac{x_s^{*\alpha}(\alpha)}{x_r^{*\alpha}(\alpha)} + (x_s - x_s^*(\alpha)) + \sum_{i=s+1}^{n}(x_i - x_i^*(\alpha))\frac{x_s^{*\alpha}(\alpha)}{x_i^{*\alpha}(\alpha)} \le 0.$$

Since $|x_r^*(\alpha) - x_r^*| \le \epsilon/4$, we further have

$$\sum_{r=1}^{s-1}(x_r - x_r^*(\alpha))\frac{x_s^{*\alpha}(\alpha)}{x_r^{*\alpha}(\alpha)} + (x_s - x_s^*(\alpha)) - \sum_{i=s+1}^{n}|x_i - x_i^*(\alpha)|\frac{(x_s^* + \epsilon/4)^\alpha}{(x_i^* - \epsilon/4)^\alpha} \le 0.$$

Note that $x_i^* - \epsilon/4 - (x_s^* + \epsilon/4) \ge \epsilon/2$ for any $i > s$, so, by increasing α, the third term in the above expression will become negligible. Thus, if $x_s > x_s^*(\alpha)$, the allocation for at least one user whose rate satisfies $x_r^*(\alpha) < x_s^*(\alpha)$ will decrease. The argument can be made rigorous and extended to the case $x_r^* = x_s^*$ for some r and s. Therefore, as $\alpha \to \infty$, the α-fair allocation approaches max-min fairness.

2.3 Mathematical background: stability of dynamical systems

Consider a dynamical system defined by the following differential equation:

$$\dot{x} = f(x), \qquad f : \mathcal{R}^n \to \mathcal{R}^n, \qquad (2.18)$$

where $\dot{x}$ is the derivative of x with respect to the time t. The time variable t has been omitted when no confusion is caused. Assume that $x(0)$ is given. Throughout, we will assume that f is a continuous function and that it also satisfies other appropriate conditions to ensure that the differential equation has a unique solution $x(t)$, for $t \ge 0$.

A point $x_e \in \mathcal{R}^n$ is said to be the equilibrium point of the dynamical system if $f(x_e) = 0$. We assume that $x_e = 0$ is the *unique* equilibrium point of this dynamical system.

Definition 2.3.1 (Globally, asymptotically stable) $x_e = 0$ is said to be a globally asymptotically stable equilibrium point if

$$\lim_{t \to \infty} x(t) = 0$$

for any $x(0) \in \mathcal{R}^n$.

We first introduce the Lyapunov boundedness theorem.

Theorem 2.3.1 (Lyapunov boundedness theorem) Let $V : \mathcal{R}^n \to \mathcal{R}$ be a differentiable function with the following property:

$$V(x) \to \infty \text{ as } \|x\| \to \infty. \tag{2.19}$$

Denote by $\dot{V}(x)$ the derivative of $V(x)$ with respect to t, i.e.,

$$\dot{V}(x) = \nabla V(x)\dot{x} = \nabla V(x)f(x).$$

If $\dot{V}(x) \leq 0$ for all x, there exists a constant $B > 0$ such that $\|x(t)\| \leq B$ for all t.

Proof At any time T, we have

$$V(x(T)) = V(x(0)) + \int_0^T \dot{V}(x(t)) \, dt \leq V(x(0)).$$

Note that condition (2.19) implies that $\{x : V(x) \leq c\}$ is a bounded set for any c. Letting $c = V(x(0))$, the theorem follows. □

Theorem 2.3.2 (Lyapunov global asymptotic stability theorem) If, in addition to the conditions in the previous theorem, we assume that $V(x)$ is continuously differentiable and also satisfies the following conditions:

(1) $V(x) \geq 0 \ \forall x$ and $V(x) = 0$ if and only if $x = 0$,
(2) $\dot{V}(x) < 0$ for any $x \neq 0$ and $\dot{V}(0) = 0$,

the equilibrium point $x_e = 0$ is globally, asymptotically stable.

Proof We prove this theorem by contradiction. Suppose $x(t)$ does not converge to the equilibrium point 0 as $t \to \infty$.

Note that $V(x(t))$ is non-increasing because its derivative with respect to t is non-positive $(\dot{V}(x) \leq 0)$ for any x. Since $V(x(t))$ decreases as a function of t and is lower bounded (since $V(x) \geq 0, \forall x$), it converges as $t \to \infty$. Suppose that $V(x(t))$ converges to, say, $\epsilon > 0$. Define the set

$$\mathcal{C} \triangleq \{x : \epsilon \leq V(x) \leq V(x(0))\}.$$

The set $\mathcal{C}$ is bounded since $V(x) \to \infty$ as $\|x\| \to \infty$ and it is closed since $V(x)$ is a continuous function of x. Thus, $\mathcal{C}$ is a compact set.

Let

$$-a = \sup_{x \in \mathcal{C}} \dot{V}(x),$$

where $a > 0$ is finite because $\dot{V}(x)$ is continuous in x and $\mathcal{C}$ is a compact set. Now we write $V(x(t))$ as

$$V(x(t)) = V(x(0)) + \int_0^t \dot{V}(x(s)) \, ds$$

$$\leq V(x(0)) - at,$$

which implies that

$$V(x(t)) = 0, \quad \forall\, t \geq \frac{V(x(0))}{a},$$

and

$$x(t) = 0, \quad \forall\, t \geq \frac{V(x(0))}{a}.$$

This contradicts the assumption that $x(t)$ does not converge to 0. $\qquad\square$

The Lyapunov global asymptotic stability theorem requires that $\dot{V}(x) \neq 0$ for any $x \neq 0$. In the case $\dot{V}(x) = 0$ for some $x \neq 0$, global asymptotic stability can be studied using Lasalle's invariance principle. The proof of the theorem is omitted in this book.

Theorem 2.3.3 (Lasalle's invariance principle) Replace condition (2) of Theorem 2.32 by

$$\dot{V}(x) \leq 0, \quad \forall x,$$

and suppose that the only trajectory $x(t)$ that satisfies

$$\dot{x}(t) = f(x(t)) \quad \text{and} \quad \dot{V}(x(t)) = 0, \quad \forall t,$$

is $x(t) = 0$, $\forall t$. Then $x = 0$ is globally, asymptotically stable. $\qquad\square$

2.4   Distributed algorithms: primal solution

In Section 2.2, we formulated the resource allocation problem as a convex optimization problem. However, the technique used to solve the optimization problem in Example 2.2.1 assumed that we had complete knowledge of the topology and routes. Clearly this is infeasible in a giant network such as the Internet. In this section and the next, we will study distributed algorithms which only require limited information exchange among the sources and the network for implementation.

The approach in this section is called the primal solution. Instead of imposing a strict capacity constraint on each link, we append a cost to the sum network utility:

$$W(x) = \sum_{r \in \mathcal{S}} U_r(x_r) - \sum_{l \in \mathcal{L}} B_l\left(\sum_{s:l \in s} x_s\right). \tag{2.20}$$

Here, x is the vector of rates of all sources and $B_l(\cdot)$ is the cost or price of sending data on link l. Thus, $W(x)$ represents a tradeoff: increasing the data rates x results in increased utility, but there is a price to be paid for the increased data rates at the links. If B_l is interpreted as a "barrier" function associated with link l, it should be chosen so that it increases to infinity when the arrival rate on link l approaches the link capacity c_l. Thus, it will ensure that the arrival rate is smaller than the capacity of a link. If B_l is interpreted as a "penalty" function which penalizes the arrival rate for exceeding the link capacity, rates slightly larger than the link capacity may be allowable, but this will result in packet losses over the link. One can also interpret c_l as a virtual capacity of the link which is smaller than the real capacity, in which case, even if the arrival rate exceeds c_l, one may still be operating within the link capacity. While it is not apparent in the deterministic formulation here, later in the book we will see that, even when the arrival rate on a link is less than its capacity, due to randomness in the arrival process, packets in the network will experience delay or packet loss. The function $B_l(\cdot)$ may thus be used to represent average delay, packet loss rate, etc.

We assume that B_l is a continuously differentiable convex function so that it can be written equivalently as

$$B_l\left(\sum_{s:l\in s} x_s\right) = \int_0^{\sum_{s:l\in s} x_s} f_l(y)dy, \tag{2.21}$$

where $f_l(\cdot)$ is an increasing, continuous function. We call $f_l(y)$ the *congestion price function*, or simply the price function at link l, since it associates a price with the level of congestion on the link. It is straightforward to see that B_l defined in the above fashion is convex, since integrating an increasing function results in a convex function (see Result 2.1.1). Note also that the convexity of B_l ensures that the function (2.20) is a strictly concave function.

We will assume that U_r and f_l are such that the maximization of (2.20) results in a solution with $x_r > 0$, $\forall r \in \mathcal{S}$. Then, the first-order condition for optimality states that the maximizer of (2.20) must satisfy

$$U'_r(x_r) - \sum_{l:l\in r} f_l\left(\sum_{s:l\in s} x_s\right) = 0, \qquad r \in \mathcal{S}. \tag{2.22}$$

Clearly, it is not practically feasible to solve (2.22) offline and implement the resulting data rates in the network since, as mentioned earlier, the topology of the network is unknown. Therefore, we will develop a decentralized algorithm under which each user can collect limited information from the network and solve for its own optimal data rate. A natural candidate for such an algorithm is the so-called gradient ascent algorithm from optimization theory. The basic idea behind the gradient ascent algorithm is intuitive, especially if the concave function is a function of one variable: since a concave function has a derivative which is a decreasing function and the optimal solution is obtained at the point where the derivative is zero, it makes sense to seek a solution by moving in the direction of the derivative. More generally, for a function of many variables, the gradient ascent algorithm suggests moving in the direction of the gradient.

Consider the algorithm

$$\dot{x}_r = k_r(x_r)\left(U'_r(x_r) - \sum_{l:l\in r} f_l\left(\sum_{s:l\in s} x_s\right)\right). \tag{2.23}$$

The right-hand side of the above differential equation is simply the derivative of (2.20) with respect to x_r, while $k_r(\cdot)$ is simply a step-size parameter which determines how far one moves in the direction of the gradient. The scaling function $k_r(\cdot)$ must be chosen such that the equilibrium of the differential equation is the same as the optimal solution to the resource allocation problem. For example, if $k_r(x_r) > 0$, setting $\dot{x}_r = 0$ for all r yields the same set of equations as (2.22). Algorithm (2.23) is called a *primal algorithm* since it arises from the primal formulation of the utility maximization problem. Note that the primal algorithm is a congestion control algorithm for the following reasons: when the route price $q_r = \sum_{l:l\in r} f_l(\sum_{s:l\in s} x_s)$ is large, the congestion controller decreases its transmission rate. Further, if x_r is large, $U'(x_r)$ is small (since $U_r(x_r)$ is concave) and thus the rate of increase is small. Thus, the network can be viewed as a control system with the network providing the feedback to allow the sources to adjust their rates. Control systems are typically viewed as block diagrams, and to visualize our congestion control algorithm as a block diagram, we introduce some notation.

Let R be the routing matrix of the network, i.e., the (l, r) element of this matrix is given by

$$R_{lr} = \begin{cases} 1, & \text{if route } r \text{ uses link } l, \\ 0, & \text{otherwise.} \end{cases}$$

Let

$$y_l = \sum_{s:l\in s} x_s \qquad (2.24)$$

be the load on link l. Thus,

$$y_l = \sum_s R_{ls} x_s.$$

Letting y be the vector of all y_l ($l \in \mathcal{L}$), we have

$$y = Rx. \qquad (2.25)$$

Let $p_l(t)$ denote the price of link l at time t, i.e.,

$$p_l(t) = f_l\left(\sum_{s:l\in s} x_s(t)\right)$$

$$= f_l(y_l(t)). \qquad (2.26)$$

Then the price of a route is just the sum of link prices p_l of all the links in the route. So we define the price of route r to be

$$q_r = \sum_{l:l\in r} p_l. \qquad (2.27)$$

Also let p be the vector of all link prices and q be the vector of all route prices. We thus have

$$q = R^T p. \qquad (2.28)$$

This notation allows us to depict the primal algorithm as a block diagram, as shown in Figure 2.8.

We now establish the global asymptotic stability of (2.23) using the Lyapunov technique described in Section 2.3. Since $W(x)$ is a strictly concave function, it has a unique minimizer $\hat{x}$. Further, $W(\hat{x}) - W(x)$ is non-negative and is equal to zero only at $x = \hat{x}$. Thus, $W(\hat{x}) - W(x)$ is a natural candidate Lyapunov function for the system (2.23). We use this Lyapunov function to prove the following theorem.

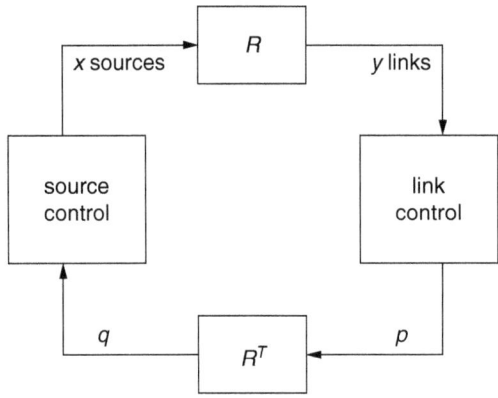

Figure 2.8 A block diagram view of the congestion control algorithm. The controller at the source uses congestion feedback from the links to perform its action.

Theorem 2.4.1 Consider a network in which all sources adjust their data rates according to the primal control algorithm (2.23). Define $V(x) = W(\hat{x}) - W(x)$, where $W(x)$ is given by (2.20). Assume that the functions $U_r(\cdot)$, $k_r(\cdot)$ and $f_l(\cdot)$ are such that $V(x) \to \infty$ as $||x|| \to \infty$, $\hat{x}_i > 0$ for all i, and the equilibrium point of (2.23) is the maximizer of (2.20). Then, the controller in (2.23) is globally asymptotically stable.

Proof Differentiating $V(\cdot)$, we get

$$\dot{V} = -\sum_{r \in \mathcal{S}} \frac{\partial V}{\partial x_r} \dot{x}_r = -\sum_{r \in \mathcal{S}} k_r(x_r) \left(U_r'(x_r) - q_r \right)^2 < 0, \quad \forall x \neq \hat{x}, \qquad (2.29)$$

and $\dot{V} = 0$ if $x = \hat{x}$. Thus, all the conditions of the Lyapunov theorem are satisfied, and so the system state will converge to $\hat{x}$, starting from any initial condition. $\qquad \square$

In the proof of Theorem 2.4.1, we have assumed that the utility, price, and scaling functions are such that $W(x)$ satisfies the conditions required to apply the Lyapunov stability theorem. It is easy to find functions that satisfy these properties. For example, if $U_r(x_r) = w_r \log(x_r)$ and $k_r(x_r) = x_r$, the primal congestion control algorithm for source r becomes

$$\dot{x}_r = w_r - x_r \sum_{l:l \in r} f_l(y_l),$$

and thus the unique equilibrium point can be obtained by solving $w_r/x_r = \sum_{l:l \in r} f_l(y_l)$. Further, if $f_l(\cdot)$ is such that $B_l(\cdot)$ is a polynomial function, $W(x)$ goes to $-\infty$, as $||x|| \to \infty$, and thus $V(x) \to \infty$ as $||x|| \to \infty$.

2.4.1 Congestion feedback and distributed implementation

For a congestion control algorithm to be useful in practice, it should be amenable to decentralized implementation. We now present one possible manner in which the primal algorithm could be implemented in a distributed fashion. We first note that each source simply needs to know the sum of the link prices on its route to adjust its data rate as suggested by the algorithm. Suppose that every packet has a field (a certain number of bits) set aside in its header to collect the price of its route. When the source releases a packet into the network, the price field can be set to zero. Then, each link on the route can add its price to the price field so that, by the time the packet reaches its destination, the price field will contain the route price. This information can then be fed back to the source to implement the congestion control algorithm. A noteworthy feature of the congestion control algorithm is that the link prices depend only on the total arrival rate to the link, and not on the individual arrival rates of each source using the link. Thus, each link has only to keep track of the total arrival rate to compute the link price. If the algorithm required each link to keep track of individual source arrival rates, it would be infeasible to implement since the number of sources using high-capacity links could be prohibitively large. Thus, the primal algorithm is both amenable to a distributed implementation and has low overhead requirements.

Packet headers in the Internet are already crowded with a lot of other information, such as source/destination addresses to facilitate routing, so Internet practitioners do not like to add another field in the packet header to collect congestion information. In view of this, the overhead required to collect congestion information can be further reduced to accommodate practical realities. Consider the extreme case where there is only one bit available in the packet header to collect congestion information. Suppose that each packet is *marked* with probability $1 - e^{-p_l}$ when the packet passes through link l. Marking simply means that a bit in the packet header is flipped from a 0 to a 1 to indicate congestion. Then, along a route r, a packet is marked with probability

$$1 - e^{\sum_{l:l\in r} p_l}.$$

If the acknowledgement for each packet contains one bit of information to indicate whether a packet is marked or not, then, by computing the fraction of marked packets, the source can compute the route price $\sum_{l:l\in r} p_l$. The assumption here is that the x_r's change slowly so that each p_l remains roughly constant over many packets. Thus, one can estimate p_l approximately.

While marking, as mentioned above, has been widely studied in the literature, the predominant mechanism used for congestion feedback in the Internet today is through packet drops. Buffers used to store packets at a link have finite capacity, and therefore a packet that arrives at a link when its buffer is full is dropped immediately. If a packet is dropped, the destination of the packet will not receive it. So, if the destination then provides feedback to the source that a packet was not received, this provides an indication of congestion. Clearly, such a scheme does not require even a single bit in the packet header to collect congestion information. However, strictly speaking, this type of congestion feedback cannot be modeled using our framework since we assume that a source's data rate is seen by all links on the route, whereas, if packet dropping is allowed, some packets will not reach all links on a source's route. However, if we assume that the packet loss rate at each link is small, we can approximate the rate at which a link receives a source's packet by the rate at which the source is transmitting packets. Further, the end-to-end drop probability on a route can be approximated by the sum of the drop probabilities on the links along the route if the drop probability at each link is small. Thus, the optimization formulation approximates reality under these assumptions. To complete the connection to the optimization framework, we have to specify the price function at each link. A crude approximation to the drop probability (also known as packet loss rate) at link l is $((y_l - c_l)/y_l)^+$, which is non-zero only if $y_l = \sum_{r:l\in r} x_r$ is larger than c_l. This approximate formula for the packet loss rate can serve as the price function for each link.

2.5 Distributed algorithms: dual solution

In this section we consider another distributed algorithm based on the dual formulation of the utility maximization problem. Consider the resource allocation problem that we would like to solve,

$$\max_{x_r} \sum_{r\in\mathcal{S}} U_r(x_r), \tag{2.30}$$

subject to the constraints

$$\sum_{r:l\in r} x_r \leq c_l, \quad \forall l \in \mathcal{L}, \tag{2.31}$$

$$x_r \geq 0, \quad \forall r \in \mathcal{S}. \tag{2.32}$$

The Lagrange dual of the above problem is obtained by incorporating the constraints into the maximization by means of Lagrange multipliers as follows:

$$D(p) = \max_{\{x_r \geq 0\}} \sum_r U_r(x_r) - \sum_l p_l \left(\sum_{s:l\in s} x_s - c_l \right). \tag{2.33}$$

Here the p_l's are the Lagrange multipliers that we saw in Section 2.1. The dual problem may then be stated as

$$\min_{p\geq 0} D(p).$$

As in the case of the primal problem, we would like to design an algorithm that ensures that all the source rates converge to the optimal solution. Note that, in this case, we are looking for a gradient descent (rather than the gradient ascent we saw in the primal formulation), since we would like to minimize $D(p)$. To find the direction of the gradient, we need to know $\partial D/\partial p_l$.

We first observe that, in order to achieve the maximum in (2.33), x_r must satisfy

$$U_r'(x_r) = q_r, \tag{2.34}$$

where, as usual, $q_r = \sum_{l:l\in r} p_l$ is the price of a particular route r. Note that we have assumed that $x_r > 0$ in writing down (2.34). This would be true, for example, if the utility function is an α-utility function with $\alpha > 0$. Now,

$$\frac{\partial D}{\partial p_l} = \sum_r U_r'(x_r)\frac{\partial x_r}{\partial p_l} - (y_l - c_l) - \sum_k p_k \frac{\partial y_k}{\partial p_l}$$

$$= \sum_r U_r'(x_r)\frac{\partial x_r}{\partial p_l} - (y_l - c_l) - \sum_k p_k \sum_{r:k\in r} \frac{\partial x_r}{\partial p_l}$$

$$= \sum_r U_r'(x_r) \frac{\partial x_r}{\partial p_l} - (y_l - c_l) - \sum_r \frac{\partial x_r}{\partial p_l} \sum_{k:k \in r} p_k$$

$$= \sum_r U_r'(x_r) \frac{\partial x_r}{\partial p_l} - (y_l - c_l) - \sum_r \frac{\partial x_r}{\partial p_l} q_r.$$

Thus, using (2.34), we have

$$\frac{\partial D}{\partial p_l} = -(y_l - c_l). \tag{2.35}$$

Recalling that to minimize $D(p)$ we have to *descend* in the direction of the gradient, from (2.34) and (2.35) we have the following dual control algorithm:

$$x_r = U_r'^{-1}(q_r) \tag{2.36}$$

and

$$\dot{p}_l = h_l(y_l - c_l)_{p_l}^+, \tag{2.37}$$

where $h_l > 0$ is a constant and $(g(x))_y^+$ denotes

$$(g(x))_y^+ = \begin{cases} g(x), & y > 0, \\ \max(g(x), 0), & y = 0. \end{cases}$$

Note that p_l is not allowed to become negative because the KKT conditions impose such a condition on the Lagrange multipliers. Also, we remark that p_l has the same dynamics of the queue length at the link when $h_l \equiv 1$. It increases at rate y_l, which is the arrival rate, and decreases at rate c_l, the capacity of the link. Thus, the links do not have to compute their dual variables explicitly when $h_l \equiv 1$; the dual variables are simply the queue lengths.

The stability of this algorithm follows in a manner similar to that of the primal algorithm by considering $D(p)$ as the Lyapunov function, since the dual algorithm is simply a gradient algorithm for finding the minimum of $D(p)$.

In Chapter 7, we will discuss practical TCP protocols based on the primal and dual formulations. When we discuss these protocols, we will see that the price functions and congestion control mechanisms obtained from the two formulations have different interpretations.

2.6   Feedback delay and stability

We have seen in Sections 2.4 and 2.5 that the primal and dual algorithms are globally, asymptotically stable if link prices were fed back *instantaneously* to sources and rate adjustments at sources were reflected *instantaneously* at links. Both assumptions are not true in reality due to delays. In this section, to illustrate how to analyze congestion control algorithms in the presence of delays, we will consider a simple one-link network and study the impact of feedback delay on the stability of a primal congestion controller.

Consider a simple system with one link of capacity c and one source with utility function $\log x$. In this case, the proportionally fair congestion controller derived in Section 2.4 becomes

$$\dot{x} = k(x)\left(\frac{1}{x} - f(x)\right).$$

Choosing $k(x) = \kappa x$ for some $\kappa > 0$, so that when x is close to zero the rate of increase is bounded, this controller can be written as

$$\dot{x} = \kappa(1 - xf(x)). \tag{2.38}$$

Suppose that $f(x)$ is the loss probability or marking probability when the arrival rate at the link is x. Then, the congestion control algorithm can be interpreted as follows: increase x at rate κ and decrease it proportional to the rate at which packets are marked, with κ being the proportionality constant.

In reality, there is a delay from the time at which a packet is released at the source to the time at which it reaches the link, called the forward delay, and a delay for any feedback to reach the source from the link, called the backward delay. So, one cannot implement the above congestion controller in the form (2.38). Let us denote the forward delay by T_f and the backward delay by T_b. Taking both delays into consideration, we have

$$y(t) = x(t - T_f)$$

and

$$\begin{aligned} q(t) &= p(t - T_b) \\ &= f(y(t - T_b)) \\ &= f(x(t - T_f - T_b)) \\ &= f(x(t - T)), \end{aligned}$$

where $T = T_f + T_b$ is called the Round-Trip Time (RTT). The congestion controller becomes the following *delay differential equation*:

$$\dot{x} = \kappa(1 - x(t - T)f(x(t - T))). \tag{2.39}$$

Let $\hat{x}$ be the unique equilibrium point of this delay differential equation, i.e.,

$$1 - \hat{x}f(\hat{x}) = 0.$$

To analyze this system, we will assume that $x(t)$ is close to the equilibrium and derive conditions under which the system is asymptotically stable, i.e.,

$$z(t) \triangleq x(t) - \hat{x}$$

converges to zero as $t \to \infty$. To this end, we will linearize the system around the equilibrium point and derive conditions under which the resulting linear delay differential equation is asymptotically stable.

Clearly, we would like to study the global, asymptotic stability of the system instead of just the asymptotic stability assuming that the system is close to its equilibrium, but typically studying the global, asymptotic stability is a much harder problem when there are many links and many sources.

2.6.1   Linearization

To study the asymptotic stability of (2.39), we substitute $x(t) = \hat{x} + z(t)$ and linearize the delay differential equation as follows:

$$\dot{z} = \kappa \left(1 - (\hat{x} + z(t-T))f(\hat{x} + z(t-T))\right)$$
$$\approx \kappa \left(1 - (\hat{x} + z(t-T))(f(\hat{x}) + f'(\hat{x})z(t-T))\right),$$

where in the second line we only retain terms linear in z in Taylor's series. The rationale is that, since x is close to $\hat{x}$, z is close to zero and $z = x - \hat{x}$ dominates z^2, z^3, etc. Now, using the equilibrium condition

$$1 - \hat{x}f(\hat{x}) = 0,$$

we obtain

$$\dot{z} = -\kappa(f(\hat{x}) + \hat{x}f'(\hat{x}))z(t-T), \tag{2.40}$$

where we have again dropped z^2 terms for the same reasons as before.

To understand the stability of the linear delay differential equation (2.40), we introduce the following theorem, which characterizes the necessary and sufficient condition for a general linear delay differential equation to be asymptotically stable.

Theorem 2.6.1 Consider a system governed by the following linear delay differential equation:

$$\dot{x}(t) = -ax(t) - bx(t-T), \tag{2.41}$$

where the initial condition $x(t)$, $-T \le t \le 0$, is specified. For any choice of the initial condition, $x(t)$ converges to zero (the unique equilibrium of the system) as $t \to \infty$ if and only if there exists a χ such that

$$0 < \chi < \frac{\pi}{T}, \quad a = -c\cos\chi T, \quad c = \frac{\chi}{\sin\chi T}, \quad \text{and} \quad -a \le b \le c. \tag{2.42}$$

If $a = 0$, the condition simplifies to

$$0 \le b \le \frac{\pi}{2T}. \qquad \qquad \square$$

The shaded area in Figure 2.9 is the set of a and b that satisfies condition (2.42). Theorem 2.6.1 is a well-known result in the field of delay differential equations, and we will not prove it here. The following proposition is an immediate result of the theorem.

Proposition 2.6.2 The linearized version of the proportionally fair controller, given in (2.40), is asymptotically stable if and only if

$$\kappa T(f(\hat{x}) + \hat{x}f'(\hat{x})) \le \frac{\pi}{2}. \tag{2.43}$$

$$\square$$

Equation (2.43) suggests that the parameter κ should be chosen inversely proportional to T. This means that the congestion control algorithm should react more slowly when the feedback delay is large. This is very intuitive since, when T is large, the algorithm is

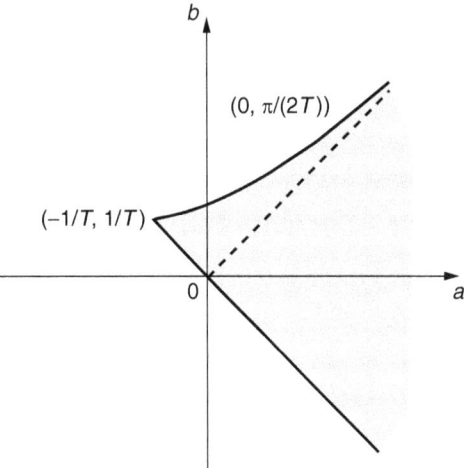

Figure 2.9 The set of a and b that satisfies condition (2.42).

reacting to events that occurred a long time ago and thus should change the source rate x very slowly in response to such old information.

In this section, we considered the problem of deriving delay differential equation models for congestion control protocols operating on a single link accessed by a single source. The modeling part can be easily extended to general networks with an arbitrary number of links and sources. However, the stability analysis of the resulting equations is considerably more challenging than in the case of the simple single-source, single-link model. We will not address this problem in this book, but we provide references in the Notes section at the end of the chapter.

2.7 Game-theoretic view of utility maximization

In earlier sections, we used utility maximization as a tool to understand network archi-tecture and algorithms. In doing so, it was assumed that all users will act as a team to maximize total network utility. It is interesting to relax this assumption and understand what happens if each user attempts selfishly to maximize its own utility minus any price incurred for transmitting at a certain data rate. In this section, we will first introduce a pric-ing scheme called the *VCG mechanism*, under which no user has an incentive to lie about their utility function. Therefore, the network can learn the true utility functions and thus solve the network utility maximization problem. However, the VCG mechanism requires each user to convey its utility function to the network. We will show that this communica-tion burden can be lessened under a reasonable assumption. Specifically, it will be shown that, in a large network such as the Internet, the selfish goals of the users coincide with the system-wide goal if users are *price-takers*, i.e., they take the price given by the network without attempting to infer the impact of their actions on the network. This is a reason-able assumption for the Internet since no one user can impact the network significantly and hence cannot use any reasonable inference algorithm to determine the impact of its own actions on the network price. On the other hand, it is also of theoretical interest to understand how far away we are from the team-optimal solution if users are strategic, i.e., the users are *not price-takers*. We will address this issue in this section and present a lower bound on the total network utility if the users are strategic.

2.7.1   The Vickrey–Clarke–Groves mechanism

Recall that the goal of the utility maximization problem introduced in Section 2.2 was to solve the following optimization problem:

$$\max_{x \geq 0} \sum_r U_r(x_r)$$

subject to

$$\sum_{r:l \in r} x_r \leq c_l, \qquad \forall l,$$

where x_r is the rate allocated to user r, U_r is the user's utility function, and c_l is the capacity of link l.

In a large network such as the Internet, it is not feasible to solve the utility maximization problem in a centralized fashion. However, to understand the game-theoretic implications of the solution, let us suppose that the network is small and that a central network planner wants to solve the utility maximization problem. To solve the problem, the network planner first has to ask each user to reveal its utility function. However, there may be an incentive for a user to lie about its utility function; for example, by incorrectly reporting its utility function, a user may receive a larger data rate. Suppose that user r reveals its utility function as $\tilde{U}_r(x_r)$, which may or may not be the same as $U_r(x_r)$. Then, the network will solve the maximization problem

$$\max_{x \geq 0} \sum_r \tilde{U}_r(x_r)$$

subject to

$$\sum_{r:l \in r} x_r \leq c_l, \qquad \forall l,$$

and allocate the optimal rate $\tilde{x}_r$ to user r. Suppose that the user is charged a price q_r by the network for receiving this rate allocation. The goal is to select the price such that there is no incentive for each user to lie about its utility function. Surprisingly, the following simple idea works: user r is charged an amount equal to the reduction in the sum utility of the other users in the network due to the presence of user r. Specifically, the network first obtains the optimal solution $\{\bar{x}_s\}$ to the following problem, which assumes that user r is not present in the network:

$$\max_{x \geq 0} \sum_{s \neq r} \tilde{U}_s(x_s)$$

subject to

$$\sum_{s \neq r:l \in s} x_s \leq c_l, \qquad \forall l.$$

The price q_r is then computed as

$$q_r = \sum_{s \neq r} \tilde{U}(\bar{x}_s) - \sum_{s \neq r} \tilde{U}(\tilde{x}_s),$$

which represents the increase in the sum utility of other users due to the absence of user r. This pricing mechanism is called the Vickrey–Clarke–Groves (VCG) mechanism.

Let us now consider how users will report their utilities knowing that the network plans to implement the VCG mechanism. Clearly, each user r will announce a utility function such that its *payoff*, i.e., the utility minus the price, is maximized:

$$U_r(\tilde{x}_r) - q_r.$$

We will now see that the payoff obtained by user r when lying about its utility function is always less than or equal to the payoff obtained when it truthfully reports its utility function. In other words, we will show that there is no incentive to lie.

If user r reveals its utility function truthfully, its payoff is given by

$$\mathcal{U}^t = U_r(\tilde{x}_r^t) - \left(\sum_{s \neq r} \tilde{U}_s(\tilde{x}_s^t) - \sum_{s \neq r} \tilde{U}_s(\bar{x}_s^t) \right),$$

where $\{\tilde{x}_s^t\}$ is the allocation given to the users by the network and $\{\bar{x}_s^t\}$ is the solution of the network utility maximization problem when user r is excluded from the network, both being computed using the submitted utility functions. The superscript t indicates that user r has revealed its utility function truthfully. Next, suppose that user r lies about its utility function and denote the network planner's allocation by $\tilde{x}^l$, where superscript l indicates that user r has lied. Now, the payoff for user r is given by

$$\mathcal{U}^l = U_r(\tilde{x}_r^l) - \left(\sum_{s \neq r} \tilde{U}_s(\tilde{x}_s^l) - \sum_{s \neq r} \tilde{U}_s(\bar{x}_s^l) \right).$$

Since $\{\bar{x}_s^l\}$ is the allocation excluding user r, it is independent of the utility function submitted by user r.

If truth-telling were not optimal, $\mathcal{U}^l > \mathcal{U}^t$ for some incorrect utility function of user r, which would imply

$$U_r(\tilde{x}_r^l) + \sum_{s \neq r} \tilde{U}_s(\tilde{x}_s^l) > U_r(\tilde{x}_r^t) + \sum_{s \neq r} \tilde{U}_s(\tilde{x}_s^t).$$

The above expression contradicts the fact that $\tilde{x}^t$ is the optimal solution to

$$\max_{x \geq 0} U_r(x_r) + \sum_{s \neq r} \tilde{U}_s(x_s)$$

subject to the capacity constraints. It is worth noting that we have not made any assumptions about the strategies of other users in showing that user r has no incentive to lie. A strategy that is optimal for a user, independent of the strategies of others, is called a *dominant* strategy in game theory parlance. Thus, *truth-telling* is a dominant strategy under the VCG mechanism.

Example 2.7.1

Consider a single-link network shared by three users. The link has unit capacity, and the utility functions associated with the three users are: $U_0(x_0) = \log x_0$, $U_1(x_1) = 3 \log x_1$, and $U_2(x_2) = 4 \log x_2$. If all users reveal their true utility functions, the network planner solves

$$\max_{x_0 + x_1 + x_2 \leq 1} \log x_0 + 3 \log x_1 + 4 \log x_2,$$

which yields allocation

$$x_0^t = 1/8, \quad x_1^t = 3/8, \quad x_2^t = 1/2.$$

The price user 2 has to pay is given by

$$\left(\max_{x_0 + x_1 \leq 1} \log x_0 + 3 \log x_1 \right) - \left(\log x_0^t + 3 \log x_1^t \right)$$

$$= \log \frac{1}{4} + 3 \log \frac{3}{4} - \left(\log \frac{1}{8} + 3 \log \frac{3}{8} \right)$$

$$= 4 \log 2.$$

The payoff for user 2 is

$$4 \log \frac{1}{2} - 4 \log 2 = -8 \log 2.$$

Now, if user 2 lies about its utility and submits $U_2(x_2) = 2 \log x_2$, the network planner decides the resource allocation by solving

$$\max_{x_0 + x_1 + x_2 \leq 1} \log x_0 + 3 \log x_1 + 2 \log x_2,$$

which yields

$$x_0^l = 1/6, \quad x_1^l = 1/2, \quad x_2^l = 1/3.$$

The price user 2 has to pay is now given by

$$\left(\max_{x_0 + x_1 \leq 1} \log x_0 + 3 \log x_1 \right) - \left(\log x_0^l + 3 \log x_1^l \right)$$

$$= \log \frac{1}{4} + 3 \log \frac{3}{4} - \left(\log \frac{1}{6} + 3 \log \frac{1}{2} \right)$$

$$= 4 \log \frac{3}{2}.$$

The payoff for user 2 in this case is

$$4 \log \frac{1}{3} - 4 \log \frac{3}{2} = -4 \log \frac{9}{2},$$

which is smaller than $-8 \log 2$, the payoff when user 2 submits its true utility. □

Example 2.7.2

Consider an auction with a single item and n players. The true value of the item to player r is denoted by w_r, and the value bid by player r is denoted by $\tilde{w}_r$. A scheme called *Vickrey's second price auction* allocates the item to the bidder who has the highest bid, say bidder r^*, and charges the bidder $\max_{r \neq r^*} \tilde{w}_r$, i.e., the second highest bid. We will next show that Vickrey's second price auction is a special case of the VCG algorithm.

Define the utility function for bidder r to be $w_r x_r$, where $x_r \in \{0, 1\}$, $x_r = 1$, indicates that bidder r wins the item and $x_r = 0$ otherwise. Then, after obtaining the bids $\{\tilde{w}_r\}$, the VCG algorithm solves the utility maximization problem

$$\max_{x_r} \sum_{r=1}^{n} \tilde{w}_r x_r$$

$$\text{subject to} \quad \sum_{r=1}^{n} x_r \leq 1, \quad x_r \in \{0, 1\}, \quad \forall r.$$

It is easy to verify that the maximum is

$$\max_r \tilde{w}_r = \tilde{w}_{r^*}.$$

So the VCG algorithm allocates the item to the bidder with the highest bid ($x_{r^*}^* = 1$ and $x_r^* = 0$ for $r \neq r^*$). The price user r has to pay is

$$\left(\max \sum_{r \neq r^*} \tilde{w}_r x_r \right) - \sum_{r \neq r^*} \tilde{w}_r x_r^* = \max_{r \neq r^*} \tilde{w}_r,$$

the second highest bid. So Vickrey's second price auction is a special case of the VCG algorithm. $\square$

It is useful now to discuss the practicality of the VCG mechanism. While it is true that there is no incentive to lie under the VCG mechanism, it imposes an unreasonable computational burden on the network, since the network has to solve several optimization problems to compute the price for each user. In Section 2.7.2, we show that a simpler mechanism exists, under which truth-telling is optimal, provided we make an assumption about users' behaviors. The mechanism we will develop is related to the theory developed in Section 2.5.

2.7.2 The price-taking assumption

Suppose that the users do not announce their utility functions and simply react to the price announced by the network in a myopic fashion. We will further assume that users cannot anticipate the impact of their actions on the price set by the network. Such users are called *price-taking* users. Price-taking is a reasonable assumption in a network such as the Internet, where the number of users is in the millions, and hence any one user cannot possibly estimate its impact on the network.

Suppose the network charges q dollars for unit rate (bits per second) transferred by any user in the network. For price-taking users, each user will attempt to maximize its net utility assuming that q is a constant, i.e.,

$$\max_{x_r} U_r(x_r) - x_r q.$$

Thus, if $U_r'(x_r) \to \infty$ as $x_r \to 0$, the optimal value of x_r is strictly greater than zero and is given by

$$U_r'(x_r) = q.$$

The network then may use the following differential equation to compute q as a function of time t:

$$\dot{q} = \left(\sum_r x_r - c \right)^+_q.$$

So we have the following algorithm

$$x_r = U_r'^{-1}(q),$$

$$\dot{q} = \left(\sum_r x_r - c \right)^+_q.$$

Recall that these equations represent the dual algorithm for the network utility maximization problem, and we have already shown that the dual algorithm converges to the solution of the optimization problem under reasonable conditions. So the selfish goals of the users coincide with the system-wide goal if the users are price-taking.

2.7.3 Strategic or price-anticipating users

Although in the context of the Internet it is reasonable to assume that users are price-taking, it is nevertheless interesting to understand the impact of removing this assumption. In particular, in the case of price-taking users, we noted that the solution to the users' selfish net utility maximization problems coincided with the solution to the network utility maximization problem. On the other hand, if users anticipate the impact of their actions on the price of the resource, individual user problems' solutions need not coincide with the global optimum of the network utility maximization problem. Therefore, it is of interest to characterize the ratio of the sum utility obtained by strategic (price-anticipating) users to the sum utility obtained by network utility maximization. This ratio is called the *Price of Anarchy* (PoA) as it compares anarchy (a situation in which users are competing against each other) with a tranquil central optimization solution.

Again, we assume that user r has a utility function $U_r(x_r)$, where $U_r(0) \geq 0$ and $U_r(\cdot)$ is twice differentiable, concave, and increasing in $[0, \infty)$.

Each user r bids w_r (dollars per second) that it is willing to pay instead of submitting its entire utility function. Since the network does not know the utility functions, it uses the bids as weights and allocates $\{x_r\}$ according to weighted proportional fairness:

$$\max_{x_r \geq 0} \sum_r w_r \log x_r$$

$$\text{subject to} \quad \sum_r x_r \leq c.$$

The solution to the above optimization problem is

$$x_r^*(w_r, w_{-r}) = \frac{w_r c}{\sum_s w_s},$$

assuming at least one $w_r > 0$, where $w_{-r} = (w_1, \ldots, w_{r-1}, w_{r+1}, \ldots, w_R)$ and R is the number of users. If $w_r = 0$, $\forall r$, let us say $x_r^*(w_r, w_{-r}) = 0$. Therefore, the network charges a price of

$$q = \frac{\sum_r w_r}{c}$$

dollars for unit transmission rate (bits per second), and so user r can transmit with rate $x_r = w_r/q$ when paying w_r. This pricing mechanism is called the *Kelly mechanism*. The net utility that user r realizes is given by

$$U_r\left(\frac{w_r}{q}\right) - w_r. \tag{2.44}$$

This pricing mechanism is of particular interest because, when the users are price-taking, the equilibrium point of the system coincides with the network-wide optimal solution. Recall that if the users are price-taking, each user will attempt to maximize its net utility assuming that q is a constant. The optimal value of w_r is given by

$$U_r'\left(\frac{w_r}{q}\right) = q. \tag{2.45}$$

The equilibrium point of the Kelly mechanism with price-taking users, therefore, is $\{\hat{x}, \hat{q}\}$, such that

$$U_r'\left(\frac{\hat{w}_r}{\hat{q}}\right) = \hat{q}, \tag{2.46}$$

$$\hat{q} = \frac{\sum_r \hat{w}_r}{c}. \tag{2.47}$$

We claim that the resulted resource allocation $\{\hat{x}_r = \hat{w}_r/q\}$ solves the network utility maximization problem $\sum_r U_r(x_r)$. To see this, note that the KKT optimality conditions for the network utility maximization problem are given by

$$U_r'(x_r) = q, \tag{2.48}$$

$$\sum_r x_r = c. \tag{2.49}$$

Note that (2.48) and (2.49) are satisfied by (2.46) and (2.47), respectively, by recalling that $x_r = w_r/q$. Therefore, the Kelly mechanism chooses prices in such a manner that the optimal bid by a price-taking user results in data rates that solve the network utility maximization problem. We comment that, in this subsection, we focus on the efficiency of equilibrium points, and will not discuss how to reach these equilibrium points.

Now consider strategic users who anticipate the impact of their actions on the price of the resource to see whether the selfish goals again coincide with the system goal. Given the Kelly mechanism, how much should each user bid? If the users are rational, we can expect users to settle on bids $\{w_r\}$ such that no user has an incentive to deviate from its bid, i.e., bids $\{\hat{w}_r\}$ such that

$$\hat{w}_r \in \arg\max_{w_r \geq 0} U_r(x_r^*(w_r, \hat{w}_{-r})) - w_r. \tag{2.50}$$

Such a set of $\{\hat{w}_r\}$ is called a Nash Equilibrium (NE). It is not obvious whether a Nash equilibrium exists, and, if it exists, it may not be unique.

Lemma 2.7.1 There exists a $\{\hat{w}_r\}$ satisfying (2.50), i.e., there exists a unique NE.

Proof The definition of NE in (2.50) implies that no user has an incentive to deviate from its bid. Using this property, we will first show that, at a NE, at least two users must have non-zero bids. If no user has a non-zero bid, then a particular user r can increase its bid by a small amount ϵ and get the entire resource. So the all-zero bid cannot be an equilibrium. If only one user has a non-zero bid, then it has an incentive to decrease its bid arbitrarily close to zero, while still getting the entire resource. So this cannot be a NE either. In view of this,

$$x_r^*(w_r, \hat{w}_{-r}) = \frac{w_r c}{\sum_{s \neq r} \hat{w}_s + w_r}$$

is well defined since the denominator cannot be equal to zero. Assuming U_r is differentiable, a necessary condition for $\hat{w}_r$ to satisfy (2.50) is given by

$$U_r'\left(x_r^*(\hat{w}_r, \hat{w}_{-r})\right) \left(\frac{c}{\sum_s \hat{w}_s} - \frac{\hat{w}_r c}{\left(\sum_s \hat{w}_s\right)^2}\right) - 1 = 0, \quad \text{if} \quad \hat{w}_r > 0,$$

and

$$U_r'(0) \left(\frac{c}{\sum_s \hat{w}_s}\right) - 1 \leq 0, \quad \text{if} \quad \hat{w}_r = 0,$$

which implies

$$\begin{cases} U_r'(x_r^*(\hat{w}_r, \hat{w}_{-r})) \left(1 - \frac{\hat{x}_r^*(\hat{w}_r, \hat{w}_{-r})}{c}\right) = \frac{\sum_s \hat{w}_s}{c}, & \text{if } \hat{w}_r > 0, \\ U_r'(0) \leq \frac{\sum_s \hat{w}_s}{c}, & \text{if } \hat{w}_r = 0. \end{cases} \tag{2.51}$$

If we can show that there exists a solution to (2.51), then we have proved the lemma.

Consider the following optimization problem:

$$\max_{x_r \geq 0} \quad \sum_r \int_0^{x_r} U_r'(y) \left(1 - \frac{y}{c}\right) dy \tag{2.52}$$

$$\text{subject to} \quad \sum_r x_r \leq c.$$

The objective $\int_0^{x_r} U_r'(y) (1 - y/c) \, dy$ is concave since $U_r'(y) (1 - y/c)$ is a decreasing function. Thus, by the KKT theorem, there exists $\lambda \geq 0$ such that the optimal solution $\{z_r\}$ to (2.52) satisfies

$$\sum_r z_r \leq c,$$

and

$$\begin{cases} U_r'(z_r)\left(1 - \frac{z_r}{c}\right) - \lambda = 0, & \text{if } z_r > 0, \\ U_r'(0) \leq \lambda, & \text{if } z_r = 0. \end{cases} \tag{2.53}$$

We will prove the existence of a NE by showing that a solution to (2.53) can be used to generate a NE. Since (2.53) is simply the solution to (2.52), and a solution to (2.52) exists, this will complete the proof of the lemma.

First, we note that $\sum_r z_r = c$ since the objective in (2.52) is increasing in each x_r. Let $\lambda = \sum_s \hat{w}_s/c$, and solve (2.53) to obtain $\{z_r\}$. It is easy to see that (2.51) is satisfied with $x_r^* = z_r$. It is also not difficult to see that $\lambda z_r > 0$ for at least two users by examining (2.53) and using the condition

$$\sum_r z_r = c.$$

To see this, note that all z_r's cannot be zero since $\sum_r z_r = c$. Further if only one $z_r > 0$, then it must be equal to c, in which case $\lambda = 0$ from the first equation in (2.53), which makes all z_r's equal to zero.   □

Next we prove the main theorem, which shows that the network utility under strategic users is at least 75% of the maximum network utility under price-taking users.

Theorem 2.7.2 Let $\{x_r^*\}$ be the solution to

$$\max_{x_r \geq 0} \quad \sum_r U_r(x_r)$$

$$\text{subject to:} \quad \sum_r x_r \leq c.$$

Let $\hat{x}_r$ be the allocation under a NE. If $U_r(0) \geq 0$ for all r,

$$\frac{\sum_r U_r(\hat{x}_r)}{\sum_r U_r(x_r^*)} \geq \frac{3}{4}.$$

Note: The ratio $\sum_r U_r(\hat{x}_r) / \sum_r U_r(x_r^*)$ is the price of anarchy. The reason for this terminology is that we are comparing "anarchy," where everyone is competing with everyone else, to a cooperative optimal solution $\{\hat{x}_r\}$.

Proof The key idea behind the proof is to show that linear utility functions lower bound the PoA, and 3/4 is the lower bound on the PoA of linear utility functions.

By the concavity of U_r, we have

$$U_r(x_r^*) \leq U_r(\hat{x}_r) + U_r'(\hat{x}_r)(x_r^* - \hat{x}_r)$$

$$= (U_r(\hat{x}_r) - \hat{x}_r U_r'(\hat{x}_r)) + U_r'(\hat{x}_r)x_r^*$$

$$\leq (U_r(\hat{x}_r) - \hat{x}_r U_r'(\hat{x}_r)) + x_r^* \max_r U_r'(\hat{x}_r),$$

which implies that

$$\sum_r U_r(x_r^*) \leq \left(\sum_r (U_r(\hat{x}_r) - \hat{x}_r U_r'(\hat{x}_r))\right) + c \max_r U_r'(\hat{x}_r), \tag{2.54}$$

where we have used the fact that

$$\sum_r x_r^* = c.$$

Next, by adding and subtracting $\sum_r \hat{x}_r U_r'(\hat{x}_r)$, we have

$$\sum_r U_r(\hat{x}_r) = \sum_r (U_r(\hat{x}_r) - \hat{x}_r U_r'(\hat{x}_r)) + \sum_r \hat{x}_r U_r'(\hat{x}_r). \qquad (2.55)$$

However, by the concavity of U_r and the assumption that $U_r(0) \geq 0$,

$$U_r(0) \leq U_r(\hat{x}_r) + (0 - \hat{x}_r) U_r'(\hat{x}_r),$$

which yields

$$U_r(\hat{x}_r) - \hat{x}_r U_r'(\hat{x}_r) \geq 0, \qquad (2.56)$$

since we have assumed $U_r(0) \geq 0$, $\forall r$. By (2.54), (2.55), and (2.56), we conclude that

$$\text{PoA} \geq \sum_r \left(\frac{U_r'(\hat{x}_r)}{\max_{\hat{x}_r} U_r'(\hat{x}_r)} \right) \frac{\hat{x}_r}{c},$$

where the inequality becomes equality when the utility functions are all linear and $U_r(0) = 0$ for all r.

We recall that $\hat{x}_r$ satisfies

$$U_r'(\hat{x}_r) \left(1 - \frac{\hat{x}_r}{c} \right) = \frac{\sum_s \hat{w}_s}{c}, \quad \text{if } \hat{x}_r > 0, \qquad (2.57)$$

$$U_r'(0) < \frac{\sum_s \hat{w}_s}{c}, \quad \text{if } \hat{x}_r = 0. \qquad (2.58)$$

Without loss of generality, let us assume

$$\max_{\hat{x}_r} U_r'(\hat{x}_r) = U_1'(\hat{x}_1).$$

If $\hat{x}_r > 0$, then $\hat{x}_1 \geq \hat{x}_r > 0$ from (2.57). Since $\hat{x}_r > 0$ for at least one r, we obtain

$$\hat{x}_1 > 0,$$

which further implies that

$$U_1'(x_1) \left(1 - \frac{\hat{x}_1}{c} \right) = \frac{\sum_s \hat{w}_s}{c}.$$

Hence, (2.57) and (2.58) can be written as

$$U_r'(\hat{x}_r) \left(1 - \frac{\hat{x}_r}{c} \right) = U_1'(\hat{x}_1) \left(1 - \frac{\hat{x}_1}{c} \right), \quad \text{if } \hat{x}_r > 0, \qquad (2.59)$$

and

$$U_r'(0) \leq U_1'(\hat{x}_1) \left(1 - \frac{\hat{x}_1}{c} \right), \quad \text{if } \hat{x}_r = 0. \qquad (2.60)$$

Replacing $U_r'(\hat{x}_r)$ using (2.59) for any r for which $\hat{x}_r \neq 0$, we obtain

$$\text{PoA} \geq \frac{\hat{x}_1}{c} + \sum_{\substack{r:r\neq 1 \\ \hat{x}_r \neq 0}} \frac{\left(1 - \frac{\hat{x}_1}{c}\right)\hat{x}_r}{\left(1 - \frac{\hat{x}_r}{c}\right)c},$$

where we have the following constraints:

$$\hat{x}_r \leq \hat{x}_1, \quad \forall r,$$

$$\sum_r \frac{\hat{x}_r}{c} = 1,$$

$$\hat{x}_r \geq 0.$$

Now define $y_r = \hat{x}_r/c$, so we can rewrite the PoA as

$$\text{PoA} \geq \min \quad y_1 + \sum_{r\neq 1} \frac{1 - y_1}{1 - y_r}y_r$$

$$\text{subject to:} \quad y_r \leq y_1, \quad \forall r, \tag{2.61}$$

$$\sum_{r=1}^{R} y_r = 1, \quad y_r \geq 0 \quad r = 1, 2, \ldots, R. \tag{2.62}$$

From (2.61) and (2.62), $y_1 \geq 1/R$. If we fix the value of y_1, the optimization problem becomes

$$\min \sum_{r\neq 1} \frac{y_r}{1 - y_r}, \quad \text{subject to} \sum_{r\neq 1} y_r = 1 - y_1.$$

The solution is given by $y_r = (1 - y_1)/(R - 1)$, so that

$$\text{PoA} \geq \min_{\frac{1}{R}\leq y_1 \leq 1} y_1 + \frac{(1 - y_1)^2}{R - 1} \sum_{r\neq 1} \frac{1}{1 - y_r}.$$

Since

$$1 - y_r = 1 - \frac{1 - y_1}{R - 1} = \frac{R - 2 + y_1}{R - 1},$$

we have

$$\sum_{r\neq 1} \frac{1}{1 - y_r} = \frac{(R - 1)^2}{R - 2 + y_1}$$

and

$$\text{PoA} \geq \min_{\frac{1}{R}\leq y_1 \leq 1} y_1 + \frac{(R - 1)(1 - y_1)^2}{R - 2 + y_1}.$$

Since $y_1 \leq 1$, we further obtain

$$\frac{R - 1}{R - 2 + y_1} \geq 1,$$

and

$$\text{PoA} \geq \min_{\frac{1}{R} \leq y_1 \leq 1} y_1 + (1 - y_1)^2$$

$$\geq \min_{0 \leq y_1 \leq 1} y_1 + (1 - y_1)^2$$

$$= \frac{3}{4}.$$

The worst PoA happens when the utilities functions are all linear functions, i.e., $U_r(x_r) = w_r x_r$. Assume the utility functions are $U_1(x_1) = 2x_1$, and $U_r(x_r) = x_r$ for $r = 2, \ldots, R$. The solution to the NUM problem is $x_1^* = 1$ and $x_r^* = 0$ for $r = 2, \ldots, R$, which results in a net utility of 2. The NE, according to condition (2.57), is $\hat{x}_1 = R/(2R-1)$, and $\hat{x}_r = 1/(2R - 1)$ for $r = 2, \ldots, R$, which results in a net utility of

$$\frac{2R}{2R - 1} + \frac{R}{2R - 1} = \frac{3R}{2R - 1}.$$

So the PoA is $3R/(4R - 2)$, which converges to $3/4$ as $R \to \infty$. $\qquad\square$

2.8   Summary

- **Network utility maximization** Suppose we have a network with a set of traffic sources S and a set of links $\mathcal{L}$. Each link l has a capacity c_l, and each source in S transmits data at some rate x_r, along a fixed route, where a route is simply a collection of links. The source derives a utility $U_r(x_r)$ when transmitting at rate x_r. The network utility maximization problem is to allocate rates to sources to maximize the net utility $\sum_{r \in S} U_r(x_r)$ subject to $\sum_{r:l \in r} x_r \leq c_l$ for all link l and $x_r \geq 0$ for all r.

- **Fairness** By associating different utility functions with the users, different notions of fairness can be obtained. Resource allocation obtained by using the following utility function (with $\alpha > 0$) is called α-fair:

$$\max \sum_r \frac{x_r^{1-\alpha}}{1 - \alpha}.$$

When $\alpha \to 1$, the resulting allocation is called proportionally fair, and when $\alpha \to \infty$ the resulting allocation is called max-min fair.

- **Primal congestion control algorithm** One way to solve the network utility maximization problem is to relax the capacity constraint and maximize the following objective:

$$W(x) = \sum_{r \in S} U_r(x_r) - \sum_{l \in \mathcal{L}} B_l \left(\sum_{s:l \in s} x_s \right),$$

where $B_l(y)$ is the cost of sending data at rate y on link l. The primal congestion control algorithm is a gradient ascent algorithm for solving the relaxed optimization problem:

$$\dot{x}_r = k_r(x_r) \left(U_r'(x_r) - \sum_{l:l \in r} f_l \left(\sum_{s:l \in s} x_s \right) \right),$$

where $f_l(y) = B_l'(y)$ can be regarded as the congestion price on link l.

- **Dual congestion control algorithm** The dual solution is to consider the dual of the network utility maximization problem:

$$\min_{p \geq 0} \max_{\{x_r \geq 0\}} \sum_r U_r(x_r) - \sum_l p_l \left(\sum_{s:l \in s} x_s - c_l \right),$$

where the p_l's are the Lagrange multipliers. The following dual congestion control algorithm is the gradient descent solution of the dual problem:

$$x_r = U_r'^{-1}(q_r) \quad \text{and} \quad \dot{p}_l = h_l \left(\sum_{s:l \in s} x_s - c_l \right)^+_{p_l}.$$

- **The Vickrey–Clarke–Groves (VCG) mechanism** The VCG mechanism is a pricing scheme to ensure that users do not have an incentive to lie about their true utility functions. Suppose that user r reveals its utility function as $\tilde{U}_r(\cdot)$, which may or may not be its true utility function $U_r(\cdot)$. The network planner allocates the optimal solution of $\max_{x \geq 0} \sum_r \tilde{U}_r(x_r)$ as the rates to the users. Then, it charges user r a price

$$q_r = \sum_{s \neq r} \tilde{U}(\bar{x}_s) - \sum_{s \neq r} \tilde{U}(\tilde{x}_s),$$

where $\tilde{x}_s$ is the optimal solution to $\max_{x \geq 0} \sum_{s:s \neq r} \tilde{U}_s(x_s)$. The price represents the decrease in the sum utility of other users due to the presence of user r. It can be shown that an optimal strategy for each user to maximize its payoff is to reveal its true utility function.

2.9   Exercises

Exercise 2.1 (Bottleneck links and max-min fair rate allocation) Let x_r be the rate allocated to user r in a network where users' routes are fixed. Link l is called a bottleneck link for user r if $l \in r$ and

$$y_l = c_l \quad \text{and} \quad x_s \leq x_r \quad \forall s \quad \text{such that} \quad l \in s,$$

i.e., link l is fully utilized and user r has the highest transmission rate among all users using link l. Show that $\{x_r\}$ is a max-min fair rate allocation *if and only if* every source has at least one bottleneck link.

Exercise 2.2 (A max-min fair resource allocation algorithm) Show that the allocation $\{x_r\}$ obtained from the algorithm below is a max-min fair allocation.

(1) Let $\mathcal{S}^0$ be the set of all sources in the network, and let $c_l^0 = c_l$, i.e., the capacity of link l.

(2) Set $k = 0$.

(3) Let $\mathcal{S}_l^k \subseteq \mathcal{S}^k$ be the set of sources whose routes include link l, and let $|\mathcal{S}_l^k|$ be the cardinality of this set. Define $f_l^k = c_l^k / |\mathcal{S}_l^k|$, which is called the fair share on link l at the kth iteration.

(4) For each source $r \in \mathcal{S}^k$, let $z_r^k = \min_{l:l \in r} f_l^k$, which is the minimum of the fair shares on its route.

(5) Let $\mathcal{T}^k$ be the set of sources such that $z_r^k = \min_{s \in \mathcal{S}^k} z_s^k$, and set $x_r = z_r^k$, $\forall r \in \mathcal{T}^k$. The sources in $\mathcal{T}^k$ are permanently allocated rate z_r^k.

(6) Set $\mathcal{S}^{k+1} = \mathcal{S}^k \setminus \mathcal{T}^k$ and

$$c_l^{k+1} = c_l^k - \sum_{r:l \in r \text{ and } \in \mathcal{T}^k} z_r^k$$

for all l. In other words, sources whose rate allocations are finalized are removed from the set of sources under consideration and the capacity of each link is reduced by the total rate allocated to such sources.

(7) Go to step (3).

Hint: Use the result in Exercise 2.1.

Exercise 2.3 (Different notions of fairness in a simple network) Consider a two-link, three-source network as shown in Figure 2.10. Link A has a capacity of 2 packets/time slot and link B has a capacity of 1 packet/time slot. The route of source 0 consists of both links A and B, the route of source 1 consists of only link A, and the route of source 2 consists of only link B. Compute the resource allocations under the proportional fairness, minimum potential delay fairness, and max-min fairness. Hint: For the max-min fair rate allocation, consider the algorithm in Exercise 2.2; for the other two resource allocations, use Lagrange multipliers and the KKT theorem.

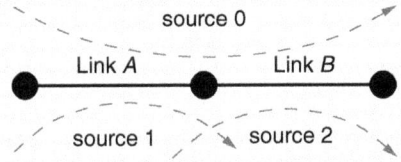

Figure 2.10 Two-link, three-source network.

Exercise 2.4 (NUM in a simple network) Consider again the same two-link, three-user network shown in Figure 2.10. Now assume that the link capacities are $C_A = C_B = 1$. Suppose that the utility functions of the users are given as follows:

$$U_0(x_0) = \log(x_0),$$
$$U_1(x_1) = \log(1 + x_1),$$
$$U_2(x_2) = \log(1 + x_2).$$

Compute the data transmission rates of the three users, x_0, x_1, and x_2, which maximize the sum network utility.

Exercise 2.5 (The utility function of a primal congestion controller) Consider the following primal congestion control algorithm:

$$\dot{x}_r = k_r \left[(1 - q_r) - q_r x_r \right],$$

where q_r is the sum of the link prices on route r, x_r is the transmission rate of user r and $k_r > 0$ is some constant. Identify the utility function of user r. Hint: Recall the form of the primal congestion control algorithm, and compare it to the above differential equation.

Exercise 2.6 (An alternative proof of the stability of the primal algorithm) Consider the primal congestion controller with $\kappa_r(x) = 1 \ \forall x$ and $\forall r$. Use the Lyapunov function

$$\sum_r (x_r - \hat{x}_r)^2$$

to prove that the controller is globally, asymptotically stable, where $\hat{x}$ is the global maximizer of $W(x) = \sum_r U_r(x_r) - \sum_l B_l(y_l)$.

Hint: Since $\hat{x}$ is the global maximizer, it has the properties presented in Result 2.1.9.

Exercise 2.7 (The primal congestion controller with non-negligible p_l) Assume link prices $p_l(y_l) \in [0,1]$ and $q_r = 1 - \prod_{l:l \in r}(1 - p_l)$. For example, if p_l is the probability that a packet is *marked* on link l, then q_r is the probability that a packet is marked on route r. In this exercise, you will be asked to prove that the primal congestion controller is globally, asymptotically stable under this model, without the assumption that p_l's are really small.

(1) First, show that the primal congestion controller in this case can be rewritten as

$$\dot{x}_r = k_r(x_r)\left(\prod_{l:l \in r}(1 - p_l) - (1 - U_r'(x_r))\right).$$

(2) Show that $W(x)$, given by

$$W(x) = \sum_l \int_0^{y_l} \log(1 - p_l(y))dy - \sum_r \int_0^{x_r} \log(1 - U_r'(x))dx,$$

is strictly concave. Hint: Assume $1 - U_r'(x_r) > 0$ so that $\log(1 - U_r'(x))$ is well defined. The following fact may be useful: log is an increasing function.

(3) Use $W(\hat{x}) - W(x)$ as the Lyapunov function to show the global, asymptotic stability of the primal congestion controller, where $\hat{x}$ is the global maximizer of $W(x)$. Hint: Assume that U_r, k_r, and p_l are such that $W(x) \to \infty$ as $||x|| \to \infty$, there exists a unique equilibrium point, and $x_r(t)$ can never go below zero.

Exercise 2.8 (Multi-path routing) In this problem, we expand the scope of the utility maximization problem to include adaptive, multi-path routing. Let s denote a source and let $\mathcal{R}(s)$ denote the set of routes used by source s. Each source s is allowed to split its packets along multiple routes. Let z_s denote the rate at which source s generates data and let x_r denote the rate on route r. Thus, the penalty function formulation of the utility maximization problem becomes

$$\max_x \ \sum_s U_s(z_s) - \sum_l \int_0^{y_l} f_l(y)dy + \epsilon \sum_r \log x_r,$$

where

$$z_s = \sum_{r \in \mathcal{R}(s)} x_r, \qquad y_l = \sum_{r:l \in r} x_r,$$

and $\epsilon > 0$ is a small number.

(1) Even when U_S is a strictly concave function, argue that the above objective need not be strictly concave if $\epsilon = 0$. (Thus, we have introduced the ϵ term only to ensure strict concavity. But the impact of this term on the optimal solution will be small if ϵ is chosen to be small.)

(2) Derive a congestion control (and rate-splitting across routes) algorithm and prove that it asymptotically achieves the optimal rates which solve the above utility maximization problem. Hint: Use the approach used to derive the primal congestion control algorithm.

Exercise 2.9 (The global stability of the dual algorithm) Recall the dual algorithm

$$x_r = U_r'^{-1}(q_r),$$
$$\dot{p}_l = h_l(y_l - c_l)_{p_l}^+.$$

Prove that the dual algorithm is globally, asymptotically stable when the routing matrix R has full row rank, i.e., given q, there exists a unique p that satisfies $q = R^T p$.

Exercise 2.10 (The primal-dual algorithm for congestion control) Consider the following congestion control algorithm:

$$\dot{x}_r = \kappa_r \left(\frac{w_r}{x_r} - q_r \right),$$
$$\dot{p}_l = h_l (y_l - c_l)_{p_l}^+,$$

where $q_r = \sum_{l:l \in r} p_l$, $y_l = \sum_{r:l \in r} x_r$, and κ_r and h_l are positive constants. This algorithm is called the primal-dual algorithm for congestion control.

(1) Show that the equilibrium point of the above congestion control algorithm solves a utility maximization problem, which allocates rates in a weighted proportionally fair manner.

(2) Assume that the equilibrium point is unique, and show that the congestion controller is globally, asymptotically stable by using the Lyapunov function

$$V(x,p) = \sum_r \frac{(x_r - \hat{x}_r)^2}{\kappa_r} + \frac{\sum_l (p_l - \hat{p}_l)^2}{h_l},$$

where $(\hat{x}, \hat{p})$ denotes the equilibrium point. To do this, show that (i) $\dot{V} \leq 0$ and (ii) $\dot{V} = 0$ implies $(x(t), p(t)) = (\hat{x}, \hat{p})$. The result then follows from LaSalle's invariance principle (see Section 2.3, Theorem 2.3.3).

Note: In this problem, we have derived a third type of congestion control algorithm, called the primal-dual algorithm. The question of which one of these algorithms is best is debatable. Clearly all of the algorithms lead to the same steady-state rate allocation.

Exercise 2.11 (A discrete-time version of the dual algorithm) Consider the following discrete-time version of the dual congestion control algorithm: at each time slot k, each source chooses a transmission rate $x_r(k)$, which is the solution to

$$\max_{0 \leq x_r \leq X_{\max}} U_r(x_r) - q_r(k)x_r,$$

where $X_{\max}$ is the maximum rate at which any user can transmit. Each link l computes its price $p_l(k)$ according to the following update rule, which is a discretization of the continuous-time algorithm:

$$p_l(k+1) = (p_l(k) + \epsilon(y_l - c_l))^+,$$

where $\epsilon > 0$ is a small step-size parameter. The variables y_l and q_r are defined as usual:

$$q_r(k) = \sum_{l:l\in r} p_l(k), \qquad y_l(k) = \sum_{r:l\in r} x_r(k).$$

We will show that, on average, the above discrete-time algorithm is nearly optimal in the sense that it approximately solves the utility maximization problem.

(1) Consider the Lyapunov function

$$V(k) = \frac{1}{2} \sum_l p_l^2(k).$$

Show that

$$V(k+1) - V(k) \le K\epsilon^2 + \epsilon \sum_r q_r(k)(x_r(k) - \hat{x}_r),$$

for some constant $K > 0$, where $\hat{x}$ is an optimal solution to the utility maximization problem

$$\max_{x\ge 0} \sum_r U_r(x_r), \qquad \text{subject to} \qquad \sum_{r:l\in r} x_r \le c_l.$$

Assume that $X_{\max} > \max_r \hat{x}_r$.

(2) Next, show that

$$V(k+1) - V(k) \le K\epsilon^2 + \epsilon \sum_r (U_r(x_r) - U_r(\hat{x}_r)).$$

(3) Finally, show that

$$\sum_r U_r(\hat{x}_r) \le \sum_r U_r(\bar{x}_r) + K\epsilon,$$

where

$$\bar{x}_r := \lim_{N\to\infty} \frac{1}{N} \sum_{k=1}^N x_r(k).$$

Note: For this problem, we assume U_r is concave, but it does not have to be strictly concave for the results of this problem to hold. If U_r is not strictly concave, there may be multiple optimal solutions $\hat{x}$. In this case, $X_{\max}$ is assumed to be greater than $\max_r \hat{x}_r$ for all possible $\hat{x}$.

Exercise 2.12 (An example illustrating the VCG algorithm) Consider the network shown in Figure 2.11, where the four links are owned by four different players. Suppose that the network wants to establish a communication path from node 1 to node 3. If a link is selected for the transmission, it incurs a cost of p_l, where $l \in \{a, b, c, d\}$. The overall system goal, also known as the social welfare problem, is to find the minimum-cost path. Define x_l to be a variable such that $x_l = 0$ if link l is selected and $x_l = 1$ otherwise. Therefore, the utility function associated with player l is

$$U_l(x_l) = \begin{cases} -p_l, & \text{if } x_l = 0, \quad \text{i.e., link } l \text{ is selected,} \\ 0, & \text{if } x_l = 1, \quad \text{i.e., link } l \text{ is not selected.} \end{cases}$$

The utility function revealed by player l is

$$\tilde{U}_l(x_l) = \begin{cases} -\tilde{p}_l, & \text{if } x_l = 0, \\ 0, & \text{if } x_l = 1, \end{cases}$$

where $\tilde{p}_l$ is the cost claimed by player l. The objective of the network is

$$\max \sum_l U_l(x_l),$$

$$\text{subject to} \quad x_a + x_c = 1, \tag{2.63}$$

$$x_b + x_d = 1, \tag{2.64}$$

$$x_a, x_b, x_c, x_d \in \{0, 1\},$$

where equalities (2.63) and (2.64) guarantee that there is a feasible path from node 1 to node 3.

Assume $p_a > p_c$ and $p_b > p_d$. Let w_l denote the price charged to link l. Write the value of w_l under the VCG pricing mechanism. *Note: You will find that the prices are non-positive. In fact, $-w_l$ can be interpreted as the payment link l receives when it is selected by the network.*

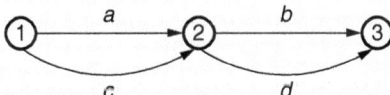

Figure 2.11 A simple network.

Exercise 2.13 (An example illustrating the PoA) Consider a network with two strategic users sharing the same link with capacity $c = 1$. The utility function of user i is $\alpha_i - (\alpha_i / (x_i + 1))$, where x_i is the rate allocated to user i. Each user bids an amount that it is willing to pay, say w_i for user i, and user i is allocated a data rate given by $x_i = w_i / (w_1 + w_2)$. Thus, the payoff to user i is $\alpha_i - (\alpha_i / x_i) - w_i$. We assume $\alpha_1 = 1$ and $\alpha_2 = 2$.

(1) Write down the NE for the bids (w_1, w_2) for these strategic users.

(2) Compute the PoA.

2.10   Notes

The utility maximization framework for studying resource allocation in communication networks was introduced in [73]. Max-min fairness was originally developed in the context of political philosophy [144], and was extensively studied in the context of communication networks in [10, 61]. Log utilities were introduced in the solution of a game where players bargain over the allocation of a common resource [129]. It is called the Nash bargaining solution in economics. It was studied under the name proportional fairness in the context of communication networks in [73]. Minimum potential delay fairness was introduced in [114]. The α-fair utility functions have been studied by economists under the name isoelastic utility functions [140]. They were proposed as a common framework to study many notions of fairness in communication networks in [124, 125].

The primal and dual algorithms for solving the network utility maximization problem were presented in [77]. The version of the dual algorithm presented in this chapter is a continuous-time version of the algorithms proposed in [107, 176]. The multi-path algorithm was also proposed in [77], while the addition of the ϵ term was proposed in [54].

The primal-dual algorithm for Internet congestion control was introduced in [93], and its convergence was proved in [94], although the algorithm at the nodes to compute the Lagrange multipliers is different from the computations presented in the Problems section in this chapter. The version of the primal-dual algorithm presented here is in [2, 173]. The block diagram view of the relationships between the primal and dual variables was suggested in [108].

One-bit feedback for congestion control was proposed in [25, 143]. A idea of using exponential functions to convert price information to one-bit information is in [107]. A crude approximation for the drop probability (also called the DropTail price function) was proposed in [93], as a limit of more accurate queueing-theoretic models.

A game-theoretic view of network resource allocation was presented in [73]; see also [155]. The price of anarchy result presented here is due to [66]. A survey of game theory in networks can be found in [118].

Several surveys of resource allocation in the Internet using the utility function framework are available in [24, 75, 150, 155].

Excellent sources for the background material on optimization include [9, 17, 110]. An introduction to differential equation models of dynamical systems and their stability can be found in [83]. Delay differential equations and stability are treated in [53].

3 Links: statistical multiplexing and queues

In Chapter 2, we assumed that the transmission rates x_r are positive, and we derived fair and stable resource allocation algorithms. In reality, since data are transmitted in the form of packets, the rates x_r are converted to discrete window sizes, which results in bursty (non-smooth) arrival rates at the links in the network. In addition, many flows in the Internet are very short (consisting of only a few packets), for which the convergence analysis in the previous chapter does not apply. Further, there may also be flows which are not congestion controlled. Because of these deviations, the number of incoming packets at a link varies over time and may exceed the link capacity occasionally even if the mean arrival rate is less than the link capacity. So buffers are needed to absorb bursty arrivals and to reduce packet losses. To understand the effect of bursty arrivals and the role of buffering in communication networks, in this chapter we model packet arrivals at links as random processes and study the packet level performance at a link using discrete-time queueing theory. This chapter is devoted to answering the following questions.

- *How large should the buffer size be to store bursty packet arrivals temporarily before transmission over a link?*
- *What is the relationship between buffer overflow probabilities, delays, and the burstiness of the arrival processes?*
- *How do we provide isolation among flows so that each flow is guaranteed a minimum rate at a link, independent of the burstiness of the other flows sharing the link?*

3.1 Mathematical background: the Chernoff bound

In this section, we present the Chernoff bound, which provides a bound on the tail distribution of the sum of independent random variables.

Lemma 3.1.1 (Markov's inequality) For a positive random variable X, the following inequality holds for any $\epsilon > 0$:

$$\Pr(X \geq \epsilon) \leq \frac{E(X)}{\epsilon}.$$

Proof Define a random variable Y such that $Y = \epsilon$ if $X \geq \epsilon$ and $Y = 0$ otherwise. So

$$E[X] \geq E[Y] = \epsilon \Pr(X \geq \epsilon). \qquad \square$$

Theorem 3.1.2 (The Chernoff bound) Consider a sequence of independently and identically distributed (i.i.d.) random variables $\{X_i\}$ with mean $\mu = E[X_i]$. For any constant x, the following inequality holds:

$$\Pr\left(\sum_{i=1}^{n} X_i \geq nx\right) \leq \exp\left(-n \sup_{\theta \geq 0}\{\theta x - \log M(\theta)\}\right), \tag{3.1}$$

where $M(\theta) \triangleq E\left[e^{\theta X_1}\right]$ is the moment generating function of X_1.

If $\{X_i\}$ are Bernoulli random variables with parameter μ, and $1 \geq x > \mu$, then

$$\Pr\left(\sum_{i=1}^{n} X_i \geq nx\right) \leq \exp\left(-nD(x\|\mu)\right), \tag{3.2}$$

where

$$D(x\|\mu) = x \log \frac{x}{\mu} + (1 - x) \log \frac{1 - x}{1 - \mu}$$

is the Kullback–Leibler distance between Bernoulli random variables with parameter x and parameter μ.

Proof For any $\theta \geq 0$, we have

$$\Pr\left(\sum_{i=1}^{n} X_i \geq nx\right) \leq \Pr(e^{\theta \sum_{i=1}^{n} X_i} \geq e^{\theta nx})$$

$$\leq \frac{E\left[e^{\theta \sum_{i=1}^{n} X_i}\right]}{e^{\theta nx}}, \tag{3.3}$$

where the first inequality becomes an equality if $\theta > 0$ and the second inequality follows from the Markov inequality. Since inequality (3.3) holds for all $\theta \geq 0$, we further obtain

$$\Pr\left(\sum_{i=1}^{n} X_i \geq nx\right) \leq \inf_{\theta \geq 0} \frac{E\left[e^{\theta \sum_{i=1}^{n} X_i}\right]}{e^{\theta nx}}$$

$$= e^{-n \sup_{\theta \geq 0}\{\theta x - \log M(\theta)\}}. \tag{3.4}$$

Recall that $M(\theta) \triangleq E\left[e^{\theta X_i}\right]$ is the moment generating function of X_i. Inequality (3.4) is called the Chernoff bound.

Inequality (3.2) holds because

$$\sup_{\theta \geq 0}\{\theta x - \log M(\theta)\} = D(x\|\mu)$$

when $\{X_i\}$ are Bernoulli random variables and $x > 0$. $\square$

3.2 Statistical multiplexing and packet buffering

When multiple data sources share the same link, the bandwidth of the link needs to be allocated properly. To guarantee no packet loss and small transmission latencies, one may allocate the bandwidth according to the data sources' peak transmission rates. For example, if a data source has a peak rate of R bps, then a bandwidth of R bps of the link is reserved for that source. This approach will provide very good Quality of Service (QoS) to data sources in terms of bandwidth, delay, and jitter, but could cause the link to be under-utilized since the typical total data rate of the sources may be much smaller than the sum of their peak rates.

The bandwidth allocated to sources is often much smaller than the sum of their peak rates, and is slightly larger than the sum of the average rates of the sources. The link then relies on the fact that the probability that the sum of rates exceeds the sum of the average rates is small. This type of resource allocation is called *statistical multiplexing*. Compared to bandwidth allocation based on peak rates, statistical multiplexing allows a link to support a larger number of data sources, as shown in Example 3.2.1.

Example 3.2.1

Consider a link with bandwidth 10 Mbps, which is shared by multiple data sources. At any given time, a source is active with a probability of 0.1, and transmits at a rate of 100 kbps when active.

If the link bandwidth is allocated according to the peak rate, the link needs to reserve 100 kbps for each source. In this case, the maximum number of sources that can be allowed is given by

$$\frac{10\,\text{Mbps}}{100\,\text{kbps}} = 100.$$

Now consider statistical multiplexing and assume there are n sources using the link. Define X_i to be a random variable such that $X_i = 1$ if source i is active and $X_i = 0$ otherwise. We apply the Chernoff bound for Bernoulli random variables to bound the probability that the aggregated rate of active sources exceeds the link capacity, i.e., the following overflow probability:

$$\Pr\left(\sum_{i=1}^{n} X_i \geq 100\right). \tag{3.5}$$

The result is illustrated in Figure 3.1.

We observe that the link can accommodate up to 750 sources if the overflow probability is allowed to be 0.01%. We can increase the number of sources to 800 if the overflow probability is allowed to be 1%. Thus, statistical multiplexing can dramatically increase network capacity at the cost of very small loss probabilities. □

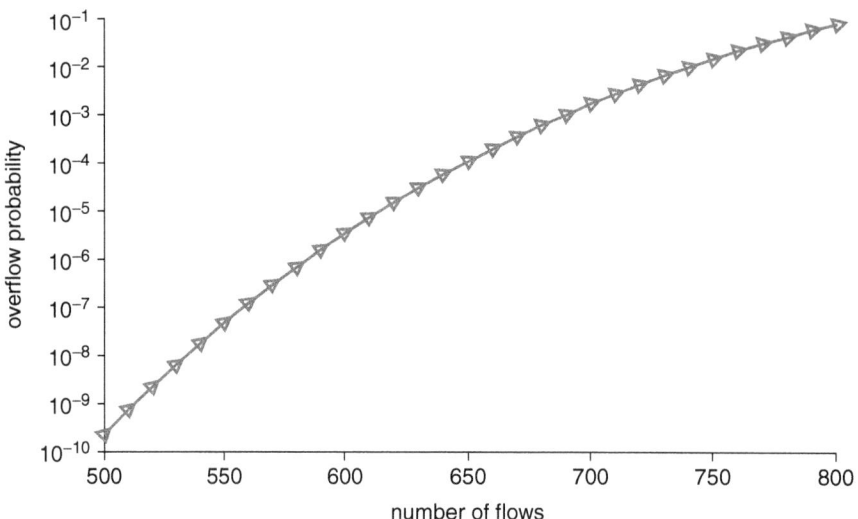

Figure 3.1 The overflow probabilities versus the number of sources in the network.

3.2.1   Queue overflow

In Example 3.2.1, we have seen that statistical multiplexing performs well even without buffer. In practice, when the number of arrivals exceeds the link capacity, the packets will first be stored in a buffer, instead of being dropped immediately. The focus of the rest of this chapter is to understand the behavior of the buffer under various arrival processes and buffer models. As a starting point, we assume the buffer is of infinite size and calculate the probability that the amount of buffered packets exceeds a certain threshold B.

We model a single link shared by n sources as a discrete-time queueing system with a single server and infinite buffer space, as shown in Figure 3.2. The server represents the link and can serve c packets per time slot. We assume packets are of the same size here; we will study varying packet sizes later. We define $a_i(t)$ to be the number of packets injected by source i in time slot t, and assume $a_i(t)$ are i.i.d. across time and sources. Further, assume $\lambda \triangleq E[a_i(t)] < c/n$, so the overall arrival rate is less than the link capacity.

Let $q(t)$ denote the number of buffered packets (queue length) at the beginning of time slot t. We assume packets arrive at the beginning of each time slot and depart at the end of each time slot. Thus, the queue evolution can be described as follows:

$$q(t+1) = (q(t) + a(t) - c)^+,$$

where $a(t) = \sum_{i=1}^{n} a_i(t)$ and $x^+ = \max\{0, x\}$. Assuming the system starts from $t = 0$ with empty queue, i.e., $q(0) = 0$, we next consider the queue length at time k. According to the definition of x^+, we first have

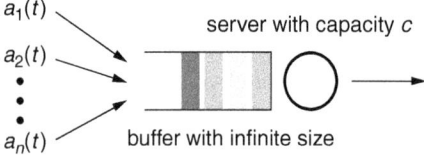

Figure 3.2 Queue shared by n sources.

$$q(k) = (q(k-1) + a(k-1) - c)^+ = \max\{q(k-1) + a(k-1) - c, 0\},$$

and

$$q(k-1) = \max\{q(k-2) + a(k-2) - c, 0\}.$$

Substituting the second equation into the first one, we obtain

$$q(k) = \max\{\max\{q(k-2) + a(k-2) - c, 0\} + a(k-1) - c, 0\}$$
$$= \max\{q(k-2) + a(k-1) + a(k-2) - 2c, a(k-1) - c, 0\}.$$

Recursively applying the procedure above, we obtain

$$q(t) = \max\left\{\max_{t \geq k \geq 1}\left(\sum_{s=1}^{k} a(t-s) - kc\right), 0\right\}. \tag{3.6}$$

Next we compute the probability that the queue length at time t exceeds B for some constant B. From (3.6), we have

$$\Pr(q(t) \geq B) = \Pr\left(\max_{t \geq k \geq 1}\left(\sum_{s=1}^{k} a(t-s) - kc\right) \geq B\right).$$

Note that

$$\Pr\left(\max_{t \geq k \geq 1}\left(\sum_{s=1}^{k} a(t-s) - kc\right) \geq B\right) = \Pr\left(\bigcup_{k=1}^{t}\left(\sum_{s=1}^{k} a(t-s) - kc \geq B\right)\right).$$

Using the union bound,

$$\Pr\left(\max_{t \geq k \geq 1}\left(\sum_{s=1}^{k} a(t-s) - kc\right) \geq B\right) \leq \sum_{k=1}^{t}\Pr\left(\sum_{s=1}^{k} a(t-s) - kc \geq B\right)$$

$$= \sum_{k=1}^{t}\Pr\left(e^{\theta\sum_{s=1}^{k} a(t-s)} \geq e^{\theta(kc+B)}\right)$$

$$\leq \sum_{k=1}^{t}E\left[e^{\theta\sum_{s=1}^{k} a(t-s)}\right]e^{-\theta(kc+B)}, \tag{3.7}$$

where the last inequality follows from the Markov inequality.

Since $a_i(t)$ are i.i.d. across time slots and sources,

$$E\left[e^{\theta\sum_{s=1}^{k} a(t-s)}\right] = \prod_{s=1}^{k}\prod_{i=1}^{n}E\left[e^{\theta a_i(t-s)}\right] = M(\theta)^{nk},$$

where $M(\theta) = E[e^{\theta a_i(t)}]$ is the moment generating function. Defining $\Lambda(\theta) = \log M(\theta)$, we obtain the following upper bound:

$$\Pr(q(t) \geq B) \leq \sum_{k=1}^{t} e^{nk\Lambda(\theta)}e^{-\theta(kc+B)}, \tag{3.8}$$

which leads to the following theorem, which shows that the probability of queue overflow decreases exponentially as B increases.

Theorem 3.2.1 For all $\theta > 0$ such that $\frac{\Lambda(\theta)}{\theta} < \frac{c}{n}$, the queue overflow probability satisfies

$$\Pr(q(t) \geq B) \leq \frac{e^{n\Lambda(\theta)-\theta c}}{1 - e^{n\Lambda(\theta)-\theta c}} e^{-\theta B}, \quad \forall t \geq 0. \tag{3.9}$$

Proof Note that upper bound (3.8) can be rewritten as

$$\Pr(q(t) \geq B) \leq \sum_{k=1}^{t} e^{-k(\theta c - n\Lambda(\theta))} e^{-\theta B} \leq e^{-\theta B} \sum_{k=1}^{\infty} e^{k(n\Lambda(\theta)-\theta c)}. \tag{3.10}$$

When $\frac{\Lambda(\theta)}{\theta} < \frac{c}{n}$ holds, we have $e^{(n\Lambda(\theta)-\theta c)} < 1$, which implies that

$$\sum_{k=1}^{\infty} e^{k(n\Lambda(\theta)-\theta c)} = \frac{e^{n\Lambda(\theta)-\theta c}}{1 - e^{n\Lambda(\theta)-\theta c}}.$$

So the theorem holds. $\qquad\qquad\qquad\qquad\qquad\qquad\qquad\qquad\qquad\qquad\qquad\qquad\quad \square$

The above estimate of the overflow probability in (3.9) is an upper bound and may be loose. We will obtain tighter estimates in Chapter 10 on large deviations in Part II of the book. For now, using the above result, we see that the overflow probability decreases exponentially with B with decay rate θ provided that

$$\frac{c}{n} > \frac{\Lambda(\theta)}{\theta}.$$

The quantity $\frac{\Lambda(\theta)}{\theta}$ is called the *effective bandwidth of a source.*

We now look at the range of this effective bandwidth. Suppose that the $a_i(t)$ take values in a finite set such that

$$a_i(t) \in \{a_1, a_2, \ldots, a_m\} \quad \text{and} \quad a_j < a_{j+1}.$$

Define

$$p_j \triangleq \Pr(a_i(t) = a_j) > 0, \quad \forall j \in \{1, 2, \ldots, m\}.$$

Note that the larger the θ, the more stringent is the QoS requirement. Consider the following two extreme cases.

- If θ is close to 0, then

$$\frac{\Lambda(\theta)}{\theta} = \frac{\log E\left[e^{\theta a_i(t)}\right]}{\theta}$$

$$\approx_{(a)} \frac{\log E[1 + \theta a_i(t)]}{\theta}$$

$$= \frac{\log(1 + \theta\lambda)}{\theta}$$

$$\to_{(b)} \lambda, \quad \text{as } \theta \to 0,$$

where both (a) and (b) can be precisely justified using Taylor's theorem.

- If θ is very large, then

$$\frac{\Lambda(\theta)}{\theta} = \frac{\log\left(\sum_{j=1}^{m} e^{\theta a_j} p_j\right)}{\theta}$$

$$\approx_{(c)} \frac{\log\left(e^{\theta a_m} p_m\right)}{\theta}$$

$$\to a_m, \qquad \text{as } \theta \to \infty,$$

where approximation (c) holds because $e^{\theta a_j}/e^{\theta a_m} \to 0$ as $\theta \to \infty$ for any $a_j < a_m$.

Thus, the effective bandwidth of the source increases from the mean arrival rate λ to the maximum arrival rate a_m, as the QoS parameter θ becomes more and more stringent.

Besides the overflow probability in Theorem 3.2.1, other performance metrics, such as expected queue length and queueing delay, and the probability of packet loss when the buffer size is finite, are also important in practice. Markov chains and queueing theory will be introduced next for the purpose of quantitatively understanding these performance metrics, at least for simple arrival processes.

3.3　Mathematical background: discrete-time Markov chains

Let $\{X_k\}$ be a discrete-time stochastic process that takes on values in a countable set $\mathcal{S}$ called the state space. Here, k is the time index. $\{X_k\}$ is called a Discrete-Time Markov Chain (DTMC, or simply a Markov chain, when the discrete nature of the time index is clear) if

$$\Pr(X_k = i_k \mid X_{k-1} = i_{k-1}, X_{k-2} = i_{k-2}, \ldots) = \Pr(X_k = i_k \mid X_{k-1} = i_{k-1}),$$

where $i_j \in \mathcal{S}$.

A Markov chain is said to be *time homogeneous* if $\Pr(X_k = j \mid X_{k-1} = i)$ is independent of k. We will only consider time-homogeneous Markov chains here. Associated with each Markov chain is a matrix called the transition probability matrix, denoted by $\mathbf{P}$, whose (i,j)th element is given by $P_{ij} = \Pr(X_k = j \mid X_{k-1} = i)$. Let $p[k]$ denote a row vector of probabilities with $p_j[k] = \Pr(X_k = j)$. This vector of probabilities evolves according to the equation

$$p[k] = p[k-1]\,\mathbf{P}.$$

Thus, $p[0]$ and $\mathbf{P}$ capture all the relevant information about the dynamics of the Markov chain.

The following questions are important in the study of Markov chains.

- Does there exist a π so that $\pi = \pi \mathbf{P}$? If such a π exists, it is called a stationary distribution.
- If there exists a unique stationary distribution, does $\lim_{k \to \infty} p[k] = \pi$ for all $p[0]$? In other words, does the distribution of the Markov chain converge to the stationary distribution starting from any initial state?

While the existence of a unique stationary distribution is desirable in the applications studied in this book, not all Markov chains have a unique steady-state distribution. We will first present an example of a Markov chain that does not have a stationary distribution, and then we give another example where a stationary distribution exists but the probability distribution over the states does not converge to the stationary distribution in steady state. Motivated by these examples, we will impose some conditions to guarantee the existence of a stationary distribution to which the Markov chain converges in steady state.

Example 3.3.1

Consider a trivial two-state Markov chain with states a and b such that the chain remains in the initial state for time slots $k \geq 0$. Thus, the transition probability matrix for this Markov chain is given by

$$\mathbf{P} = \begin{pmatrix} 1 & 0 \\ 0 & 1 \end{pmatrix}.$$

Therefore, $\pi\mathbf{P} = \pi$ is true for any distribution π and the stationary distribution is not unique. The Markov chain in this example is such that if it started in one state, it remained in the same state forever. In general, Markov chains in which the state space can be divided into two disconnected parts will not possess a unique stationary distribution. $\square$

The above example motivates the following definitions.

Definition 3.3.1 Let $P_{ij}^{(n)} = \Pr(X_{k+n} = j \mid X_k = i)$.

(1) State j is said to be reachable from state i if there exists $n \geq 1$ so that $P_{ij}^{(n)} > 0$.
(2) A Markov chain is said to be *irreducible* if any state i is reachable from any other state j. $\square$

In this chapter, we will mostly consider Markov chains that are irreducible. The Markov chain in Example 3.3.1 was not irreducible.

Example 3.3.2

Again let us consider a two-state Markov chain with two states a and b. The Markov chain behaves as follows: if it is in state a at the current time slot, it jumps to b at the next time slot, and vice versa. Thus,

$$\mathbf{P} = \begin{pmatrix} 0 & 1 \\ 1 & 0 \end{pmatrix}.$$

The stationary distribution is obtained by solving $\pi\mathbf{P} = \pi$, which gives $\pi = (1/2 \quad 1/2)$. However, the system does not converge to this stationary distribution starting from any initial condition. To see this, note that if $p[0] = (1 \quad 0)$, then $p[1] = (0 \quad 1)$, $p[2] = (1 \quad 0)$, $p[3] = (0 \quad 1)$, $p[4] = (1 \quad 0)$, ... Therefore $\lim_{k \to \infty} p[k] \neq \pi$. The reason that this Markov chain does not converge to the stationary distribution is due to the fact that the state periodically alternated between a and b. $\quad\square$

Motivated by the above example, the following definitions lead up to the classification of a Markov chain as being either periodic or aperiodic.

Definition 3.3.2 The following definitions classify Markov chains and their states as periodic or aperiodic.

(1) State i is said to have a period $d_i \geq 1$ if $d_i = \gcd \left\{ n : P_{ii}^{(n)} > 0 \right\}$, where gcd denotes the greatest common divisor. If $P_{ii}^{(n)} = 0$, $\forall n$, we say that $d_i = \infty$.
(2) State i is said to be *aperiodic* if $d_i = 1$.
(3) A Markov chain is said to be *aperiodic* if all states are aperiodic. $\quad\square$

Next, we state the following useful lemma, which will be useful later to identify whether a Markov chain is aperiodic or not.

Lemma 3.3.1 Every state in an irreducible Markov chain has the same period. Thus, in an irreducible Markov chain, if one state is aperiodic, the Markov chain is aperiodic. $\quad\square$

The Markov chain in Example 3.3.2 was not aperiodic. The following theorem states that Markov chains which do not exhibit the type of behavior illustrated in the examples possess a stationary distribution to which the distribution converges, starting from any initial state.

Theorem 3.3.2 A finite-state-space, irreducible Markov chain has a unique stationary distribution π, and, if it is aperiodic, $\lim_{k \to \infty} p[k] = \pi$, $\forall p[0]$. $\quad\square$

Example 3.3.3

The following example illustrates the computation of the stationary distribution of a Markov chain. Consider a three-state Markov chain with the state space $\{a, b, c\}$, as shown in Figure 3.3. If the Markov chain is in state a, it switches from the current state to one of the other two states, each with probability $1/4$, or remains in the same state. If it is in state b, it switches to state c with probability $1/2$ or remains in the same state. If it is in state c, it switches to state a with probability 1. Thus,

$$\mathbf{P} = \begin{pmatrix} 1/2 & 1/4 & 1/4 \\ 0 & 1/2 & 1/2 \\ 1 & 0 & 0 \end{pmatrix}.$$

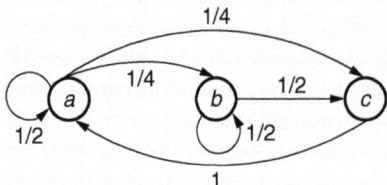

Figure 3.3 Three-state Markov chain.

This Markov chain is irreducible since it can go from any state to any other state in finite time with non-zero probability. Next, note that there is a non-zero probability of remaining in state a if the Markov chain starts in state a. Therefore, $P_{aa}^{(n)} > 0$ for all n, and state a is aperiodic. Since the Markov chain is irreducible, this implies that all the states are aperiodic. Thus, the finite-state Markov chain is irreducible and aperiodic, which implies the existence of a stationary distribution to which the probability distribution converges, starting from any initial distribution. To compute the stationary distribution π, we solve the equation

$$\pi = \pi \mathbf{P},$$

where $\pi = (\pi_a \quad \pi_b \quad \pi_c)$, subject to the constraints $\pi_a + \pi_b + \pi_c = 1$ and $\pi_a, \pi_b, \pi_c \geq 0$, to obtain $\pi = (1/2 \quad 1/4 \quad 1/4)$. $\qquad\square$

If the state space is infinite, the existence of a stationary distribution is not guaranteed, even if the Markov chain is irreducible, as the following example illustrates.

Example 3.3.4

Let the state space be the set of integers, and define the Markov chain as follows:

$$X_{k+1} = X_k + 1, \quad \text{with probability } 1/3,$$
$$= X_k - 1, \quad \text{with probability } 1/3,$$
$$= X_k, \qquad \text{with probability } 1/3.$$

It is easy to verify that this Markov chain is irreducible and aperiodic with transition probability matrix given by

$$\mathbf{P} = \begin{pmatrix} & & & \cdots & & & \\ 0 & 1/3 & 1/3 & 1/3 & 0 & 0 & 0 \\ 0 & 0 & 1/3 & 1/3 & 1/3 & 0 & 0 \\ 0 & 0 & 0 & 1/3 & 1/3 & 1/3 & 0 \\ & & & \cdots & & & \end{pmatrix}.$$

If a stationary distribution π exists, it has to satisfy $\pi = \pi \mathbf{P}$, which can be rewritten as

$$\pi_k = \frac{1}{3}\pi_{k-1} + \frac{1}{3}\pi_k + \frac{1}{3}\pi_{k+1}, \qquad \forall k.$$

Thus, we have to solve for the $\{\pi_k\}$ that satisfy

$$2\pi_k = \pi_{k-1} + \pi_{k+1}, \qquad \forall k,$$

$$\sum_{k=-\infty}^{\infty} \pi_k = 1,$$

$$\pi_k \geq 0, \qquad \forall k.$$

We will now show that we cannot find a distribution π that satisfies the above set of equations. Note that

$$\pi_2 = 2\pi_1 - \pi_0,$$

$$\pi_3 = 2\pi_2 - \pi_1 = 2(2\pi_1 - \pi_0) - \pi_1$$

$$= 3\pi_1 - 2\pi_0,$$

$$\pi_4 = 6\pi_1 - 4\pi_0 - 2\pi_1 + \pi_0$$

$$= 4\pi_1 - 3\pi_0,$$

$$\vdots$$

$$\pi_k = k\pi_1 - (k-1)\pi_0$$

$$= (k-1)(\pi_1 - \pi_0) + \pi_1, \quad k \geq 2.$$

Thus,

$$\text{if } \pi_1 = \pi_0 > 0, \quad \text{then } \pi_k = \pi_1, \quad \forall k \geq 2 \quad \text{and} \quad \sum_{k=0}^{\infty} \pi_k > 1,$$

$$\text{if } \pi_1 > \pi_0, \quad \text{then } \pi_k \to \infty,$$

$$\text{if } \pi_1 < \pi_0, \quad \text{then } \pi_k \to -\infty,$$

$$\text{if } \pi_1 = \pi_0 = 0, \quad \text{then } \pi_k = 0, \quad \forall k \geq 0.$$

A little thought shows that the last statement is also true for $k < 0$. Thus, a stationary distribution cannot exist. $\qquad\qquad\square$

Example 3.3.4 illustrates the need for more conditions beyond irreducibility to ensure the existence of stationary distributions in countable-state-space Markov chains. To this end, we introduce the notion of recurrence and related concepts.

Definition 3.3.3 The following definitions classify the states of a Markov chain as recurrent or transient.

(1) The recurrence time T_i of state i of a Markov chain is defined as

$$T_i = \min\{n \geq 1 : X_n = i \text{ given } X_0 = i\}.$$

(Note that T_i is a random variable.)

(2) A state i is said to be *recurrent* if $\Pr(T_i < \infty) = 1$. Otherwise, it is called *transient*.

(3) The mean *recurrence time* M_i of state i is defined as $M_i = E[T_i]$.

(4) A recurrent state i is called *positive recurrent* if $M_i < \infty$. Otherwise, it is called *null recurrent*.

(5) A Markov chain is called *positive recurrent* if all of its states are positive recurrent. ☐

The next two lemmas and theorem are stated without proof.

Lemma 3.3.3 Suppose $\{X_k\}$ is irreducible and that one of its states is positive recurrent, then all of its states are positive recurrent. (The same statement holds if we replace positive recurrent by null recurrent or transient.) ☐

Lemma 3.3.4 If state i of a Markov chain is aperiodic, then $\lim_{k\to\infty} p_i[k] = 1/M_i$. (This is true whether or not $M_i < \infty$, and even for transient states by defining $M_i = \infty$ when state i is transient.) ☐

Theorem 3.3.5 Consider a time-homogeneous Markov chain which is irreducible and aperiodic. Then, the following results hold.

- If the Markov chain is positive recurrent, there exists a unique π such that $\pi = \pi\mathbf{P}$ and $\lim_{k\to\infty} p[k] = \pi$. Further, $\pi_i = 1/M_i$.
- If there exists a positive vector π such that $\pi = \pi\mathbf{P}$ and $\sum_i \pi_i = 1$, it must be the stationary distribution and $\lim_{k\to\infty} p[k] = \pi$. (From Lemma 3.3.4, this also means that the Markov chain is positive recurrent.)
- If there exists a positive vector π such that $\pi = \pi\mathbf{P}$, and $\sum_i \pi_i$ is infinite, a stationary distribution does not exist, and $\lim_{k\to\infty} p_i[k] = 0$ for all i. ☐

The following example is an illustration of the application of the above theorem.

Example 3.3.5

Consider a simple model of a wireless link where, due to channel conditions, either one packet or no packet can be served in each time slot. Let $s(k)$ denote the number of packets served in time slot k and suppose that $s(k)$ are i.i.d. Bernoulli random variables with mean μ. Further, suppose that packets arrive to this wireless link according to a Bernoulli process with mean λ, i.e., $a(k)$ is Bernoulli with mean λ, where $a(k)$ is the number of arrivals in time slot k and $a(k)$ are i.i.d. across time slots. Assume that $a(k)$ and $s(k)$ are independent processes. We specify the following order in which events occur in each time slot:

- we assume that any packet arrival occurs first in the time slot, followed by any packet departure;

- packets that are not served in a time slot are queued in a buffer for service in a future time slot.

Let $q(k)$ be the number of packets in the queue at the beginning of time slot k. Then $q(k)$ is a Markov chain and evolves according to the equation

$$q(k+1) = (q(k) + a(k) - s(k))^+.$$

We are interested in the steady-state distribution of this Markov chain. The Markov chain can be pictorially depicted as in Figure 3.4, where the circles denote the states of the Markov chain (the number of packets in the queue) and the arcs denote the possible transitions with the number on an arc denoting the probability of that transition occurring. For example, $P_{i,i+1}$, the probability that the number of packets in the queue increases from i to $i+1$ from one time slot to the next, is equal to the probability that there was an arrival but no departure in the time slot. Thus,

$$P_{i,i+1} = \lambda(1 - \mu).$$

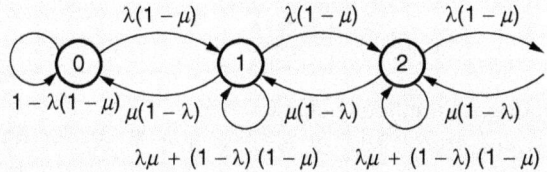

Figure 3.4 The discrete-time Markov chain for the queue.

Similarly, P_{ii} for $i > 0$ is equal to the probability of no arrival and no departure or one arrival and one departure in a time slot, and thus

$$P_{ii} = \lambda\mu + (1 - \lambda)(1 - \mu).$$

On the other hand, P_{00} is simply the probability of no arrival or one arrival and one departure, which is equal to $1 - \lambda + \lambda\mu$. If $i > 0$, then $P_{i,i-1} = 1 - P_{ii} - P_{i,i+1}$. First, we note that it is easy to see that this Markov chain is irreducible and aperiodic. To compute π, we have to solve for $\pi = \pi\mathbf{P}$, which can be written explicitly as

$$\pi_i = \pi_{i-1}P_{i-1,i} + \pi_i P_{ii} + \pi_{i+1}P_{i+1,i}, \quad i > 0, \tag{3.11}$$

$$\pi_0 = \pi_0 P_{00} + \pi_1 P_{10}.$$

The above equations have a simple interpretation: the stationary probability of being in i is equal to the sum of the probability of being in state i in the previous time slot multiplied by the probability of continuing in the same state and the probability being in another state and making a transition to state i. The above set of equations should be augmented with the constraint $\sum_i \pi_i = 1$ to solve for π.

Using the fact that $P_{ii} + P_{i,i-1} + P_{i,i+1} = 1$, a little thought shows that if we find π that satisfies

$$\pi_i P_{i,i+1} = \pi_{i+1} P_{i+1,i}, \quad \forall i,$$

it also solves (3.11). Thus,

$$\pi_{i+1} = \frac{(1-\mu)\lambda}{(1-\lambda)\mu}\pi_i,$$

which implies

$$\pi_i = \left(\frac{(1-\mu)\lambda}{(1-\lambda)\mu}\right)^i \pi_0. \tag{3.12}$$

Since $\sum_{i\geq 0} \pi_i = 1$, we obtain

$$\pi_0 \sum_{i=0}^{\infty} \left(\frac{(1-\mu)\lambda}{(1-\lambda)\mu}\right)^i = 1.$$

If we assume $\lambda < \mu$, then

$$\frac{(1-\mu)\lambda}{(1-\lambda)\mu} < 1,$$

and

$$\pi_0 = 1 - \frac{(1-\mu)\lambda}{(1-\lambda)\mu}.$$

Thus, the stationary distribution is completely characterized.

If $\lambda \geq \mu$, let $\pi_0 = 1$, and from (3.12) it is clear that $\sum_i \pi_i = \infty$. Thus, from Theorem 3.3.5, the Markov chain is not positive recurrent if $\lambda \geq \mu$. $\qquad\square$

As noted in the above example, the following theorem provides a sufficient condition for verifying the stationary distribution of a DTMC.

Theorem 3.3.6 Consider a time-homogeneous Markov chain which is irreducible and aperiodic. If there exists a positive vector π such that $\pi_i P_{ij} = \pi_j P_{ji}$ and $\sum_i \pi_i = 1$, it must be the stationary distribution. $\qquad\square$

Note that $\pi_i P_{ij} = \pi_j P_{ji}$ is called the *local balance equation*.

Unlike the above example, there are also many instances where one cannot easily find the stationary distribution by solving $\pi = \pi \mathbf{P}$. But we would still like to know if the stationary distribution exists. It is often easy to verify the irreducibility and aperiodicity of a Markov chain, but, in general, it is difficult to verify directly whether a Markov chain is positive recurrent from the definitions given earlier. Instead, there is a convenient test called Foster's test or the Foster–Lyapunov test to check for positive recurrence, and we state this next.

Theorem 3.3.7 (Foster–Lyapunov theorem) Let $\{X_k\}$ be an irreducible Markov chain with a state space $\mathcal{S}$. Suppose that there exist a function $V : \mathcal{S} \to \mathcal{R}^+$ and a finite set $\mathcal{B} \subseteq \mathcal{S}$ satisfying the following conditions:

(1) $E[V(X_{k+1}) - V(x) \mid X_k = x] \leq -\epsilon$ if $x \in \mathcal{B}^c$ for some $\epsilon > 0$, and
(2) $E[V(X_{k+1}) - V(x) \mid X_k = x] \leq A$ if $x \in \mathcal{B}$ for some $A < \infty$.

Then the Markov chain $\{X_k\}$ is positive recurrent.

Proof We will prove the result under the further assumption that $\{X_k\}$ is aperiodic. Note that the theorem itself does not require the Markov chain to be aperiodic. Following the two conditions stated in the theorem, we have

$$E[V(X_{k+1}) - V(X_k) \mid X_k = x] \leq -\epsilon \mathbb{I}_{x \in \mathcal{B}^c} + A \mathbb{I}_{x \in \mathcal{B}}$$

$$= -\epsilon \mathbb{I}_{x \in \mathcal{B}^c} + A - A \mathbb{I}_{x \in \mathcal{B}^c}$$

$$= -(A + \epsilon) \mathbb{I}_{x \in \mathcal{B}^c} + A.$$

Taking expectations on both sides, we obtain

$$E[V(X_{k+1})] - E[V(X_k)] = -(A + \epsilon) \Pr(X_k \in \mathcal{B}^c) + A,$$

$$\sum_{k=0}^{N} (E[V(X_{k+1})] - E[V(X_k)]) = -(A + \epsilon) \sum_{k=0}^{N} \Pr(X_k \in \mathcal{B}^c) + A(N + 1).$$

Thus,

$$E[V(X_{N+1})] - E[V(X_0)] = -(A + \epsilon) \sum_{k=0}^{N} \Pr(X_k \in \mathcal{B}^c) + A(N + 1),$$

$$(A + \epsilon) \sum_{k=0}^{N} \Pr(X_k \in \mathcal{B}^c) = A(N + 1) + E[V(X_0)] - E[V(X_{N+1})]$$

$$\leq A(N + 1) + E[V(X_0)],$$

$$\frac{A + \epsilon}{N + 1} \sum_{k=0}^{N} \Pr(X_k \in \mathcal{B}^c) \leq A + \frac{1}{N + 1} E[V(X_0)],$$

$$\lim_{N \to \infty} \sup \frac{1}{N + 1} \sum_{k=0}^{N} \Pr(X_k \in \mathcal{B}^c) \leq \frac{A}{A + \epsilon},$$

$$\lim_{N \to \infty} \inf \frac{1}{N + 1} \sum_{k=0}^{N} \Pr(X_k \in \mathcal{B}) \geq \frac{\epsilon}{A + \epsilon}.$$

Suppose that any state of the Markov chain is not positive recurrent, then all states are not positive recurrent since the Markov chain is irreducible. Thus, $M_i = \infty$, $\forall i$, and $\lim_{k \to \infty} p_i[k] = 0$, $\forall i$. Thus,

$$\lim_{k \to \infty} \Pr(X_k \in \mathcal{B}) = 0 \quad \text{or} \quad \lim_{N \to \infty} \inf \frac{1}{N + 1} \sum_{k=0}^{N} \Pr(X_k \in \mathcal{B}) = 0,$$

which contradicts the fact that it is $\geq \epsilon / (A + \epsilon)$.   □

Next, we present two extensions of the Foster–Lyapunov theorem without proof.

Theorem 3.3.8 An irreducible Markov chain $\{X_k\}$ is positive recurrent if there exists a function $V : \mathcal{S} \to \mathcal{R}^+$, a positive integer $L \geq 1$, and a finite set $\mathcal{B} \subseteq \mathcal{S}$ satisfying the following conditions:

$$E[V(X_{k+L}) - V(x) \mid X_k = x] \leq -\epsilon \mathbb{I}_{x \in \mathcal{B}^c} + A \mathbb{I}_{x \in \mathcal{B}}$$

for some $\epsilon > 0$ and $A < \infty$.   □

Theorem 3.3.9 An irreducible Markov chain $\{X_k\}$ is positive recurrent if there exists a function $V : \mathcal{S} \to \mathcal{R}^+$, a function $\eta : \mathcal{S} \to \mathcal{R}^+$, and a finite set $\mathcal{B} \subseteq \mathcal{S}$ satisfying the following conditions:

$$E[V(X_{k+\eta(x)}) - V(x) \mid X_k = x] \leq -\epsilon \eta(x) \mathbb{I}_{x \in \mathcal{B}^c} + A \mathbb{I}_{x \in \mathcal{B}}$$

for some $\epsilon > 0$ and $A < \infty$.   □

The following theorem provides conditions under which a Markov chain is not positive recurrent.

Theorem 3.3.10 An irreducible Markov chain $\{X_k\}$ is either transient or null recurrent if there exists a function $V : \mathcal{S} \to \mathcal{R}^+$ and a finite set $\mathcal{B} \subseteq \mathcal{S}$ satisfying the following conditions:

- $E[V(X_{k+1}) - V(X_k) \mid X_k = x] \geq 0, \forall x \in \mathcal{B}^c$;
- there exists some $x \in \mathcal{B}^c$ such that $V(x) > V(y)$ for all $y \in \mathcal{B}$, and
- $E[|V(X_{k+1}) - V(X_k)| \mid X_k = x] \leq A$ for some $A < \infty$ and $\forall x \in \mathcal{S}$.   □

3.4   Delay and packet loss analysis in queues

3.4.1   Little's law

We start this section with the famous Little's law, which states that the mean queue length in queueing system is equal to the product of the mean arrival rate and the expected waiting time. Little's law holds for very general arrival processes and service disciplines, and for both discrete-time and continuous-time queueing systems. In this book, we only derive Little's law for discrete-time queueing systems. The derivation for continuous-time systems is similar.

We assume that packets arrive at the beginning of a time slot and are served at the end of a time slot. The queue length at time slot t, denoted by $q(t)$, is the number of packets remaining in the system at the beginning of time slot t, *before packet arrivals occur.*

Let $A(t)$ denote the number of packet arrivals up to and including time slot t, and let $\mathbb{I}_i(t)$ be an indicator of the presence of packet i in the queue at time t, i.e.,

$$\mathbb{I}_i(t) = \begin{cases} 1, & \text{if packet } i \text{ arrived in a time slot} < t \text{ and departed in a time slot} \geq t, \\ 0, & \text{otherwise,} \end{cases}$$

where packets are indexed according to arrival times and ties are broken arbitrarily. Note that $\mathbb{I}_i(t) = 0$ if packet i arrives in time slot t. Since $\mathbb{I}_i(t) = 1$ means that packet i remains in the system at the beginning of time slot t, $q(t)$ can be written as

$$q(t) = \sum_{i=1}^{A(t-1)} \mathbb{I}_i(t).$$

Further, the waiting time of packet i, denoted by w_i, is defined to be

$$w_i = \sum_{t=1}^{\infty} \mathbb{I}_i(t).$$

Note that, according to this definition, the waiting time of a packet is zero if the packet arrives and departs in the same time slot.

We define $\lambda(T)$ to be the average arrival rate by time slot T, i.e.,

$$\lambda(T) = \frac{A(T)}{T},$$

and $L(T)$ to be the average queue length by time slot T, i.e.,

$$L(T) = \frac{\sum_{t=1}^{T} q(t)}{T}.$$

Further, we define $W(n)$ to be the average waiting time of the first n packets, i.e.,

$$W(n) = \frac{1}{n} \sum_{k=1}^{n} w_k.$$

We further define the following three limits:

$$\lambda = \lim_{T \to \infty} \lambda(T), \quad L = \lim_{T \to \infty} L(T), \quad W = \lim_{n \to \infty} W(n).$$

So λ is the average arrival rate, L is the average queue length, and W is the average waiting time.

Theorem 3.4.1 (Little's law) Assuming that λ and W exist and are finite, L exists and $L = \lambda W$.

Proof According to the definition of $L(T)$, we have

$$L(T) = \frac{1}{T} \sum_{t=1}^{T} q(t) = \frac{1}{T} \sum_{t=1}^{T} \sum_{i=1}^{A(t-1)} \mathbb{I}_i(t) = \frac{1}{T} \sum_{i=1}^{A(T-1)} \sum_{t=1}^{T} \mathbb{I}_i(t).$$

We first derive an upper bound on L. Note that

$$\sum_{t=1}^{T} \mathbb{I}_i(t) \le \sum_{t=1}^{\infty} \mathbb{I}_i(t) = w_i,$$

so

$$L(T) = \frac{1}{T} \sum_{i=1}^{A(T-1)} \sum_{t=1}^{T} \mathbb{I}_i(t) \le \frac{1}{T} \sum_{i=1}^{A(T-1)} w_i.$$

When λ and W exist and are finite, we have

$$\lim_{T \to \infty} \frac{1}{T} \sum_{i=1}^{A(T-1)} w_i = \lim_{T \to \infty} \frac{T-1}{T} \frac{A(T-1)}{T-1} \frac{\sum_{i=1}^{A(T-1)} w_i}{A(T-1)} = \lambda W,$$

so

$$L = \lim_{T \to \infty} L(T) \le \lambda W. \qquad (3.13)$$

Next we show that λW is also a lower bound on L. To prove this, we denote the set of packets that have departed up to and including time slot T by $\mathcal{D}(T)$. Note that the packets can be served in an arbitrary order, so $\mathcal{D}(T)$ can be an arbitrary subset of $\{1,\ldots,A(T)\}$.

Since any packet i ($i \in \mathcal{D}(T-1)$) departed from the buffer before time slot T, we have $w_i = \sum_{t=1}^{T-1} \mathbb{I}_i(t)$, which leads to the following lower bound on $L(T)$:

$$L(T) = \frac{1}{T} \sum_{i=1}^{A(T-1)} \sum_{t=1}^{T} \mathbb{I}_i(t) \geq \frac{1}{T} \sum_{i\in\mathcal{D}(T-1)} w_i = \frac{|\mathcal{D}(T-1)|}{T} \times \frac{1}{|\mathcal{D}(T-1)|} \sum_{i\in\mathcal{D}(T-1)} w_i.$$

Now, if the following two inequalities hold:

$$\lim_{T\to\infty} \frac{|\mathcal{D}(T)|}{T} \geq \lambda \quad \text{and} \quad \lim_{T\to\infty} \frac{1}{|\mathcal{D}(T)|} \sum_{i\in\mathcal{D}(T)} w_i \geq W, \tag{3.14}$$

we have $L \geq \lambda W$, so the theorem holds. The interested reader can read the rest of the proof, which establishes (3.14). However, this part can be skipped without compromising the understanding of the rest of this chapter.

To prove (3.14), we will establish the following claim: given any $\epsilon > 0$, there exist constants a_ϵ, b_ϵ, and δ_ϵ such that (i) $\delta_\epsilon \to 0$ as $\epsilon \to 0$ and (ii) for any $T \geq b_\epsilon$,

$$\{a_\epsilon,\ldots,A((1-\delta_\epsilon)T)\} \subseteq \mathcal{D}(T).$$

This claim indicates that, for sufficiently large T, most packets that arrived up to time slot T must have departed by time slot T. From this claim, it is straightforward to show (3.14).

We now verify the claim to complete the proof. Since $\lim_{n\to\infty} W(n)/n$ exists and is finite,

$$\lim_{n\to\infty} \frac{w_n}{n} = \lim_{n\to\infty} \frac{\sum_{i=1}^n w_i}{n} - \lim_{n\to\infty} \frac{n-1}{n} \frac{\sum_{i=1}^{n-1} w_i}{n-1}$$
$$= W - W$$
$$= 0.$$

So, given any $\epsilon > 0$, there exists n_ϵ such that $w_n \leq \epsilon n$ for all $n \geq n_\epsilon$. Further, $\lim_{T\to\infty}(A(T)/T) = \lambda$ implies that, given $\epsilon > 0$, there exists T_ϵ such that

$$(\lambda - \epsilon)T \leq A(T) \leq (\lambda + \epsilon)T \tag{3.15}$$

for any $T \geq T_\epsilon$.

Given ϵ, we choose T large enough such that

$$T \geq \frac{T_\epsilon}{1 - \epsilon(\lambda + \epsilon)}, \tag{3.16}$$

$$T \geq \frac{n_\epsilon}{(\lambda - \epsilon)(1 - \epsilon(\lambda + \epsilon))}. \tag{3.17}$$

Since $(1 - \epsilon(\lambda + \epsilon))T \geq T_\epsilon$ according to (3.16),

$$A((1 - \epsilon(\lambda + \epsilon))T) \geq (\lambda - \epsilon)(1 - \epsilon(\lambda + \epsilon))T \geq n_\epsilon,$$

where the first and second inequalities follow from condition (3.15) and (3.17), respectively. Now, for any packet i, such that $n_\epsilon \leq i \leq A((1 - \epsilon(\lambda + \epsilon))T)$, according to the definition of n_ϵ

$$w_i \leq \epsilon i \leq \epsilon A((1 - \epsilon(\lambda + \epsilon))T) \leq \epsilon(\lambda + \epsilon)(1 - \epsilon(\lambda + \epsilon))T \leq \epsilon(\lambda + \epsilon)T,$$

where the third inequality holds due to conditions (3.15) and (3.16). To that end, packet i for $n_\epsilon \leq i \leq A((1 - \epsilon(\lambda + \epsilon))T)$ must depart by

$$(1 - \epsilon(\lambda + \epsilon))T - 1 + w_i \leq (1 - \epsilon(\lambda + \epsilon))T - 1 + \epsilon(\lambda + \epsilon)T = T - 1,$$

so

$$\{n_\epsilon, \ldots, A((1 - \epsilon(\lambda + \epsilon))T)\} \subseteq \mathcal{D}(T).$$

The claim therefore holds with

$$a_\epsilon = n_\epsilon, \quad b_\epsilon = \max\left\{\frac{T_\epsilon}{1 - \epsilon(\lambda + \epsilon)}, \frac{n_\epsilon}{(\lambda - \epsilon)(1 - \epsilon(\lambda + \epsilon))}\right\}, \quad \text{and} \quad \delta_\epsilon = \epsilon(\lambda + \epsilon),$$

which goes to 0 as $\epsilon \to 0$.   □

The above derivation of Little's law assumes that L, λ, and W are sample path averages. In applications, we will apply Little's law to steady-state expected queue lengths, steady-state expected arrival rates, and steady-state expected waiting times. Thus, we will make an implicit assumption throughout the book that the stochastic processes that we consider are ergodic, i.e., processes for which steady-state expectations and sample-path averages are equal. We will also assume that λ and W exist as required by Little's law. We further remark that Little's law is applicable when departures occur followed by arrivals in the discrete-time model, with appropriate definitions of queue length and waiting time.

Next we will consider single-server queueing systems under various traffic and buffer models. The results help us understand queueing delays and packet drop probabilities over a link.

3.4.2   The Geo/Geo/1 queue

We consider a single-server queue with infinite buffer space. Packets arrive to this queue according to an i.i.d. Bernoulli process with parameter λ. In each time slot, either one packet is served with probability μ or no packet is served with probability $1 - \mu$. Equivalently, we can assume that the server serves one unit of data per time slot and packet sizes are geometrically distributed with mean $1/\mu$. Under these assumptions, the inter-arrival and departure times are geometrically distributed, so the queue is called a Geo/Geo/1 queue.

This simple queueing system can be viewed as a birth–death process, as shown in Figure 3.5, where the state of the Markov chain is the queue length. The parameter

$$\alpha = \Pr(1 \text{ arrival, no departure}) = \lambda(1 - \mu)$$

is the probability that the queue length increases by one, and

$$\beta = \Pr(\text{no arrival, 1 departure}) = (1 - \lambda)\mu$$

is the probability that the queue length decreases by one.

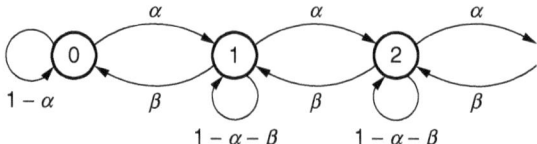

Figure 3.5 The birth–death process of a Geo/Geo/1 queue.

Let π denote the steady-state distribution of the Markov chain. We will attempt to find π by solving the local balance equation

$$\beta \pi_{i+1} = \alpha \pi_i.$$

Dividing by β on both sides yields

$$\pi_{i+1} = \rho \pi_i, \tag{3.18}$$

where $\rho = \alpha/\beta = \lambda(1-\mu)/\mu(1-\lambda)$. Since equality (3.18) holds for all i, we can further obtain

$$\pi_i = \rho^i \pi_0. \tag{3.19}$$

Normalizing to make the sum of the probabilities equal to 1, we have

$$\sum_{i=0}^{\infty} \pi_i = \pi_0 \sum_{i=0}^{\infty} \rho^i = 1. \tag{3.20}$$

If $\rho < 1$, i.e., $\lambda/\mu < 1$, then from (3.20) we get $\pi_0/(1-\rho) = 1$ and $\pi_0 = 1 - \rho$. According to (3.19),

$$\pi_i = \rho^i(1-\rho),$$

and the Markov chain is positive recurrent. If $\rho \geq 1$, $\sum_i \pi_i = \infty$. So the Markov chain is not positive recurrent.

Assume that $\rho < 1$, then the average queue length

$$E[q] = \sum_{i=0}^{\infty} \rho^i(1-\rho)i$$

$$= (1-\rho)\rho \sum_{i=1}^{\infty} i\rho^{i-1}$$

$$= (1-\rho)\rho \frac{1}{(1-\rho)^2}$$

$$= \frac{\rho}{1-\rho}.$$

By Little's law, the average waiting time of a packet is given by

$$W = \frac{L}{\lambda} = \frac{\rho}{\lambda(1-\rho)}.$$

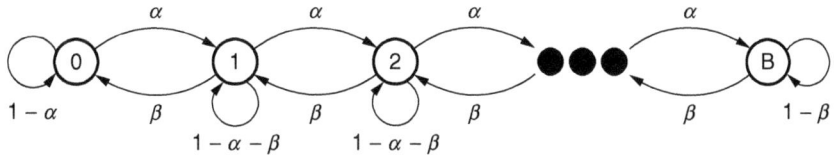

Figure 3.6 The birth–death process of a Geo/Geo/1/B queue.

3.4.3   The Geo/Geo/1/B queue

We now consider the same queueing model as in Section 3.4.2, but we further assume that the buffer size is finite. We let the maximum buffer size be denoted by B, i.e., B is the maximum number of packets allowed in the queue. When the buffer is full, newly arriving packets are dropped. It is easy to see that $q(t)$ is again a Markov chain, as shown in Figure 3.6, and that the steady-state distribution satisfies

$$\beta\pi_{i+1} = \alpha\pi_i$$

for $0 \le i \le B - 1$. Defining $\rho = \alpha/\beta$, we have

$$\pi_{i+1} = \rho\pi_i$$

for $0 \le i \le B - 1$. Normalizing the probabilities, we have $\pi_0 \sum_{i=0}^{B} \rho^i = 1$, which implies

$$\pi_0 \frac{1 - \rho^{B+1}}{1 - \rho} = 1.$$

Hence, we obtain

$$\pi_0 = \frac{1 - \rho}{1 - \rho^{B+1}},$$

$$\pi_i = \frac{(1 - \rho)\rho^i}{1 - \rho^{B+1}}, \quad i = 0, 1, 2, \dots, B.$$

Note that the Markov chain is positive recurrent for any ρ because this is a finite-state Markov chain. In the Geo/Geo/1/B queue model, the quantity of interest is the *fraction of arriving packets that are dropped*. We present an outline of the calculations involved in obtaining an expression for this quantity.

Denote by p_d the fraction of arriving packets that are dropped, i.e.,

$$p_d = \lim_{T\to\infty} \frac{\sum_{t=0}^{T} \mathbb{I}_{\{q(t)=B, a(t)=1\}}}{\sum_{t=0}^{T} \mathbb{I}_{\{a(t)=1\}}},$$

where $a(t) = 1$ if there is a packet arrival at time t, and $a(t) = 0$ otherwise. In other words, p_d is the fraction of time slots that the buffer is full when there is an arrival.

Assuming ergodicity and that $q(t)$ is in steady state, we obtain

$$p_d = \lim_{T\to\infty} \frac{\frac{1}{T}\sum_{t=1}^{T} \mathbb{I}_{\{q(t)=B, a(t)=1\}}}{\frac{1}{T}\sum_{t=1}^{T} \mathbb{I}_{\{a(t)=1\}}}$$

$$= \frac{\Pr(a(t) = 1, q(t) = B)}{\Pr(a(t) = 1)}$$

$$= \Pr(q(t) = B | a(t) = 1).$$

Since the arrivals are i.i.d. across time slots, and independent of the services, $q(t)$ is independent of $a(t)$, which implies that

$$p_d = \Pr(q(t) = B) = \pi_B. \tag{3.21}$$

Thus, we have used the fact that the next arrival is independent of past arrivals and services. This observation may not be true for other arrivals, but it holds for Bernoulli arrival processes. Equation (3.21) is a special case of the more general result called the BASTA property: Bernoulli Arrivals See Time Averages, since p_d (what the arrivals see) is equal to π_B (the time average).

3.4.4   The discrete-time G/G/1 queue

We now consider a G/G/1 queue where the first G refers to the fact that the arrival process is general and the second G refers to the fact that the service process is general. We assume that the arrivals (and potential departures) are i.i.d. across time and that the arrival process is independent of the potential departure process. We also assume that both the arrival process and the potential departure process have finite second moments.

The queue dynamics can be described as follows:

$$q(t+1) = (q(t) + a(t) - s(t))^+,$$

where $a(t)$ is the number of packet arrivals at the beginning of time slot t and $s(t)$ is the number of packet departures at the end of time slot t. Assume the first and second moments of $a(t)$ and $s(t)$ exist, and

$$E[a(t)] = \lambda \quad \text{and} \quad E[a^2(t)] = m_{2a},$$

$$E[s(t)] = \mu \quad \text{and} \quad E[s^2(t)] = m_{2s}.$$

The following theorem states that the queueing system is positive recurrent when $\lambda < \mu$, i.e., mean arrival rate is strictly less than mean service rate.

Theorem 3.4.2 The discrete-time G/G/1 queue is positive recurrent when $\lambda < \mu$. Further, using $q(\infty)$ to denote informally the queue in steady state, the following inequality holds:

$$E[q(\infty)] \leq \frac{m_{2a} + m_{2s} - 2\lambda\mu}{2(\mu - \lambda)}.$$

Proof Considering the Lyapunov function

$$V(t) \triangleq \frac{1}{2}q^2(t),$$

the drift of the Lyapunov function is given by

$$E[V(t+1) - V(t)|q(t) = q] = \frac{1}{2}E\left[((q + a(t) - s(t))^+)^2 - q^2\right]$$

$$\leq \frac{1}{2}E\left[(q + a(t) - s(t))^2 - q^2\right]$$

$$= \frac{1}{2}E\left[(a(t) - s(t))^2 + 2q(a(t) - s(t))\right]$$

$$= \frac{1}{2}[m_{2a} + m_{2s} - 2\lambda\mu + 2q(\lambda - \mu)]$$

$$= \frac{m_{2a} + m_{2s} - 2\lambda\mu}{2} + q(\lambda - \mu). \tag{3.22}$$

Fix $\epsilon > 0$. If

$$\frac{m_{2a} + m_{2s} - 2\lambda\mu}{2} + q(\lambda - \mu) \le -\epsilon$$

or, in other words,

$$q \ge \left(\epsilon + \frac{m_{2a} + m_{2s} - 2\lambda\mu}{2}\right)\frac{1}{\mu - \lambda},$$

then

$$E[V(t+1) - V(t)|q(t) = q] \le -\epsilon.$$

Invoking the Foster–Lyapunov theorem (Theorem 3.3.7), we conclude that the Markov chain is positive recurrent and has a stationary distribution.

We will now obtain a bound on the mean queue length. Taking expectations on both sides of inequality (3.22), we have

$$E[V(t+1) - V(t)] \le \frac{m_{2a} + m_{2s} - 2\lambda\mu}{2} + E[q(t)](\lambda - \mu).$$

Supposing the system is in steady state and that $E[V(t)]$ exists in steady state, we have

$$E[V(t+1) - V(t)] = 0.$$

Using $q(\infty)$ to denote informally the steady state, we obtain

$$E[q(\infty)] \le \frac{m_{2a} + m_{2s} - 2\lambda\mu}{2(\mu - \lambda)},$$

which we call the *discrete-time Kingman bound*.   □

The discrete-time Kingman bound is actually tight when $\lambda \to \mu$. We define $\bar{q}(\lambda) \triangleq E[q(\infty)]$ to be the expected queue length when the arrival rate is λ.

Assume that $\lim_{\lambda \to \mu} m_{2a}(\lambda) = \hat{m}_{2a}$ exists. For example, for i.i.d. Bernoulli arrivals,

$$m_{2a} = E[a^2(t)] = \lambda,$$

so

$$\lim_{\lambda \to \mu} m_{2a} = \mu.$$

Then, as $\lambda \to \mu$, the discrete-time Kingman bound becomes

$$\lim_{\lambda \to \mu} (\mu - \lambda)\bar{q}(\lambda) = \frac{\hat{m}_{2a} + m_{2s} - 2\mu^2}{2}. \tag{3.23}$$

Using Little's law, we also have an upper bound on the queueing delay, given by

$$W \leq \frac{m_{2a} + m_{2s} - 2\lambda\mu}{2\lambda(\mu - \lambda)}.$$

3.5   Providing priorities: fair queueing

In Section 3.4, we looked at delay and packet loss in discrete-time queueing systems. Note that we did not specify the order in which packets are served and that we assumed that the link employs only a work-conserving policy, i.e., the link processes packets whenever the buffer is not empty. In general, when packets from different flows are waiting at a link, the link may provide different levels of priority to packets from different flows. This priority is implemented by choosing which packet to serve when there are multiple packets in the buffer. Such algorithms are called *packet scheduling algorithms*, and they determine the amount of resources allocated to a flow and hence the Quality of Service (QoS) received by the flow.

The simplest way to serve packets is to serve them in the order in which they arrive, as shown in Figure 3.7. A queue which uses this service discipline is called a First-In-First-Out (FIFO) queue. The advantage of a FIFO queue is its simplicity. The disadvantage is that packets from one flow affect the performance of all flows. For example, if a packet from flow i arrives immediately after the arrival of 100 packets from another flow j, then, in a FIFO queue, flow i's packet can only be served after the 100 packets from flow j have been served. In other words, one cannot provide any QoS guarantee to each flow at a link.

Weighted Fair Queueing (WFQ) is a packet scheduling policy that can isolate flows from each other to a large extent. In WFQ, packets are sorted into classes, e.g., each flow can be treated as a separate class; and a separate queue is maintained for each class, as shown in Figure 3.8. Each class is assigned a weight, denoting by w_i the weight associated with class i. The goal of WFQ is to allocate a bandwidth of $w_i c / \sum_{j \in \mathcal{N}(t)} w_j$ to queue i, where c is the capacity of the link and $\mathcal{N}(t)$ is the set of non-empty queues at time t. So, under WFQ, the link bandwidth is dynamically and fairly allocated to all non-empty queues, and class-i packets are served with a rate at least $w_i c / \sum_{j=1}^{N} w_j$, where N is the number of classes (queues) in the system.

Of course, packets are not fluids and one cannot allocate a rate equal to $w_i c / \sum_{i=1}^{N} w_j$ to flow i. For example, once a link starts processing flow j's packet, it cannot serve any other flow's packets till the current packet in service has departed. Thus, temporarily, all flows other than flow j will receive no service. WFQ is a scheme that closely approximates the allocation of rate $w_i c / \sum_{i=1}^{N} w_j$ to each flow i, while taking into account the fact that each packet has non-zero and even possibly varying lengths. We next explain the key idea behind WFQ and its implementation. Without loss of generality, we assume that each flow

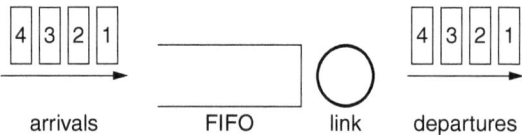

Figure 3.7 FIFO queue: the numbers are the indices of the packets.

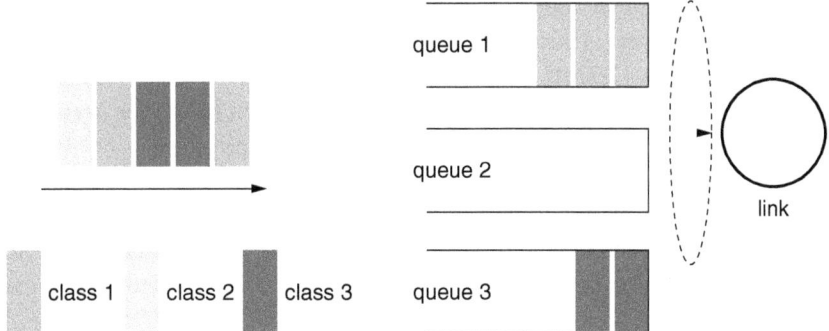

Figure 3.8 WFQ queue.

is a separate class. We first consider the case where all packets are single-bit packets and all flows have the same weight. For this special case, WFQ is simply round-robin scheduling among all non-empty queues. This ideal form of WFQ is known as Processor Sharing (PS). As in Figure 3.8, the link will transmit one packet from queue 1, skip queue 2, and then transmit one packet from queue 3. Letting $N(t)$ denote the number of non-empty queues at time t, each non-empty queue receives a transmission rate of

$$\frac{c}{N(t)}.$$

In other words, flows that have packets in the buffer share the link capacity equally.

If flows are associated with different weights, the ideal form of WFQ, i.e., where bit-level service can be provided, is called Generalized Processor Sharing (GPS), whereby the link transmits w_i bits from queue i when it is the turn of queue i, where w_i is the weight of flow i. So, queue i ($i \in \mathcal{N}(t)$) receives a transmission rate of

$$\frac{w_i}{\sum_{j \in \mathcal{N}(t)} w_j} c.$$

The link capacity is shared among non-empty queues in proportion to their weights. It is useful to think of service as proceeding in rounds. Suppose queue 1, queue 2, and queue 3 are the non-empty queues at a time, then one round of service corresponds to serving w_1 bits from queue 1, w_2 bits from queue 2, and w_3 bits from queue 3. On the other hand, if queue 1 and queue 3 are the only non-empty queues, each round would consist of w_1 bits of service to queue 1 and w_3 bits of service to queue 3. When we consider packetized versions of WFQ in the following, this concept of rounds will be useful.

In reality, it is impossible to implement PS and GPS for packet scheduling because packets having varying sizes and packet transmissions cannot be interrupted. To deal with varying packet sizes, WFQ maintains a round counter $R(t)$, called the virtual time. This virtual time evolves as follows:

$$R(t) = \int_0^t \frac{c}{\sum_{j \in \tilde{\mathcal{N}}(\tau)} w_j} \mathbb{I}_{\tilde{\mathcal{N}}(\tau) \neq \emptyset} \, d\tau, \tag{3.24}$$

where $\tilde{\mathcal{N}}(t)$ is the set of backlogged flows at time t if we had implemented GPS. The virtual time $R(t)$ is the number of rounds of service that the link has provided up to time t assuming that GPS is being implemented. Recall that the link completes one round after transmitting w_i bits in queue i for all non-empty queues. To see this, suppose that $w_i = 1$ for all i. If only one queue is non-empty, the router can make c rounds every second, so $R(t + \delta t) = R(t) + c\delta t$. If m queues are non-empty, the link can make c/m rounds every second, so $R(t + \delta t) = R(t) + (c/m)\delta t$. Thus, equation (3.24) for $R(t)$, the virtual time at time t, is simply the number of rounds served up to time t.

For a packet of size S, the link needs exactly S rounds to transmit the entire packet. Let $F_{(i,k+1)}$ denote the virtual time when the $(k + 1)$th packet of flow i is completely served, and let $S_{(i,k+1)}$ denote the size of the packet; $F_{(i,k+1)}$ is called the *finish tag* of the packet. Assuming the packet arrives at time t, the finish tag of the $(k+1)$th packet can be computed by considering the following two cases.

- If the queue for flow i is not empty when the packet arrives, the packet first waits in the queue until all packets arriving ahead of it are served. Since the finish tag of the kth packet of flow i is $F_{(i,k)}$, the $(k + 1)$th packet receives service starting from virtual time $F_{(i,k)}$, and leaves the queue at virtual time

$$F_{(i,k)} + \frac{S_{(i,k+1)}}{w_i}.$$

  The above expression can be understood by noting that when a packet is at the head of a queue, it takes $S_{(i,k+1)}/w_i$ rounds to serve the packet, independent of the states of other queues.

- If the queue for flow i is empty when the packet arrives, it can be served immediately. In this case, the packet leaves the network at virtual time

$$R(t) + \frac{S_{(i,k+1)}}{w_i}.$$

Note that $F_{(i,k)} > R(t)$ if queue i is not empty at time t, so the finish tag of the $(k + 1)$th packet of flow i can be computed recursively as follows:

$$F_{(i,k+1)} = \max\left\{F_{(i,k-1)}, R(t)\right\} + \frac{S_{(i,k+1)}}{w_i}.$$

It is straightforward to see that, under GPS, packets depart in the order of their finish tags. WFQ simply schedules packets according to their finish tags: whenever the link is idle,

---

Algorithm 1 Weighted Fair Queueing (WFQ)

1: $M_i(t)$: the finish tag of the packet at the tail of queue i at time t.
2: The link keeps track of the virtual time,

$$R(t) = \int_0^t \frac{c}{\sum_{j \in \mathcal{N}(\tau)} w_j} \mathbb{I}_{\mathcal{N}(\tau) \neq \emptyset}\, d\tau,$$

where $\tilde{\mathcal{N}}(\tau) = \{j : M_j(\tau) \geq R(\tau)\}$, i.e., the set of non-empty queues at time τ if GPS were used.
3: When the kth packet of flow i arrives at time t, a finish tag $F_{(i,k)}$ is assigned to the packet, where

$$F_{(i,k)} = \max\left\{F_{(i,k-1)}(t), R(t)\right\} + \frac{S_{(i,k)}}{w_i}.$$

4: When the link is free, it selects the packet with the smallest finish tag, among all packets in the queues, for transmission.

---

the packet with the smallest finish tag is selected for service. The algorithm is presented in Algorithm 1.

Example 3.5.1

Consider an output link with $c = 1$ bps and shared by two flows, called flow i and flow j. The link maintains one queue for each flow, and both queues are empty at the beginning ($t = 0$). As shown in Figure 3.9, flow i has two packets, one arriving at time $t = 0$ and the other arriving at time $t = 2$, and flow j has two packets, one arriving at the beginning of time slot 1 and the other arriving at time $t = 7$. The numbers in the boxes are packet sizes and the numbers above the boxes are the finish tags. The evolution of the virtual time is also illustrated in Figure 3.9. The transmission rates provided to the two queues under WFQ and GPS are shown in Figure 3.10. □

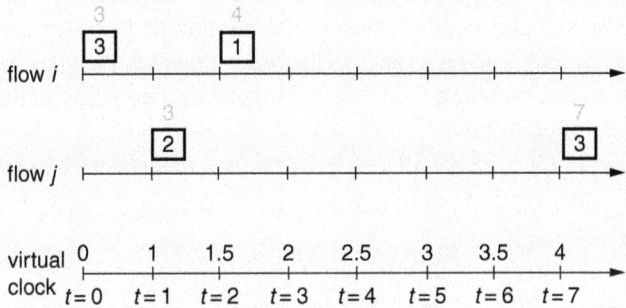

Figure 3.9 Finish tags and the virtual time.

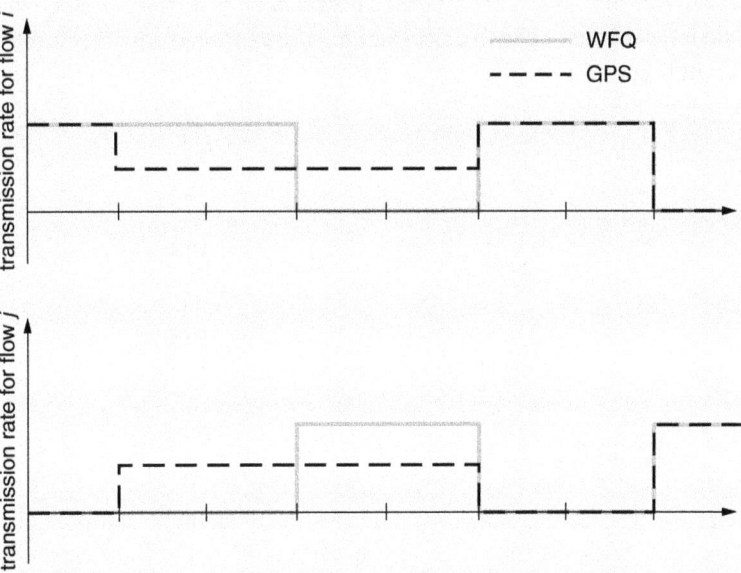

Figure 3.10 The transmission rates under WFQ and GPS. The thicker gray lines are transmission rates under WFQ and the dashed lines are transmission rates under GPS.

3.5.1 Key properties

In the following theorem, we establish a deterministic bound on the difference between the departure times of a packet under WFQ and under GPS, where the departure time of a packet is the time at which all bits of the packet have been transmitted.

Theorem 3.5.1 Denote by τ_k^w the departure time of packet k under WFQ, and by τ_k^g the departure time of packet k under GPS. The following inequality holds for any packet:

$$\tau_k^w \le \tau_k^g + \frac{S_{\max}}{c},$$

where $S_{\max}$ is the maximum packet size.

Proof Define a busy period of the system to be a time period during which the link is always transmitting packets. Since both WFQ and GPS are work-conserving policies, the busy periods under these two algorithms are identical. Consider a specific busy period, and number the packets served during this busy period according to the order in which they are served under WFQ, which is also the same as the order in which they depart under WFQ. Without loss of generality, we will let $t = 0$ be the time at the beginning of the busy period. Denote by S_n the size of packet n.

Consider packet k. If $\tau_m^g \le \tau_k^g$ for all $1 \le m \le k-1$, i.e., packets 1 to $k-1$ depart no later than packet k under GPS, the link has transmitted at least $\sum_{n=1}^{k} S_n$ bits by the time packet k departs, which implies that

$$\tau_k^g \ge \sum_{n=1}^{k} \frac{S_n}{c} = \tau_k^w.$$

Now consider the case where there is a packet m ($1 \le m \le k-1$) such that $\tau_m^g > \tau_k^g$. Let m^* be the largest index of such a packet, i.e., $\tau_{m^*}^g > \tau_k^g$ and $\tau_n^g \le \tau_k^g$ for $m^* < n \le k-1$, as shown in Figure 3.11.

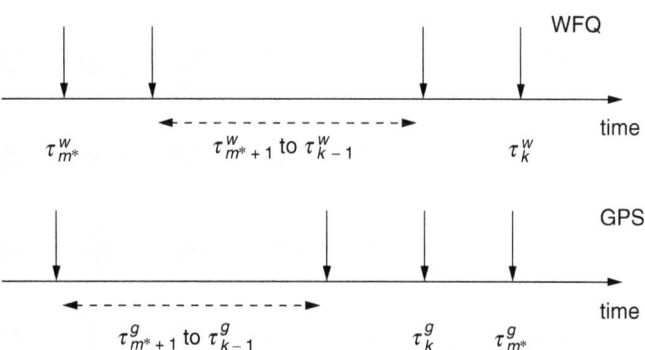

Figure 3.11 The packets are indexed according to the order of the departure times under WFQ; m^* is the largest index such that $\tau_{m^*}^g > \tau_k^g$, so packets $m^* + 1$ to $k - 1$ depart before packet k under GPS.

Let $b_{m^*}^w$ denote the time at which packet m^* starts to receive service under WFQ. We next show that packets $m^* + 1$ to k are served between $b_{m^*}^w$ and τ_k^g under GPS. This claim holds because of the following two facts:

(i) packets $m^* + 1$ to k must arrive after $b_{m^*}^w$ because, otherwise, packet m^* should not be selected under WFQ since its finish tag is larger than those of packets $m^* + 1$ to k, as shown in Figure 3.11; and

(ii) packets $m^* + 1$ to $k - 1$ depart no later than packet k under GPS, as shown in Figure 3.11.

Therefore, packets $m^* + 1$ to k arrive after $b_{m^*}^w$ and depart by τ_k^g under GPS. During this time, the link must have transmitted at least $\sum_{n=m^*+1}^{k} S_n$ bits under GPS, which takes $\sum_{n=m^*+1}^{k} S_n / c$ seconds. We conclude that

$$\tau_k^g \geq b_{m^*}^w + \frac{\sum_{n=m^*+1}^{k} S_n}{c} \geq \frac{\sum_{h=1}^{m^*-1} S_h}{c} + \frac{\sum_{n=m^*+1}^{k} S_n}{c}$$

$$= \tau_k^w - \frac{S_{m^*}}{c} \geq \tau_k^w - \frac{S_{\max}}{c},$$

where the second inequality holds because the packets are indexed according to the order they are served under WFQ, so packets 1 to $m^* - 1$ have been completely served by time $b_{m^*}^w$ under WFQ.   □

Based on Theorem 3.5.1, we can also bound the difference between the numbers of transmitted bits of flow i under WFQ and under GPS.

Theorem 3.5.2 Denote by $x_i^w(t)$ the number of flow i's bits that have been transmitted under WFQ up to time t, and by $x_i^g(t)$ the number of flow i's bits that have been transmitted under GPS up to time t. The following inequality holds for any $t \geq 0$:

$$x_i^w(t) \geq x_i^g(t) - S_{\max}.$$

Proof We prove this theorem by contradiction. Assume that

$$x_i^w(t) < x_i^g(t) - S_{\max} \tag{3.25}$$

holds for some flow i at time t. Since the per-flow queues are FIFO queues, this assumption implies that there exists a packet of flow i, say packet k, which has departed by time t under GPS but has not under WFQ. Let $L_k^w(t)$ denote the remaining size of packet k under WFQ by time t.

The departure time of packet k under WFQ satisfies

$$\tau_k^w \geq t + \frac{L_k^w(t)}{c} \tag{3.26}$$

since it has $L_k^w(t)$ bits left at time t. The departure time of the packet under GPS satisfies

$$\tau_k^g < t - \frac{S_{\max} - L_k^w(t)}{c} \tag{3.27}$$

because (i) according to assumption (3.25), by time t, GPS must have transmitted at least $S_{\max} - L_k^w(t)$ more bits belonging to flow i than WFQ, excluding the $L_k^w(t)$ remaining bits of packet k; and (ii) packets belonging to the same flows are served in FIFO order, so the additional bits are transmitted after the departure of packet k.

Combining inequalities (3.26) and (3.27), we obtain

$$\tau_k^w - \frac{L_k^w(t)}{c} \geq t > \tau_k^g + \frac{S_{\max} - L_k^w(t)}{c},$$

which implies that

$$\tau_k^w > \tau_k^g + \frac{S_{\max}}{c},$$

and contradicts Theorem 3.5.1. □

The major difficulty in implementing WFQ is the complexity of emulating GPS and maintaining the virtual time. Note that the set of non-empty queues under GPS is different from the set of non-empty queues under WFQ (i.e., the actual non-empty queues in the network). So WFQ needs to keep track of $\tilde{N}(t)$, which needs to be updated when there is a packet arrival or when the virtual time is larger than the smallest finish tag in the buffer (i.e., there would be a departure if GPS was used). Many approximations have been suggested in the literature to overcome this issue, but we do not consider them here.

WFQ guarantees that a packet departs no later than $S_{\max}/c$, but its performance can be further improved by using a scheduling policy called Worst Case Fair WFQ (WFWFQ), which has the two properties stated in Theorems 3.5.1 and 3.5.2, and, in addition, guarantees that the departure time of a packet under WFQ is no earlier than a fixed constant. The key idea behind WFWFQ is that, instead of selecting the packet with the smallest finish tag *among all packets*, it selects the packet with the smallest finish tag among *those packets that would have started service under GPS*. The proofs of the properties of WFWFQ will be considered in exercises at the end of the chapter.

3.6 Summary

- **The Chernoff bound** Consider a sequence of independently and identically distributed (i.i.d.) random variables $\{X_i\}$ with mean $\mu = E[X_i]$. For any constant x, the following inequality holds:

$$\Pr\left(\sum_{i=1}^n X_i \geq nx\right) \leq \exp\left(-n \sup_{\theta \geq 0}\{\theta x - \log M(\theta)\}\right),$$

where $M(\theta) \triangleq E\left[e^{\theta X_1}\right]$ is the moment generating function of X_1.

- **Queue overflow probability** Consider a single-server queueing system with infinite buffer space and shared by n sources. The server can serve c packets per time slot, and the number of packets injected by source i in time slot t is $a_i(t)$, which are i.i.d. across time and sources. Assume $\lambda \triangleq E[a_i(t)] < c/n$. Let $q(t)$ denote the queue length at time slot t, for all $\theta > 0$ such that $\Lambda(\theta)/\theta < c/n$, the following inequality holds:

$$\Pr(q(t) \geq B) \leq \frac{e^{n\Lambda(\theta)-\theta c}}{1 - e^{n\Lambda(\theta)-\theta c}} e^{-\theta B}, \qquad \forall t \geq 0,$$

where $\Lambda(\theta)$ is the log-moment generating function.

- **The Foster–Lyapunov theorems** Let $\{X_k\}$ be an irreducible Markov chain with a state space $\mathcal{S}$. The Markov chain $\{X_k\}$ is positive recurrent if there exist a function $V : \mathcal{S} \to \mathcal{R}^+$ and a finite set $\mathcal{B} \subseteq \mathcal{S}$ such that one of the following conditions holds:
  - $E[V(X_{k+1}) - V(x) \mid X_k = x] \leq -\epsilon \mathbb{I}_{x \in \mathcal{B}^c} + A\mathbb{I}_{x \in \mathcal{B}}$ for some $\epsilon > 0$ and $A < \infty$;
  - $E[V(X_{k+L}) - V(x) \mid X_k = x] \leq -\epsilon \mathbb{I}_{x \in \mathcal{B}^c} + A\mathbb{I}_{x \in \mathcal{B}}$ for some $\epsilon > 0, A < \infty$, and positive integer $L \geq 1$;
  - $E[V(X_{k+\eta(x)}) - V(x) \mid X_k = x] \leq -\epsilon \eta(x)\mathbb{I}_{x \in \mathcal{B}^c} + A\mathbb{I}_{x \in \mathcal{B}}$ for some $\epsilon > 0, A < \infty$, and function $\eta : \mathcal{S} \to \mathcal{R}^+$.

- **Little's law** In a queueing system, if the mean arrival rate and expected waiting time exist, the mean queue length is equal to the product of the mean arrival rate and the expected waiting time.

- **Discrete-time queueing systems**
  - **The Geo/Geo/1 queue** The mean queue length equals $\rho/(1 - \rho)$ and the expected waiting time equals $\rho/\lambda(1 - \rho)$.
  - **The Geo/Geo/1/B queue** The fraction of arriving packets that are dropped is $(1 - \rho)\rho^B/(1 - \rho^{B+1})$.
  - **The G/G/1 queue** The mean queue length satisfies

$$E[q] \leq \frac{m_{2a} + m_{2s} - 2\lambda\mu}{2(\mu - \lambda)}.$$

  This is called the discrete-time Kingman bound.

- **Fair queueing** Weighted Fair Queueing (WFQ) is a packet scheduling policy that allocates a data rate of $w_i c/\sum_{j \in \mathcal{N}(t)} w_j$ to a non-empty queue i, where w_i is the weight of queue i, c is the capacity of the link, and $\mathcal{N}(t)$ is the set of non-empty queues sharing the link at time t.

3.7   Exercises

Exercise 3.1 (The Chernoff bound for i.i.d. Bernoulli random variables) Let $\{X_i\}$ be i.i.d. Bernoulli random variables with mean μ. Prove that, for $x > 0$,

$$\Pr\left(\sum_{i=1}^{n} X_i \geq n(\mu + x)\right) \leq \exp\left(-nD((\mu + x)\|\mu)\right),$$

where

$$D((\mu + x)\|\mu) = (\mu + x)\log\frac{\mu + x}{\mu} + (1 - \mu - x)\log\frac{1 - \mu - x}{1 - \mu}.$$

Exercise 3.2 (The Chernoff bound for non-identical Bernoulli random variables) Suppose that X_i is $\text{Ber}(\mu_i)$ and that $X_1, X_2, \ldots$ are independent. In other words, the random variables are independent but not identical. Show that

$$\Pr\left(\frac{X_1 + \cdots + X_n}{n} \geq \mu + x\right) \leq e^{-nD((\mu+x)\|\mu)},$$

where

$$\mu = \frac{1}{n} \sum_{i=1}^{n} \mu_i.$$

Hint: Use the AM–GM (arithmetic mean–geometric mean) inequality:

$$\left(\prod_{i=1}^{n} x_i \right)^{1/n} \leq \frac{1}{n} \sum_{i=1}^{n} x_i$$

for any non-negative $x_1, x_2, \ldots, x_n$.

Exercise 3.3 (The Chernoff bound for non-identical bounded random variables) Let $X_i \in [0, 1]$ with mean μ_i. Show that the conclusions of Exercise 3.2 hold if X_i's are independent. (In other words, we are strengthening the result to allow X_i's which take values in $[0, 1]$ instead of just $\{0, 1\}$.) Hint: Use the convexity of $e^{\theta x}$ to note that $e^{\theta x}$ is below the line joining the points $(0, 1)$ and $(1, e^{\theta})$ when $\theta \geq 0$ and $0 \leq x \leq 1$.

Exercise 3.4 (The effective bandwidth and the burstiness of a source) Consider a single-source, single-server queue. Let $a(t)$ be the number of bits generated by the source at time t, and assume $\{a(t)\}$ are i.i.d. Bernoulli random variables such that

$$a(t) = \begin{cases} M, & \text{with probability } 1/M, \\ 0, & \text{with probability } 1 - 1/M. \end{cases} \tag{3.28}$$

Compute the effective bandwidth $\Lambda(\theta)/\theta$, and prove that $\Lambda(\theta)/\theta$ is an increasing function of M ($M > 0$) for any $\theta > 0$. This exercise shows that the effective bandwidth of a source increases as the burstiness of the source increases. Hint: It would be helpful to prove the following fact:

$$\frac{d}{dM} \frac{\Lambda(\theta)}{\theta} > 0.$$

Exercise 3.5 (Birth–death chains) Consider a Markov chain whose state space is the set of non-negative integers and whose transition matrix satisfies $P_{ij} = 0$ if $|i - j| > 1$. Such Markov chains are called *birth–death chains*. Consider a birth–death chain with $P_{i,i+1} = \lambda_i$ and $P_{i+1,i} = \mu_i$, and assume that $\lambda_i, \mu_i > 0$ for all i. Further assume that $P_{ii} > 0$ for all i.

(1) Show that the Markov chain is irreducible and aperiodic.

(2) Obtain conditions under which the Markov chain is (a) positive recurrent and (b) not positive recurrent. *Hint: Theorem 3.3.6 may be useful.*

Exercise 3.6 (Queue overflow probability with time-correlated arrivals) In   Theorem 3.2.1, we have shown that, if $a_i(t)$ are i.i.d. across time slots, then for all $\theta > 0$ such that $\Lambda(\theta)/\theta < c/n$, the queue overflow probability satisfies

$$\Pr(q(t) \geq B) \leq \frac{e^{n\Lambda(\theta) - \theta c}}{1 - e^{n\Lambda(\theta) - \theta c}} e^{-\theta B}$$

for all t. Now consider general stationary arrival processes $\{a_i(t)\}$ that may be correlated across time slots, and assume that there exists $\epsilon > 0$ such that

$$\frac{\Lambda(\theta)}{\theta} + \epsilon \leq \frac{c}{n},$$

where

$$\Lambda(\theta) = \lim_{k \to \infty} \frac{1}{k} \log E\left[e^{\theta \sum_{s=0}^{k-1} a_i(s)}\right]. \tag{3.29}$$

Prove that there exists a constant (κ) such that, for any $t \geq 0$,

$$\Pr(q(t) \geq B) \leq \kappa e^{-\theta B}.$$

Hint: Start from inequality (3.7) and use the fact that $a_i(t)$ is stationary, i.e., the distribution of $a_i(0)$, $a_i(1), \ldots, a_i(k-1)$ is the same as the distribution of $a_i(t-k)$, $a_i(t-k+1), \ldots, a_i(t-1)$.

Exercise 3.7 (The effective bandwidth of a Markovian source) Consider an ON–OFF source described as follows: the source is in one of the two states ON or OFF. When in the ON state, the source generates one packet per time slot, and when in the OFF state it does not generate any packets. The source switches between ON and OFF states according to a Markov chain. Let $a(k)$ denote the state of the Markov chain at time slot k. We denote the ON state by 1 and the OFF state by 0. Let $p_{ij} \in (0, 1)$ be the probability of switching from state i to state j. Thus, the transition probability matrix is given by

$$P = \begin{bmatrix} p_{00} & p_{01} \\ p_{10} & p_{11} \end{bmatrix}.$$

Assume that the Markov chain is in steady state, i.e., the initial probability distribution is the steady-state distribution.

(1) Define $M_i(k) = E\left[e^{\theta \sum_{l=1}^{k-1} a(l)} \mid a(0) = i\right]$. Show that

$$\begin{pmatrix} M_0(k+1) \\ M_1(k+1) \end{pmatrix} = \begin{pmatrix} p_{00} & p_{01} \\ p_{10}e^{\theta} & p_{11}e^{\theta} \end{pmatrix} \begin{pmatrix} M_0(k) \\ M_1(k) \end{pmatrix}. \tag{3.30}$$

(2) Compute the eigenvalues of

$$\begin{pmatrix} p_{00} & p_{01} \\ p_{10}e^{\theta} & p_{11}e^{\theta} \end{pmatrix}.$$

(3) Let λ_1 and λ_2 be the two eigenvalues obtained in part (2). Compute the effective bandwidth of the source, i.e., $\Lambda(\theta)/\theta$, where $\Lambda(\theta)$ is defined in (3.29), using the following basic fact from linear algebra and dynamical systems: there exist constants a_1, a_2, b_1, and b_2 such that

$$M_0(k+1) = a_1 \lambda_1^k + a_2 \lambda_2^k, \tag{3.31}$$

$$M_1(k+1) = b_1 \lambda_1^k + b_2 \lambda_2^k. \tag{3.32}$$

Exercise 3.8 (Queue-length stability in wireless channels) Consider a simple model of a discrete-time wireless channel: the channel is ON with probability μ and OFF with probability $1 - \mu$; when the channel is ON, it can serve one packet in the slot, and, when the

channel is OFF, the channel cannot serve any packets. Packets arrive at this wireless channel according to the following arrival process: a maximum of one arrival occurs in each
instant. The probability of an arrival in the current slot is 0.8 if there was an arrival in the
previous time slot. The probability of an arrival in the current slot is 0.1 if there was no
arrival in the previous time slot.

(1) For what values of μ would you expect this system to be stable?
    Hint: The arrival process is itself a Markov chain, but it is a simple two-state Markov
    chain. Find the mean arrival rate of the packet arrival process, assuming that it is in
    steady state. The mean service rate must be larger than this mean arrival rate.

(2) For these values of μ, show that an appropriate Markov chain describing the state of
    this queueing system is stable (i.e., positive recurrent) using the extended version of
    the Foster–Lyapunov theorem (Theorem 3.3.8).
    *Hint: The state of the Markov chain describing the queueing system is the number
    of packets in the queue along with the state of the arrival process. Additionally, you
    may need the fact that the expected value of the time average of the arrival process
    converges to the average steady-state arrival rate calculated in part (1), i.e.,*

$$\lim_{m \to \infty} \frac{1}{m} E\left[\sum_{i=0}^{m-1} a(k+i) \,\middle|\, a(k) = a \right] = E[a(k)]$$

for any k and $a(k)$.

Exercise 3.9 (Positive recurrence of a simple Markov chain) Consider a simple model of
a discrete-time wireless channel: the channel is ON with probability μ and OFF with probability $1 - \mu$; when the channel is ON, it can serve one packet in the slot, and, when the
channel is OFF, the channel cannot serve any packets. Assume that the number of arrivals
per time slot is Bernoulli with mean μ when the queue length is less than or equal to B
and is equal to λ otherwise. Compute the stationary distribution and the expected queue
length in steady state of this Markov chain when it is positive recurrent. Clearly identify
the conditions under which the Markov chain is positive recurrent and the conditions under
which it is not.

Exercise 3.10 (Geo/D/1 queues) Consider a single-server queue with infinite buffer. In
each time slot, half a packet is served. In other words, it takes two time slots to serve a
packet. Packets arrive to this queue according to an i.i.d. Bernoulli process with parameter
λ. This queue is called the Geo/D/1 queue. Compute the steady-state distribution π, the
mean queue length, and the mean queueing delay.

Exercise 3.11 (Geo/D/1/B queues) Consider a single-server queue with a buffer of size B
packets. In other words, a newly arriving packet is dropped if there are already B packets
in the queue. In each time slot, half a packet is served, i.e., it takes two time slots to serve
a packet. Packets arrive to this queue according to an i.i.d. Bernoulli process with parameter λ. This queue is called the Geo/D/1/B queue. Compute the steady-state distribution π
and the fraction of arriving packets that are dropped.

Exercise 3.12 (A queue with departures before arrivals in a time slot) Consider a discrete-time queue in which a departure occurs first, then arrivals occur in a time slot. The

service and arrival processes are independent Bernoulli processes, respectively, such that one packet can depart in a time slot with probability μ, and one packet can arrive with probability λ.

(1) Compute the steady-state distribution of the queue length assuming $\lambda < \mu$. Assume unlimited buffer space. Clearly draw the DTMC.

(2) Repeat (1) assuming that a newly arriving packet is dropped if there are already B packets in the system (assume $B \geq 2$). What is the probability that an arriving packet is dropped? Hint: Clearly understand when an arriving packet is dropped. You don't have to find a closed-form expression, but you do have to find answers in terms of summations.

Exercise 3.13 (The Kingman bound in the heavy-traffic regime) Consider the discrete-time G/G/1 queue used to obtain the Kingman bound:

$$q(t + 1) = (q(t) + a(t) - s(t))^+.$$

Another way to represent the above queueing dynamics is to define a non-negative random variable $u(t)$, which denotes unused service in a time slot, and rewrite the above equation as

$$q(t + 1) = q(t) + a(t) - s(t) + u(t).$$

Note that $u(t) \leq s(t)$. For the rest of this problem, we will assume that the initial probability distribution for this system is the steady-state distribution, i.e., we consider the system in steady state.

(1) Using the fact that, in steady state, $E[q(t + 1) - q(t)] = 0$, show that $E[u(t)] = \mu - \lambda$.

(2) We now show that the Kingman upper bound on $E[q(t)]$ is tight in heavy traffic under the assumption that $s(t) \leq S_{\max}$ for all t, where $S_{\max}$ is some positive constant. Prove that

$$E[q(t)] \geq \frac{1}{2(\mu - \lambda)} \left(E[(a - s)^2] - E[u^2] \right),$$

and show that the upper and lower bounds on $E[q(t)](\mu - \lambda)$ coincide when $\lambda \to \mu$. Hint: Consider the Lyapunov function $V(t) = q^2(t)/2$.

Exercise 3.14 (The Kingman bound with a general service distribution) Show that the heavy-traffic tightness of the Kingman bound holds even without the assumption $s(t) \leq S_{\max}$. In other words, show that the bound is tight in heavy traffic without assuming that there is an upper bound on the amount of packets per time slot; however, continue to assume that the number of arrivals and service per time slot have finite second moments.

Exercise 3.15 (The Kingman bound when departures occur before arrivals) In discrete-time systems, one can make different assumptions on the order in which arrivals and departures can occur. In this problem, we will assume $q(k + 1) = (q(k) - s(k))^+ + a(k)$. Thus, we assume that departures occur first, followed by arrivals. Furthermore, assume that the arrivals and departures are i.i.d. over time, the mean arrival rate (λ) is less than the mean service rate (μ), and $E[(a(k) - s(k))^2]$ is finite.

(1) Show that the Markov chain q is positive recurrent and hence that it has a stationary distribution. (We implicitly assume that the arrival and service processes are such that q is irreducible and aperiodic.)

(2) Now assume that the Markov chain is in steady state and obtain an upper bound on the expected queue length.

(3) Assume $s(k) \leq S_{\max}$ and show that the upper bound is tight in the heavy-traffic sense.

Exercise 3.16 (WFQ example) Consider a link shared by two flows, a and b. For simplicity, assume that the system starts at time $t = 0$ and that the transmission time of a packet of length S is S time units. Packets belonging to the flows arrive at the link as follows:

- flow A: packets of length 8 arrive at time slots 5, 25, 45, and 65;
- flow B: packets of length 5 arrive at time slots 0, 10, 20, 30, 40, and 50.

(1) Assume the link is operated under GPS. For each packet sent on the outgoing link, write down the time when the packet starts to receive service and the delay between the arrival and departure.

(2) Assume the link is operated under GPS. Calculate the percentage of the link used by each flow up to and including time slot 59.

(3) Repeat parts (1) and (2) assuming WFQ is used.

Exercise 3.17 (Worst Case Fair WFQ (WFWFQ)) In this exercise, we use a simple example to illustrate that the performance of WFQ can be further improved by using Worst Case Fair WFQ (WFWFQ). Consider a link shared by six flows, with capacity 1 bps. All packets are of the same size of 1 bit. Flow 1 has six packets, arriving at the beginnings of time slots 0, 1, 2, ..., 5, respectively. Flow i, for $i = 2, 3, ..., 6$, has one packet arriving at the beginning of time slot 0. Further, assume $w_1 = 5$ and $w_i = 1$ for $i = 2, 3, ..., 6$.

(1) Assuming GPS is used, calculate the departure times of the 11 packets.

(2) Assuming WFQ is used, calculate the departure times of the 11 packets.

(3) Assuming WFWFQ is used, i.e., the policy selects the packet with the smallest finish tag among those packets that would have started service under GPS, calculate the departure times of the 11 packets. *Note: The output of flow 1 is smoother under WFWFQ than it is under WFQ.*

Exercise 3.18 (The performance of WFWFQ) Recall that WFWFQ is WFQ with the following additional rule: when a link is idle, select the packet with the smallest finish tag *among those packets that would have started under GPS*. For simplicity, assume all queues are infinitely backlogged. Show the following bounds hold.

(1) $\tau_k^{ww} \leq \tau_k^{g} + S_{\max}/c$.

(2) $x_i^{ww}(t) \geq x_i^{g}(t) - S_{\max}$.
Hint: The proofs of parts (1) and (2) are similar to those of Theorems 3.5.1 and 3.5.2, respectively.

(3) $x_i^{ww}(t) \leq x_i^{g}(t) + \left(1 - \frac{\phi_i}{\sum_j \phi_j}\right) S_{i,\max}$,
where ϕ_i is the weight associated with flow i and $S_{i,\max}$ is the maximum packet size of flow i. The superscript ww indicates WFWFQ and g indicates GPS. Note that the last

bound indicates that the amount of service received by queue i under WFWFQ can be ahead of that under GPS by at most $\left(1 - \phi_i / \sum_j \phi_j\right) S_{i,\max}$. Hint: Let $x_i^{ww}(t_1, t_2)$ and $x^g(t_1, t_2)$ denote the number of flow i's bits served under WFWFQ and GPS during time interval $[t_1, t_2]$, respectively. Let $b_{(i,k)}^g$ denote the time at which the kth packet of flow i starts to receive service under GPS. The following facts are useful. Consider time $\tau \in [b_{(i,k)}^g, b_{(i,k+1)}^g)$, where we have

$$x^{ww}(b_{(i,k)}^g, \tau) \leq \min\{S_{(i,k)}, c(\tau - b_{(i,k)}^g)\}$$

because the service received under WFWFQ is constrained by the link bandwidth and the packet size (note that the $k+1$th packet of flow i will not be served before $b_{(i,k+1)}^g$); and

$$x^g(b_{(i,k)}^g, \tau) = \frac{\phi_i}{\sum_j \phi_j} c(\tau - b_{(i,k)}^g)$$

because flow i is served with rate $(\phi_i / \sum_j \phi_j)c$ under GPS.

3.8   Notes

More details on the Chernoff bound and other related inequalities can be found in many books on stochastic processes, such as [36, 146]. The concept of effective bandwidths using Chernoff bounds was proposed in [58] and developed in a series of papers (see [15, 27, 32, 34, 46, 47, 49, 70, 72, 82, 123] for some representative papers and [81] for a summary). A related approach using partial differential equations to study fluid models of traffic sources was presented in [5].

A nice introduction to Markov chains can be found in [51, 135, 146]. A detailed discussion of the Foster–Lyapunov theorem for stability and instability can be found in [6, 119, 120]. A simple proof of stability using Foster–Lyapunov type drift conditions is presented in [90]. Queueing theory is the subject of many books, such as [19, 167]; its relationship to communication networks is discussed in [10, 89]. Little's law was proved in [104], and the Kingman bound presented here is a discrete-time version of a result originally established in [87]. In deriving the Kingman bound, we have assumed that the second moment of the queue length exists. This assumption can be justified by using versions of the Foster–Lyapunov theorem presented in [120], which directly give moment bounds. We will also occasionally make an ergodicity assumption in the book, i.e., time averages converge to the corresponding expectation under the stationary distribution. Such assumptions can be justified by appealing to the ergodic theorem for Markov chains, such as the one presented in [135].

A good discussion of Weighted Fair Queueing (WFQ), an algorithm that was originally suggested in [35] and analyzed further in [138], can be found in [89]. WFWFQ was suggested in [8]. A low-complexity approximation to WFQ called Deficit Round Robin (DRR) was proposed in [153].

In this chapter, we have considered only probabilistic models of arrival processes when we discussed buffer overflow and loss behavior. An entirely different approach is to assume that the arrival processes are appropriately deterministically constrained, allowing for occasional burstiness in the arrival process. Such an approach was originated in [29, 30] and is the subject of the books [16, 21].

4 Scheduling in packet switches

In Chapter 2, we learned about routing algorithms that determine the sequence of links a packet should traverse to get to its destination. But we did not explain how a router actually moves a packet from one link to another. To understand this process, let us first look at the architecture of a router. Generally speaking, a router has four major components: the input and the output ports, which are interfaces connecting the router to input and output links, respectively, a switch fabric, and a routing processor, as shown in Figure 4.1. The routing processor maintains the routing table and makes routing decisions. *The switch fabric is the component that moves packets from one link to another link.* In this chapter, we will assume that all packets are of equal size. In reality, packets in the Internet have widely variable sizes. In the switch fabric, packets are divided into equal-sized cells and reassembled at the output, hence our assumption holds.

Earlier, we *implicitly assumed* that this switch fabric operates infinitely fast, so packets are moved from input ports to output ports immediately. This allowed us to focus on the buffers at output ports. So, all our discussions so far on buffer overflow probabilities are for output queues since an output buffer is the place where packets "enter" a link. For example, WFQ, introduced in Chapter 3, may be implemented at output port buffers to provide isolation among flows. However, in reality, the switch fabric does not really operate at infinite speed. In particular, packets encounter queueing delays at input ports before they

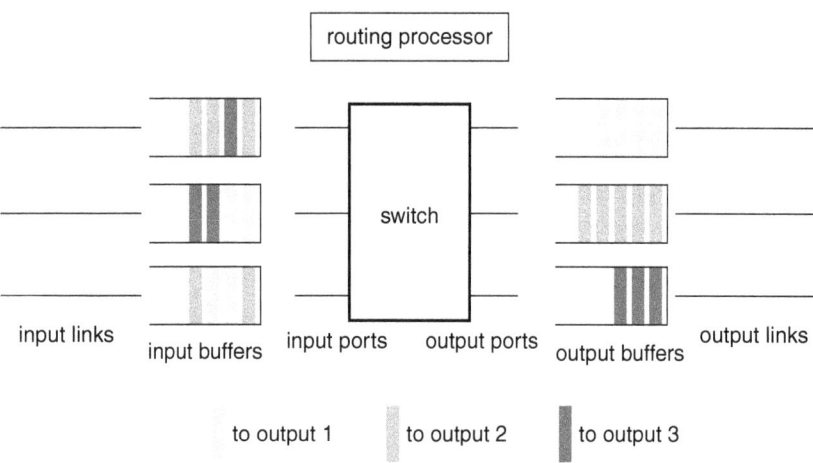

Figure 4.1 A router has four major components: input ports, output ports, switch fabric, and routing processor. Each input/output port is connected to an input/output link and maintains an input/output buffer. Packets are moved from input buffers to output buffers via the switch fabric.

are transferred to output ports. Therefore, buffers are required at both input and output ports to store packets. In this chapter, we will address the following questions.

- What is the capacity region of a switch (i.e., the set of packet arrival rates at the input region that can be processed by the switch)?
- How should the switch fabric be operated so that the queue lengths at input port buffers are finite?
- How should the complexity of switch scheduling algorithms for high-speed switches be reduced?

4.1   Switch architectures and crossbar switches

The most popular switch architecture today is the crossbar switch architecture. Figure 4.2 shows a crossbar switch with three input ports and three output ports. An $N \times N$ crossbar switch (i.e., a switch with N input ports and N output ports) consists of N^2 crosspoints. An input port is connected to an output port when the corresponding crosspoint is closed, as shown in Figure 4.2. At most one crosspoint can be closed at each row, and at most one can be closed at each column. In other words, at any given time, one input port can be connected to at most one output port, and vice versa.

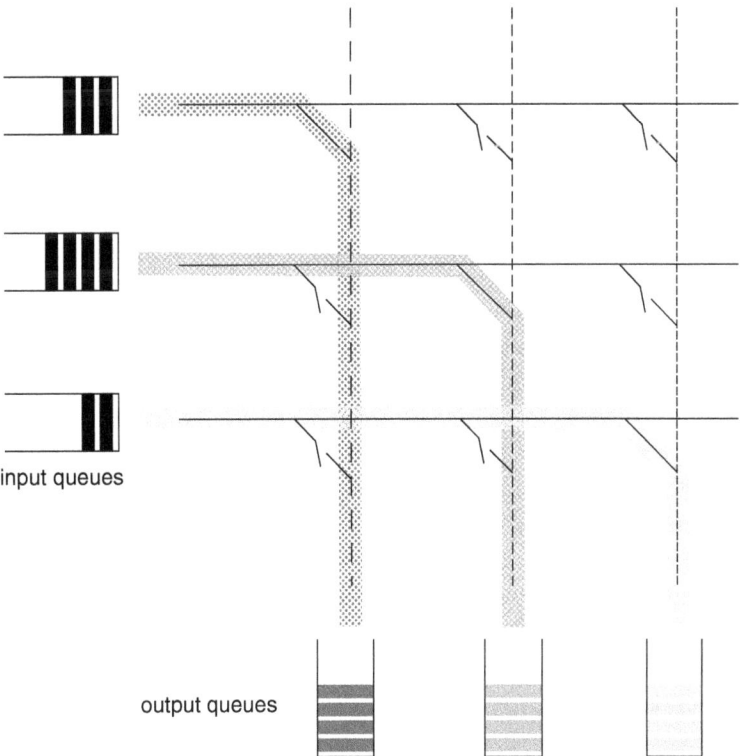

Figure 4.2 Crossbar switch with three input ports and three output ports. Three crosspoints are closed, connecting three input ports to three output ports.

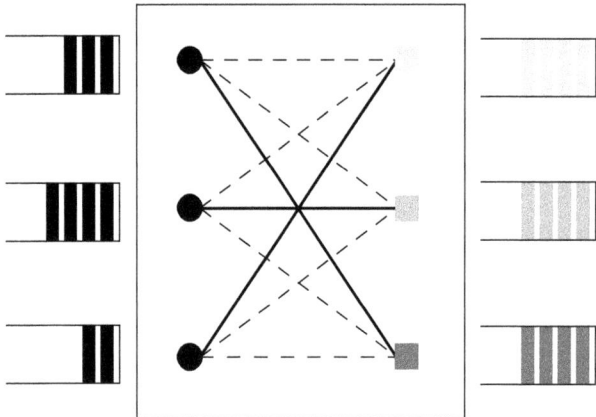

Figure 4.3 Bipartite graph representation of the switch in Figure 4.2.

A simple way to represent a crossbar switch is to use a complete bipartite graph, as shown in Figure 4.3. An $N \times N$ complete bipartite graph consists of N^2 edges, representing the N^2 crosspoints. The definitions of bipartite and complete bipartite graphs are presented in the following.

Definition 4.1.1 (Bipartite graph) A *bipartite graph* is a graph whose vertices can be partitioned into two sets $\mathcal{I}$ and $\mathcal{O}$ such that all edges connect a vertex in $\mathcal{I}$ to a vertex in $\mathcal{O}$. For a crossbar switch, $\mathcal{I}$ is the set of input ports and $\mathcal{O}$ is the set of output ports. □

Definition 4.1.2 (Complete bipartite graph) A *complete* bipartite graph is a bipartite graph where every vertex in $\mathcal{I}$ is connected to every vertex in $\mathcal{O}$. □

Definition 4.1.3 (Matching) In a graph, a matching is a set of edges such that no two share a common vertex. In other words, each node is associated with at most one edge in a matching. Figure 4.4 illustrates both a valid matching and an invalid matching. □

Definition 4.1.4 (Schedule) In a crossbar switch, a schedule is a set of connections from input ports to output ports such that no input port is connected to two output ports and vice versa. □

According to the definitions of a matching and a schedule, a schedule has to be a valid matching in the bipartite graph representing the switch. An important problem in crossbar switch design is to develop scheduling algorithms that find *good* matchings at each time instant to transfer packets from input buffers to output buffers, which is the main focus of this chapter.

4.1.1 Head-of-line blocking and virtual output queues

When the speed of a switch is not fast enough, packets will be queued at input buffers. We first consider the case where each input port is associated with a queue where packets

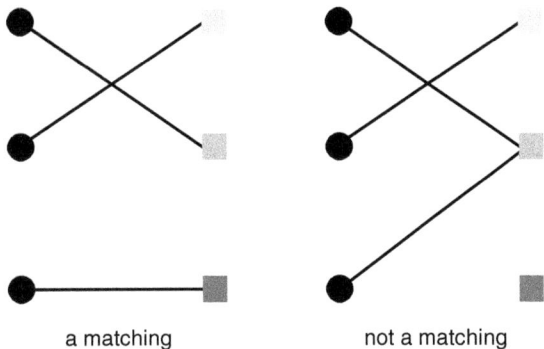

a matching   not a matching

Figure 4.4 Matching in a graph.

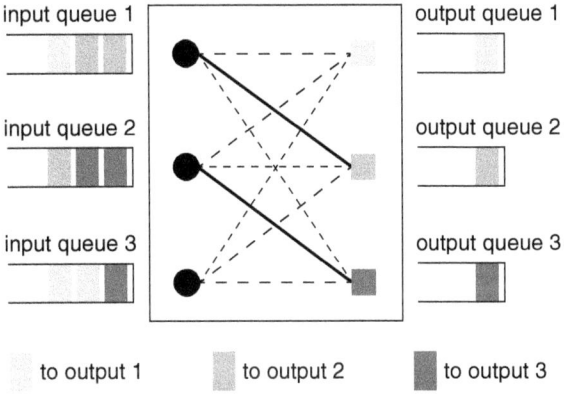

input queue 1   output queue 1

input queue 2   output queue 2

input queue 3   output queue 3

to output 1   to output 2   to output 3

Figure 4.5 Head-of-line blocking at input queues.

are transmitted in a First-In, First-Out fashion, called a FIFO queue, the switch operates in a time-slotted fashion, and at most one packet can be transferred from an input queue to an output queue during one time slot if they are connected. Assume that the status of input queues at the beginning of a time slot is as shown in Figure 4.5. If the switch can freely select any packet in an input queue to transfer, the switch can transfer one packet from input queue 1 to output queue 2, one from input queue 2 to output queue 3, and one from input queue 3 to output queue 1. However, because input queues are FIFO queues, the switch can only access the first packet in each queue (also known as the head-of-line packet). Since no head-of-line packet is destined to output port 1, even though output port 1 is free, those packets destined to output port 1 cannot be transferred. Therefore, the switch can make only two transfers, as shown in Figure 4.5. This phenomenon whereby packets behind the first packet in each queue may be blocked is called Head-of-Line (HOL) blocking, which can significantly degrade the performance of a switch.

HOL blocking occurs because the switch can only access head-of-line packets. So, each input queue can only be connected to the output queue to which its head-of-line packet is

VOQs   output queues

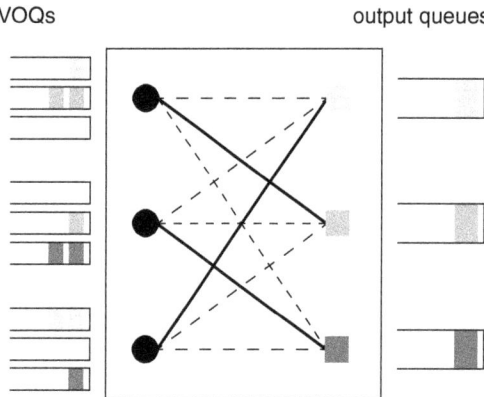

Figure 4.6 Head-of-line blocking at input queues.

destined, which significantly limits the flexibility in selecting a schedule. One approach to resolve this HOL blocking problem is to maintain a separate queue for each output port at each input port, called Virtual Output Queues (VOQs), as shown in Figure 4.6. With VOQs at each input port, the switch can access one packet for each output port (if such a packet exists). Assuming all VOQs are non-empty, an input port can transfer one packet to any output port. Consider the same situation as in Figure 4.5, but with VOQs: the switch can transfer one packet to each output queue simultaneously, as shown in Figure 4.6, so is fully utilized.

4.2   Capacity region and MaxWeight scheduling

While VOQs resolve the HOL blocking problem, it is not yet clear whether the switch can operate at its full capacity. To address this question, we first need to understand what "full capacity" means, i.e., understand the capacity region of a switch. We let $\text{VOQ}(i,j)$ denote the VOQ for output port j at input port i, and let $q_{ij}(t)$ denote the length of $\text{VOQ}(i,j)$ at time slot t. Further, denote by $a_{ij}(t)$ the number of packets arriving at input i at time slot t, and destined to output port j. We assume that $a_{ij}(t)$ is a Bernoulli random variable with parameter λ_{ij}, and that the $a_{ij}(t)$'s are independent across time and input-port and output-port pairs. We further assume that one packet can be transferred from an input port to an output port in one time slot.

Define the arrival rate matrix λ to be an $N \times N$ matrix such that the (i,j)th entry is λ_{ij}. We say that the arrival rate matrix λ is *supportable* if there exists a switch scheduling under which

$$\lim_{C \to \infty} \lim_{t \to \infty} \Pr\left(|q(t)| \geq C\right) = 0, \tag{4.1}$$

where $|q(t)|$ is the sum of the lengths of all the queues in the network at time t. In other words, λ is supportable if the probability that VOQs become infinite is zero. More simply stated, we require the queue lengths to be finite. We will first characterize the set of supportable arrival rate matrices.

Since at most one packet can be transferred from an input port in one time slot, a necessary condition for λ to be supportable is

$$\sum_{j=1}^{N} \lambda_{ij} \leq 1,$$

i.e., the aggregate arrival rate to input port i should be no more than one (packet per time slot). Similarly, each output port can accept at most one packet in each time slot, so

$$\sum_{i=1}^{N} \lambda_{ij} \leq 1,$$

i.e., the number of packets destined to output port j cannot exceed one (packet per time slot) on average. Based on the two conditions above, we define set C to be

$$C = \left\{ \lambda : \lambda \geq 0, \sum_{i=1}^{N} \lambda_{ij} \leq 1 \text{ for any } j \text{ and } \sum_{j=1}^{N} \lambda_{ij} \leq 1 \text{ for any } i \right\}. \qquad (4.2)$$

We will see in this section that C is the capacity region of the switch, and that any λ that lies strictly inside C can be supported by a scheduling algorithm, called the MaxWeight scheduling algorithm.

The first result we will present is that no scheduling algorithm can support λ if λ is not in C.

Theorem 4.2.1 If $\lambda \notin C$, no scheduling algorithm can support arrival rate matrix λ.

Proof Let $M(t)$ denote the schedule used in time slot t. So, $M(t)$ is an $N \times N$ matrix such that $M_{ij}(t) = 1$ if input port i is connected to output port j in time slot t, and $M_{ij}(t) = 0$ otherwise.

According to the definition of C, $\lambda \notin C$ implies that either $\sum_i \lambda_{ij} > 1$ for some j or $\sum_j \lambda_{ij} > 1$ for some i. Suppose the first case occurs and there exist j^* and $\epsilon > 0$ such that $\sum_i \lambda_{ij^*} \geq 1 + \epsilon$. In this case, we consider the value of $\sum_i q_{ij^*}(t+1)$. Note that

$$q_{ij}(t+1) = \left(q_{ij}(t) + a_{ij}(t) - M_{ij}(t) \right)^+ \geq q_{ij}(t) + a_{ij}(t) - M_{ij}(t),$$

so

$$\sum_i q_{ij^*}(t+1) \geq \sum_{s=1}^{t} \left(\sum_i a_{ij^*}(s) - \sum_i M_{ij^*}(s) \right).$$

According to the Strong Law of Large Numbers (SLLN), with probability 1,

$$\lim_{t \to \infty} \frac{1}{t} \sum_{s=1}^{t} \sum_i a_{ij^*}(s) = \sum_i \lambda_{ij^*} \geq 1 + \epsilon.$$

Further, since $M(t)$ is a matching, $\sum_i M_{ij^*}(t) \leq 1$ for all t. So we have

$$\frac{1}{t} \sum_{s=1}^{t} \sum_i M_{ij^*}(s) \leq 1.$$

Therefore, with probability 1,

$$\sum_i q_{ij^*}(t) \to \infty \quad \text{as} \quad t \to \infty.$$

So λ is not supportable. The same proof works if $\sum_j \lambda_{ij} \geq 1$ for some i. ☐

Theorem 4.2.1 shows that any arrival rate matrix outside of $\mathcal{C}$ cannot be supported. We next present an algorithm that supports an arrival rate matrix that lies strictly in $\mathcal{C}$. Recall that an $N \times N$ switch can be represented by an $N \times N$ complete bipartite graph. Let H denote the total number of matchings in an $N \times N$ complete bipartite graph, and let $M^{(h)}$ denote the hth matching, where we use $(\cdot)$ to indicate that the superscript is an index not a power. Note that $M^{(h)}$ is an $N \times N$ matrix such that $M_{ij}^{(h)} = 1$ if input port i is connected to output port j and $M_{ij}^{(h)} = 0$ otherwise. We now introduce the MaxWeight scheduling algorithm that can support any λ such that $(1 + \epsilon)\lambda \in \mathcal{C}$. The MaxWeight scheduling algorithm associates a weight q_{ij} with the link corresponding to VOQ(i, j) in the bipartite graph representation of the switch. It then finds a matching which maximizes the sum of the weights of the links included in the matching; hence the name MaxWeight algorithm.

MaxWeight scheduling

The switch finds a matching $M(t)$ such that

$$M(t) \in \arg \max_{M^{(h)}} \sum_{ij} q_{ij}(t) M_{ij}^{(h)},$$

and transfers a packet from VOQ(i, j) to output port j if $M_{ij}(t) = 1$ and $q_{ij}(t) + a_{ij}(t) > 0$, i.e., there is a packet in the queue.

The intuition behind the derivation of the MaxWeight algorithm is provided in Section 4.2.1.

Example 4.2.1

Consider a 3×3 crossbar switch, with the states of VOQs as shown in Figure 4.7. MaxWeight scheduling will schedule the two links shown on the left. This schedule has a sum weight of 4. The matching on the right has a sum weight equal to 3, so will not be selected by the algorithm. ☐

VOQs

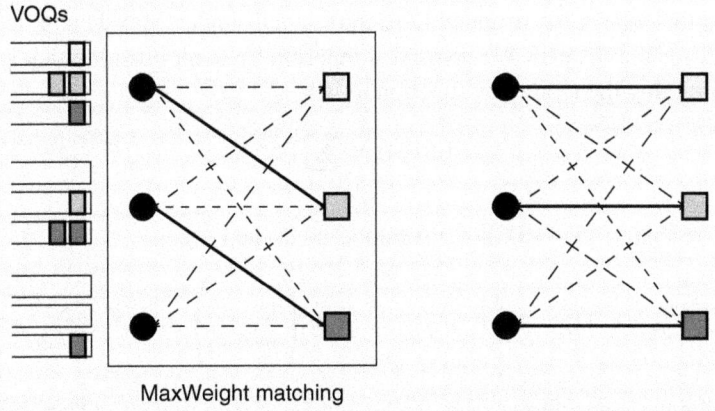

MaxWeight matching

Figure 4.7 An example of MaxWeight scheduling in a 3×3 switch.

To analyze the performance of MaxWeight scheduling, we need the Birkhoff–von Neumann theorem, which establishes the connection between doubly stochastic matrices and permutation matrices.

Definition 4.2.1 (Doubly stochastic matrix) An $N \times N$ matrix λ is a doubly stochastic matrix if $0 \leq \lambda_{ij} \leq 1$, $\sum_{i=1}^{N} \lambda_{ij} = 1$, and $\sum_{j=1}^{N} \lambda_{ij} = 1$. □

Definition 4.2.2 (Doubly substochastic matrix) An $N \times N$ matrix λ is a doubly substochastic matrix if $0 \leq \lambda_{ij} \leq 1$, $\sum_{i=1}^{N} \lambda_{ij} \leq 1$, and $\sum_{j=1}^{N} \lambda_{ij} \leq 1$. □

Definition 4.2.3 (Permutation matrix) An $N \times N$ matrix U is a permutation matrix if $\sum_{i=1}^{N} U_{ij} = 1$, $\sum_{j=1}^{N} U_{ij} = 1$, and $U_{ij} \in \{0, 1\}$. □

Lemma 4.2.2 (Birkhoff–von Neumann theorem) An $N \times N$ matrix λ is a doubly stochastic matrix if and only if it is a convex combination of permutation matrices, i.e., there exists $\beta \geq 0$ such that $\sum_{h} \beta_h = 1$ and $\lambda = \sum_{h} \beta_h U^{(h)}$, where the $U^{(h)}$'s are permutation matrices. □

We further present the following lemma that establishes the connection between doubly stochastic matrices and doubly substochastic matrices.

Lemma 4.2.3 If an $N \times N$ matrix λ is a doubly substochastic matrix, there exists a doubly stochastic matrix $\tilde{\lambda}$ such that $\lambda \leq \tilde{\lambda}$ component-wise. □

We comment that a doubly substochastic matrix λ belongs to set $\mathcal{C}$, and a permutation matrix is a valid matching.

Theorem 4.2.4 MaxWeight scheduling can support any arrival rate matrix λ such that $(1 + \epsilon)\lambda \in \mathcal{C}$.

Proof Note that $q(t)$ is an irreducible Markov chain under MaxWeight scheduling (see Exercise 4.2 for the details). We prove this theorem by demonstrating that $q(t)$ is positive recurrent. Defining the Lyapunov function to be

$$V(q(t)) = \sum_{ij} q_{ij}^2(t),$$

we next prove that, if $(1 + \epsilon)\lambda \in \mathcal{C}$ for some $\epsilon > 0$, then

$$E[V(q(t+1)) - V(q(t))| q(t) = q] \le \left(\sum_{ij} \lambda_{ij}\right) + N - 2\epsilon \sum_{i,j} \lambda_{ij}q_{ij}. \qquad (4.3)$$

Therefore, the theorem follows from the Foster–Lyapunov theorem (Theorem 3.3.7).

To prove result (4.3), we first have

$$V(q(t+1)) - V(q(t))$$

$$= \sum_{ij} \left(\left(q_{ij}(t) + a_{ij}(t) - M_{ij}(t)\right)^+\right)^2 - \sum_{ij} q_{ij}^2(t)$$

$$\le \sum_{ij} \left(q_{ij}(t) + a_{ij}(t) - M_{ij}(t)\right)^2 - \sum_{ij} q_{ij}^2(t)$$

$$= \sum_{ij} \left(2q_{ij}(t) + a_{ij}(t) - M_{ij}(t)\right)\left(a_{ij}(t) - M_{ij}(t)\right)$$

$$= \sum_{ij} 2q_{ij}(t)\left(a_{ij}(t) - M_{ij}(t)\right) + \sum_{ij} \left(a_{ij}(t) - M_{ij}(t)\right)^2.$$

Note that

$$E\left[\sum_{ij} \left(a_{ij}(t) - M_{ij}(t)\right)^2 \middle| q(t) = q\right] \le E\left[\sum_{ij} \left(a_{ij}^2(t) + M_{ij}^2(t)\right) \middle| q(t) = q\right]$$

$$\le \left(\sum_{ij} \lambda_{ij}\right) + N,$$

where $E[a_{ij}^2(t)|q(t) = q] = \lambda_{ij}$ because $a_{ij}(t)$ is Bernoulli and independent of $q(t)$, and $E[\sum_{ij} M_{ij}^2(t)|q(t) = q] \le N$ because $M(t)$ is a matching. So, taking conditional expectations on both sides of the equation above, we obtain

$$E[V(q(t+1)) - V(q(t))| q(t) = q]$$

$$\le \left(\sum_{ij} \lambda_{ij}\right) + N + 2\sum_{ij} q_{ij}\left(\lambda_{ij} - E[M_{ij}(t)|q(t) = q]\right). \qquad (4.4)$$

Recall that there exists $\epsilon > 0$ such that $(1 + \epsilon)\lambda \in \mathcal{C}$. According to Lemmas 4.2.2 and 4.2.3, there exists a vector $\beta \ge 0$ such that $\sum_h \beta_h = 1$ and $(1 + \epsilon)\lambda \le \sum_h \beta_h U^{(h)}$, where the $U^{(h)}$'s are permutation matrices and $\leq$ means element-wise inequalities here. Therefore,

$$(1+\epsilon)\sum_{ij}q_{ij}\lambda_{ij} \leq \sum_{ij}q_{ij}\left(\sum_h \beta_h U_{ij}^{(h)}\right) = \sum_h \beta_h\left(\sum_{ij}q_{ij}U_{ij}^{(h)}\right)$$

$$\leq \left(\sum_h \beta_h\right)\left(\max_h\left(\sum_{ij}q_{ij}U_{ij}^{(h)}\right)\right)$$

$$= \max_h\left(\sum_{ij}q_{ij}U_{ij}^{(h)}\right).$$

Since permutation matrices are valid matchings, we further have

$$(1+\epsilon)\sum_{ij}q_{ij}\lambda_{ij} \leq \max_h\sum_{ij}q_{ij}U_{ij}^{(h)} \leq \max_h\sum_{i,j}q_{ij}M_{ij}^{(h)} = \sum_{i,j}q_{ij}E[M_{ij}(t)|q(t)=q],$$

where the last equality arises from the definition of MaxWeight scheduling. Substituting the inequality above into inequality (4.4) yields

$$E[V(q(t+1)) - V(q(t))|q(t)=q] \leq \left(\sum_{ij}\lambda_{ij}\right) + N - 2\epsilon\sum_{i,j}\lambda_{ij}q_{ij}. \qquad (4.5)$$

Hence, inequality (4.3) holds, and the proof completes. $\qquad\square$

From Theorems 4.2.1 and 4.2.4, we conclude that any arrival rate matrix λ not in $\mathcal{C}$ cannot be supported by any switch scheduling algorithm, while any λ that is *strictly* in $\mathcal{C}$ can be supported by MaxWeight scheduling. Therefore, $\mathcal{C}$ is called the *capacity region* of an $N \times N$ switch, and MaxWeight scheduling is said to be a *throughput-optimal* algorithm.

According to the proof of Theorem 4.2.4, $q(t)$ has a well-defined stationary distribution under the MaxWeight scheduling when λ lies strictly within the capacity region. Since the matching $M(t)$ selected by the MaxWeight algorithm is determined by $q(t)$, $M(t)$ also has a well-defined stationary distribution. Let π_h denote the probability that matching $M^{(h)}$ is selected under the MaxWeight algorithm in steady state. Then, for any λ that lies strictly within the capacity region, there exists $\pi \geq 0$ such that

$$\lambda \leq \sum_h \pi_h M^{(h)}, \quad \sum_h \pi_h = 1,$$

where the inequality holds component-wise. Therefore, an alternative definition of the capacity region $\mathcal{C}$ is

$$Co\left(\left\{M^{(h)}\right\}_{h=1,\dots,H}\right),$$

i.e., the convex hull of all matchings. In Exercise 4.6, you will be asked to prove $\mathcal{C} = Co\left(\left\{M^{(h)}\right\}_{h=1,\dots,H}\right)$. This definition, using the convex hull of service rates, will be generalized to wireless networks in the next chapter, where a definition like (4.2) is hard to obtain.

4.2.1   Intuition behind the MaxWeight algorithm

In our presentation of the MaxWeight algorithm, we first presented the algorithm and then proved that it is throughput optimal. Here, we would like to point out that the proof actually contains the intuition behind how the algorithm was derived in the first place. We start the search for an appropriate scheduling algorithm by studying the drift of the Lyapunov function

$$V(q(t)) = \sum_{ij} q_{ij}^2(t).$$

As can be seen from the proof of Theorem 4.2.4, the drift of V can be upper bounded as

$$E\left[V(q(t+1)) - V(q(t))|q(t) = q\right]$$
$$\leq 2 \sum_{ij} q_{ij} \left(\lambda_{ij} - E[M_{ij}(t)|q(t) = q]\right) + \text{some constant}.$$

By the Foster–Lyapunov theorem, we would like the drift to be negative for large values of the queue lengths. The only quantity that can be controlled in the above upper bound on the drift is $M_{ij}(t)$. Therefore, we choose the schedule $\{M_{ij}(t)\}$ to maximize

$$\sum_{ij} q_{ij} M_{ij}(t),$$

hoping to make the drift negative. Then we prove that the drift indeed becomes negative for such a schedule if $\{q_{ij}\}$ lies outside a finite set. These key ideas behind the derivation of the MaxWeight scheduling algorithm will be used later to derive similar scheduling algorithms for wireless networks.

4.3   Low-complexity switch scheduling algorithms

While MaxWeight scheduling is throughput optimal, the complexity of finding the maximum weight matching is $O(N^3)$, which is not practical for high-speed switches. In this section, we introduce three low-complexity scheduling algorithms: maximal matching scheduling, pick-and-compare scheduling, and load-balanced switches. Maximal matching scheduling can be implemented with an average complexity of $O(N \log N)$, and can support any arrival rate matrix λ such that $(1 + \epsilon)2\lambda \in \mathcal{C}$ for some $\epsilon > 0$. Pick-and-compare scheduling has $O(N)$ complexity, and load-balanced switching has $O(1)$ complexity. Both pick-and-compare scheduling and load-balanced switches are throughput optimal, but have other weaknesses, which will be discussed later. Throughout this section, we assume arrivals are Bernoulli random variables, independent across time and VOQs.

4.3.1   Maximal matching scheduling

One way to reduce the complexity of the switch scheduling is to find a maximal matching at each time step rather than the maximum weight schedule. The complexity of finding a maximal matching is $O(N \log N)$, which is much lower than $O(N^3)$. We first introduce

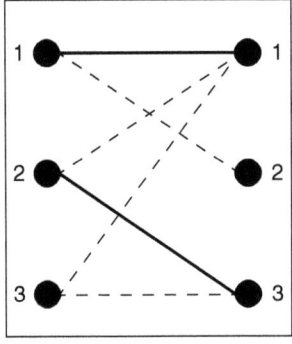 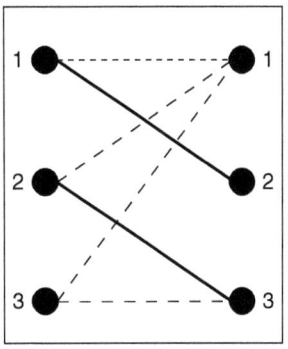

a maximal matching   a matching, but not maximal

Figure 4.8 The solid links on the left form a maximal matching because any additional link would share a common node with one of the solid links. The set of solid links on the right is a matching but not a maximal matching because link (3, 1) can be added to the set and the new set will still be a matching.

the definition of maximal matching and then present the maximal matching scheduling algorithm.

Definition 4.3.1 (Maximal matching) A maximal matching is a matching such that if any other edge is added to the set, it is not a matching.   □

Figure 4.8 illustrates a maximal matching and a matching that is not maximal.

Maximal matching scheduling

The switch first removes all edges with zero weights ($q_{ij}(t) = 0$) from the complete bipartite graph, and then finds a maximal matching $M(t)$ in the remaining bipartite graph. One packet in VOQ(i, j) is transferred to output port j if $M_{ij}(t) = 1$.

We now present an example to show that the achievable throughput under the maximal matching algorithm may be smaller than the capacity region of the switch.

Example 4.3.1

Consider a 2×2 switch. Assume $\lambda_{12} = 0$, so $q_{12}(t) = 0$ for any $t \geq 0$, and there are at most three edges we need to consider, as shown in Figure 4.9.

Consider a specific implementation of the maximal matching scheduling as follows:

(1) if $q_{11}(t) > 0$ and $q_{22}(t) > 0$, both edges (1, 1) and (2, 2) are scheduled;
(2) if $q_{11}(t) > 0$ and $q_{22}(t) = 0$, edge (1, 1) is scheduled;
(3) if $q_{11}(t) = 0$ and $q_{22}(t) > 0$, edge (2, 2) is scheduled;
(4) if $q_{11}(t) = q_{22}(t) = 0$ and $q_{21}(t) > 0$, edge (2, 1) is scheduled.

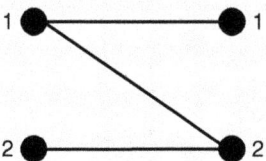

Figure 4.9 Since $q_{12}(t) = 0$ for all t, edge $(1, 2)$ is removed from the graph, and we only need to consider at most three edges.

This particular implementation of the maximal matching scheduling gives priority to VOQ$(1, 1)$ and VOQ$(2, 2)$, and serves VOQ$(2, 1)$ only if $q_{11}(t) = q_{22}(t) = 0$. Therefore, the states of VOQ$(1, 1)$ and VOQ$(2, 2)$ are independent of VOQ$(2, 1)$, and independent of each other.

Since VOQ$(1, 1)$ is served whenever it is non-empty, the Markov chain describing $q_{11}(t)$ is as shown in Figure 4.10.

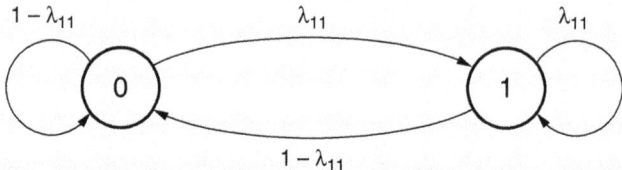

Figure 4.10 The DTMC for q_{11}. The DTMC moves from state 0 to state 1 when there is an arrival to VOQ$(1, 1)$, which occurs with probability λ_{11}. Other transition probabilities can be verified similarly.

Suppose $\lambda_{11} = 0.5$. Using the local balance equation (Theorem 3.3.6), it is easy to verify that the stationary distribution of q_{11} is $(0.5, 0.5)$. So

$$\Pr(q_{11}(t) = q_{22}(t) = 0) = (1 - \lambda_{11})(1 - \lambda_{22}) = 0.25.$$

Since VOQ$(2, 1)$ is served only if $q_{11}(t) = q_{22}(t) = 0$ under our maximal matching scheduling, we must have

$$\lambda_{21} < 0.25$$

to guarantee that $q_{21}(t)$ does not grow to infinity. On the other hand, the capacity region of the switch is given by

$$\lambda_{11} + \lambda_{21} < 1,$$
$$\lambda_{22} + \lambda_{21} < 1.$$

In the case $\lambda_{11} = \lambda_{22} = 0.5$, λ_{21} can be allowed to be up to 0.5 if the full capacity region of the switch is used. $\qquad\square$

This example suggests that there are arrival patterns and particular implementations of maximal matchings for which the throughput under the maximal matching scheduling can be much smaller than the capacity region of the switch region. The following theorem shows that the throughput under the maximal matching scheduling is at least 50% of the capacity region of the switch.

Theorem 4.3.1 Maximal matching scheduling can support any arrival rate matrix λ such that $2(1 + \epsilon)\lambda \in \mathcal{C}$ for some $\epsilon > 0$.

Proof Under maximal matching, $q(t)$ is an irreducible Markov chain. We will prove that $q(t)$ is positive recurrent based on the following two observations.

(i) There exists $\epsilon > 0$ such that

$$\sum_i 2(1 + \epsilon)\lambda_{ij} \le 1, \quad \forall j, \quad \text{and} \quad \sum_j 2(1 + \epsilon)\lambda_{ij} \le 1, \quad \forall i,$$

which implies that, for any i and j,

$$\sum_h \lambda_{ih} + \sum_k \lambda_{kj} \le \frac{1}{1 + \epsilon}. \tag{4.6}$$

(ii) Since $M(t)$ is a maximal matching, if $q_{ij}(t) \ne 0$, then input port i, output port j, or both, must appear in matching $M(t)$. Otherwise, we can add edge ij in $M(t)$ without destroying the matching property, which contradicts the fact that $M(t)$ is a maximal matching. So,

$$\sum_h M_{ih}(t) + \sum_k M_{kj}(t) \ge 1 \quad \text{if } q_{ij}(t) \ne 0. \tag{4.7}$$

These two observations indicate that the total arrival rate to the queues $\text{VOQ}(i, h)$ ($h = 1, \ldots, N$) and $\text{VOQ}(k, j)$ ($k = 1, \ldots, N$) is no more than $1/(1 + \epsilon)$, while the total service rate in each time slot is at least one. So it is reasonable to expect $q(t)$ to be positive recurrent under maximal matching scheduling, which we will prove next using the Foster–Lyapunov theorem.

Now consider the Lyapunov function

$$V(q(t)) = \sum_{ij} q_{ij}(t) \left(\sum_k q_{kj}(t) + \sum_h q_{ih}(t) \right).$$

We first compute the one-step difference of the Lyapunov function:

$V(q(t + 1)) - V(q(t))$

$$= \sum_{ij} q_{ij}(t + 1) \left(\sum_k q_{kj}(t + 1) + \sum_h q_{ih}(t + 1) \right) - q_{ij}(t) \left(\sum_k q_{kj}(t) + \sum_h q_{ih}(t) \right)$$

$$= \sum_{ij} q_{ij}(t) \left(\sum_k q_{kj}(t + 1) + \sum_h q_{ih}(t + 1) - \sum_k q_{kj}(t) - \sum_h q_{ih}(t) \right)$$

$$+ \sum_{ij} (a_{ij}(t) - M_{ij}(t)) \left(\sum_k q_{kj}(t + 1) + \sum_h q_{ih}(t + 1) \right)$$

$$= \sum_{ij} q_{ij}(t) \left(\sum_{k} a_{kj}(t) + \sum_{h} a_{ih}(t) - \sum_{k} M_{kj}(t) - \sum_{h} M_{ih}(t) \right)$$

$$+ \sum_{ij} \left(a_{ij}(t) - M_{ij}(t) \right) \left(\sum_{k} q_{kj}(t) + \sum_{h} q_{ih}(t) \right) \tag{4.8}$$

$$+ \sum_{ij} \left(a_{ij}(t) - M_{ij}(t) \right) \left(\sum_{k} a_{kj}(t) + \sum_{h} a_{ih}(t) - \sum_{k} M_{kj}(t) - \sum_{h} M_{ih}(t) \right). \tag{4.9}$$

By rearranging the summations, we can obtain

$$(4.8) = \sum_{ij} q_{ij}(t) \left(\sum_{k} a_{kj}(t) + \sum_{h} a_{ih}(t) - \sum_{k} M_{kj}(t) - \sum_{h} M_{ih}(t) \right).$$

Further,

$$(4.9) \le 2N^3$$

because $|a_{ij}(t) - M_{ij}(t)| \le 1$. Therefore, we have

$$V(q(t+1)) - V(q(t))$$

$$\le 2 \sum_{ij} q_{ij}(t) \left(\sum_{k} a_{kj}(t) + \sum_{h} a_{ih}(t) - \sum_{k} M_{kj}(t) - \sum_{h} M_{ih}(t) \right) + 2N^3.$$

Taking conditional expectations of both sides, we obtain

$$E[V(q(t+1)) - V(q(t))|q(t) = q]$$

$$\le 2 \sum_{ij} q_{ij} \left(\sum_{k} \lambda_{kj} + \sum_{h} \lambda_{ih} - \sum_{k} E[M_{kj}(t)|q(t) = q] - \sum_{h} E[M_{ih}(t)|q(t) = q] \right)$$

$$+ 2N^3$$

$$\le 2 \sum_{ij} q_{ij} \left(\frac{1}{1+\epsilon} - 1 \right) + 2N^3 \tag{4.10}$$

$$= 2N^3 - \frac{2\epsilon}{1+\epsilon} \sum_{ij} q_{ij},$$

where inequality (4.10) holds due to inequalities (4.6) and (4.7). Similar to the proof of Theorem 4.2.4, we conclude that $q(t)$ is positive recurrent using the Foster–Lyapunov theorem (Theorem 3.3.7). □

Maximal matching scheduling can be implemented with much lower complexity than maximum weighted matching. Algorithm 2 is a low-complexity implementation of maximal matching called Parallel Iterative Matching (PIM). PIM is illustrated using a 3×3 switch as an example in Figure 4.11. In this example, PIM finds a maximal matching in one iteration, but in general it may take multiple iterations for it to converge.

Algorithm 2 Parallel Iterative Matching (PIM)

1: At each time t, the switch sets $M(t) = 0$.

2: **while** $M(t)$ is not a maximal matching **do**

3:    Input port i, if not matched, sends a request message to each unmatched output port j for which $q_{ij}(t) > 0$. (An input port i is said to be matched if $M_{ij} = 1$ for any j. Similarly, an output port j is said to be matched if $M_{ij} = 1$ for any i.)

4:    Each output port j, if not matched, randomly selects one request from the received requests, and then sends a grant message to the input port that initiated the request.

5:    Input port i randomly selects a grant message (say from output port j), notifies output port j, and sets $M_{ij}(t) = 1$.

6: **end while**

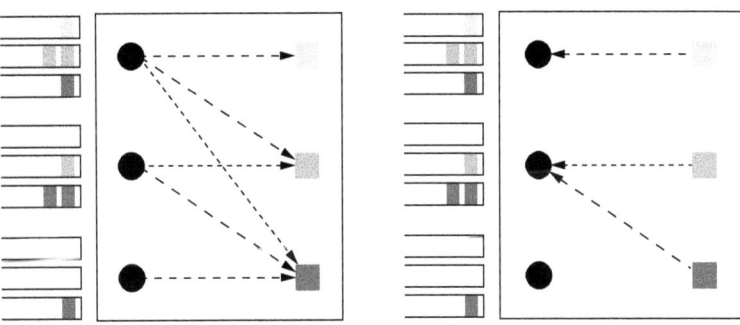

Step 1: Input ports send requests to output ports

Step 2: An output port randomly selects one request and sends a grant message back to the input port

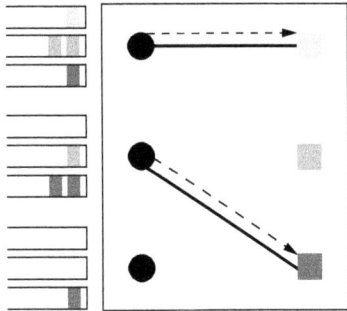

Step 3: An input port randomly selects a grant message and adds the corresponding link into the matching

Figure 4.11 The first iteration of PIM in a 3×3 switch.

4.3.2   Pick-and-compare scheduling

Maximal matching scheduling can be implemented efficiently in switches, but only supports a fraction of the capacity region. We now introduce a throughput-optimal algorithm with $O(N)$ complexity, called pick-and-compare scheduling.

The idea behind pick-and-compare scheduling is to select a maximal matching (in the complete bipartite graph representation of the switch) at the beginning of each time slot and compare it with the matching used in time slot $t-1$. The new maximal matching is accepted if its total weight is larger than the old matching. Assume queue lengths are constants, then it takes $O(H)$ time slots for the pick-and-compare algorithm to find the maximum weight matching, where H is the total number of maximal matchings in an $N \times N$ bipartite graph. When queue lengths are sufficiently large, they do not change significantly over H time slots, so the pick-and-compare algorithm works. The algorithm is presented below. You will be asked to prove the throughput optimality of this algorithm in Exercise 4.8.

Pick-and-compare scheduling

(1) The switch stores all maximal matchings, and indexes them by $h = 1, \ldots, H$.

(2) At time slot t, the switch selects the $h = (t \bmod H)$th maximal matching, and compares $W_{old} = \sum_{ij} q_{ij}(t) M_{ij}(t-1)$ and $W_{new} = \sum_{ij} q_{ij}(t) M_{ij}^{(h)}$, where $M_{ij}(t-1)$ is the matching used in time slot $t-1$.

(3) If $W_{old} \geq W_{new}$, the switch keeps using matching $M(t-1)$; otherwise, the switch uses maximal matching $M^{(h)}$ in time slot t.

The main drawback of this algorithm is that the queue lengths have to be very large for the above argument to hold. This may result is very large queue lengths in practice.

4.3.3   Load-balanced switches

Now we introduce a new switch architecture, called a load-balanced switch, which uses $O(1)$ computations at each time step, but is still throughput optimal.

The load-balanced switching architecture consists of two stages, as shown in Figure 4.12. At the first stage, one FIFO queue is maintained at each input port, and, at the second stage, VOQs are maintained at input ports. The switch stores N matchings where, in the hth matching, input port i is connected to output port $j = (i + h) \bmod N$, i.e., $M_{ij}^{(h)} = 1$ if $j = (i + k) \bmod N$ and $M_{ij}^{(h)} = 0$ otherwise. For a 3×3 switch, the three matchings are illustrated in Figure 4.13.

The key idea behind load-balanced switches is to use the N matchings at both stages in a round-robin fashion. The role of the first-stage switch is to distribute evenly the incoming traffic to the VOQs at the second-stage switch. Let $\lambda_{ij}^{(1)}$ denote the arrival rate of packets that arrive at input port i of the first switch and need to be delivered to output

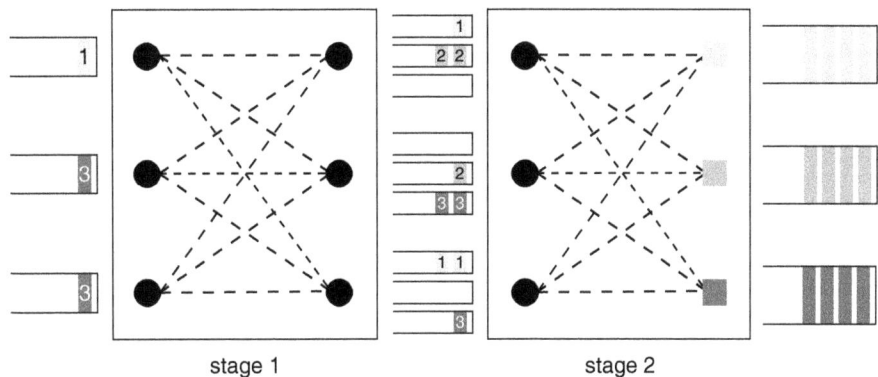

Figure 4.12 Two-stage load-balancing switch.

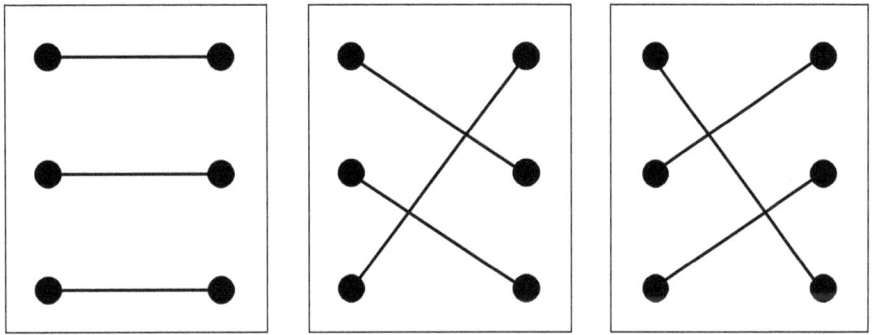

Figure 4.13 Sequence of three matchings.

port j of the second switch, and let $\lambda_{ij}^{(2)}$ denote the arrival rate into $\mathrm{VOQ}(i,j)$ at the second switch, then

$$\lambda_{ij}^{(2)} = \sum_h \frac{\lambda_{hj}^{(1)}}{N},$$

since the first stage uniformly distributes the input from each input port at stage 1 to each input port at stage 2. So, $\lambda_{ij}^{(2)} < 1/N$ if $\sum_h \lambda_{hj}^{(1)} < 1$, which holds when $\lambda^{(1)}$ lies strictly inside $\mathcal{C}$. Now, at the second stage, $\mathrm{VOQ}(i,j)$ is connected to output port j once every N time slots, and can transmit $1/N$ packet per time slot on average. Therefore, $\mathrm{VOQ}(i,j)$ is stable when $\lambda_{ij}^{(2)} < 1/N$, i.e., when $\lambda^{(1)}$ lies strictly inside $\mathcal{C}$.

Load-balanced switches

(1) At time t, set $h = t \mod N$.
(2) The hth matching is used at the first-stage switch, so input port i is connected to output port $j = (i+h) \mod N$. If the head-of-line packet at input queue i is destined to output port k, the packet is transferred to $\mathrm{VOQ}(j,k)$ at the second-stage switch.
(3) The hth matching is used at the second-stage switch. For each input port i, the switch transfers a packet from $\mathrm{VOQ}(i,j)$ to output queue j, where $j = (i+h) \mod N$.

Assume the status of queues are as in Figure 4.12 and that the first matching in Figure 4.13 is used. The packet transfers of a load-balanced switch are illustrated in Figure 4.14. □

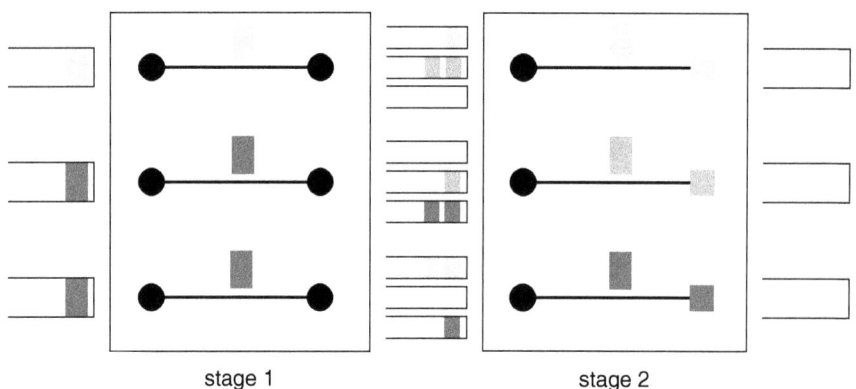

stage 1   stage 2

Figure 4.14 Packet transfers when the first matching is used.

Theorem 4.3.2 Assume that the arrivals into input queues at stage 1 are Bernoulli arrivals that are independent across time slots. Let λ_i be the probability that a packet arrives to queue i, and let p_{ij} be the probability that the arrived packet is destined to output port j. Define $\lambda_{ij} = \lambda_i p_{ij}$. The load-balanced switch architecture can support any λ such that $(1 + \epsilon)\lambda \in \mathcal{C}$ for some $\epsilon > 0$.

Proof First, each input queue at the first-stage switch can transfer one packet in each time slot, so the queue lengths at the input ports of the first stage are always equal to zero.

We now consider VOQs at the second-stage switch. Without loss of generality, we consider queue q_{1N}. The analysis for other queues is similar. We group N time slots into a super time slot, indexed by T, such that super time slot T consists of time slots $\{NT + 1, NT + 2, \ldots, (N + 1)T\}$.

During super time slot T, VOQ$(1, N)$ receives a possible arrival from each input port of the first switch. Denote by $\tilde{a}_i(T)$ the number of arrivals from input port i in super time slot T. We have

$$E[\tilde{a}_i(T)] = \lambda_i p_{ij}.$$

At the end of each super time slot, VOQ$(1, N)$ is connected to output port N, so the evolution of VOQ$(1, N)$ over super time slots can be described as

$$q_{1N}(T+1) = \left(q_{1N}(T) + \sum_{i=1}^{N} \tilde{a}_i(T) - 1 \right)^+.$$

Since $E[\sum_i \tilde{a}_i(T)] = \sum_i \lambda_i p_{ij} < 1$ when $(1 + \epsilon)\lambda \in \mathcal{C}$ for some $\epsilon > 0$, the proof that $q_{1N}(T)$ is positive recurrent is straightforward.   □

This load-balanced switch architecture achieves both low complexity and throughput optimality. The weakness of this architecture is that packets may arrive at an output port out of order. This is because packets arriving back-to-back to the same input port of the first-stage switch will be forwarded to different VOQs at the second-stage switch, and can arrive out of order when they are transferred to the output port. This problem can be solved by buffering packets at output ports. A buffer, called the reordering buffer, can be used to store packets at each output port so that a packet is only transmitted to the next router on its path only after all packets that arrived before it (at the first-stage input port) have been transmitted.

4.4   Summary

- **Capacity region of an $N \times N$ switch** Consider an $N \times N$ crossbar switch such that at most one packet can be transferred from an input port to an output port during one time slot if they are connected. Let λ_{ij} denote the mean packet arrival rate at input port i and destined to output port j. The set of rates that can be supported by the switch is called the capacity region, which is given by

$$\mathcal{C} = \left\{ \lambda : \lambda \geq 0, \sum_{i=1}^{N} \lambda_{ij} \leq 1 \text{ for any } j \text{ and } \sum_{j=1}^{N} \lambda_{ij} \leq 1 \text{ for any } i \right\}.$$

- **MaxWeight scheduling** Let $q_{ij}(t)$ denote the queue length of VOQ(i, j) at time t. The MaxWeight scheduling algorithm finds a matching $M(t)$ that maximizes $\sum_{ij} q_{ij}(t) M_{ij}$ and transfers one packet from VOQ(i, j) to output port j if there is a packet in VOQ(i, j). The MaxWeight scheduling algorithm is throughput optimal, and the complexity of finding a MaxWeight matching is $O(N^3)$.

- **Low-complexity scheduling algorithms**

  - **Maximal matching** The maximal matching scheduling algorithm finds a maximal matching at each time step rather than find the maximum weight matching. The complexity of finding a maximal matching is $O(N \log N)$, and its throughput region is half of the capacity region.

  - **Pick-and-compare scheduling** Pick-and-compare scheduling selects a maximal matching at the beginning of each time slot and compares it with the matching used in the previous time slot. The new maximal matching is accepted if its total weight is larger than the old matching. Pick-and-compare scheduling has $O(N)$ complexity, is throughput optimal, but may result in very large queue lengths.

- **Load-balanced switches** The load-balanced switching architecture consists of two stages. At the first stage, one FIFO queue is maintained at each input port, and, at the second stage, VOQs are maintained at input ports. The switch stores N matchings where, in the hth matching, input port i is connected to output port $j = (i + h) \bmod N$, and uses the N matchings at both stages in a round-robin fashion. The role of the first-stage switch is to distribute evenly the incoming traffic to the VOQs at the second-stage switch. The load-balanced switch architecture is throughput optimal and has $O(1)$ complexity, but packets may arrive at an output port out of order.

4.5   Exercises

Exercise 4.1 (HOL blocking) In this exercise, we use an example to illustrate the throughput loss induced by HOL. Consider a 2×2 switch, and assume input queues are infinitely backlogged. The destinations (output ports) of packets are randomly and uniformly chosen from $\{1, 2\}$, where 1 represents output port 1 and 2 represents output port 2. The destinations are independent across packets. Let $d_i(t)$ ($i = 1, 2$) denote the destination of the head-of-line packet of input queue i at time t. When $d_1(t) \neq d_2(t)$, both head-of-line packets can be transferred to corresponding output ports; otherwise, one of them is randomly selected and transferred to the corresponding output port.

(1) Note that $d(t)$ is a Markov chain with state space $\{11, 12, 21, 22\}$. Write down the transition matrix of this Markov chain and compute the stationary distribution of the Markov chain.

(2) Based on the stationary distribution, compute the average throughput of this 2×2 switch.

Exercise 4.2 (The irreducibility of a switch DTMC) Consider an $N \times N$ crossbar switch with VOQs and a memoryless scheduling algorithm, under which the vector of queue lengths at the VOQs, denoted by $q(t)$, is a Markov chain. Further, suppose that the algorithm has the following property: if VOQs are non-empty, then, in each time slot, at least one packet is served by the switch with a probability no less than δ. Define the state space of $q(t)$ to be

$$\mathcal{S} = \{q : q \text{ is reachable from 0 under the given scheduling algorithm}\}.$$

Prove that (i) $q(t) \in \mathcal{S}$ for any t if $q(0) \in \mathcal{S}$; and (ii) $q(t)$ is irreducible.

Exercise 4.3 (A scheduling policy which is not throughput optimal) Consider a 2×2 VOQ switch. The arrival process into VOQ(i, j) is Bernoulli with mean λ_{ij}. The arrival processes are independent across queues and time slots. Consider a scheduling policy that gives priority to edges $(1, 2)$ and $(2, 1)$, i.e., these links are scheduled if they have any packets in their queues. Given λ_{12} and λ_{21}, compute the set of $(\lambda_{11}, \lambda_{22})$ that can be supported under the priority scheduling policy above and the set of $(\lambda_{11}, \lambda_{22})$ that can be supported by the MaxWeight scheduling algorithm. Assume that $q_{12}(0) = q_{21}(0) = 0$.

Exercise 4.4 (An upper bound on the queue length under MaxWeight) Consider an $N \times N$ VOQ switch where the arrival process into queue (i,j) is Bernoulli with mean λ_{ij}. Assume that

$$\sum_i \lambda_{ik} < 1 \qquad \text{and} \qquad \sum_j \lambda_{lj} < 1, \qquad \forall k, l.$$

Compute an upper bound on the steady-state mean of the sum of the queue lengths in the network assuming that the MaxWeight scheduling algorithm is used. Try to get as tight a bound on the upper bound as possible. Hint: Proceed as in the Lyapunov stability proof, but assume that the system is already in steady state, so $E[V(q(t+1)) - V(q(t))] = 0$, as in the derivation of the Kingman bound in Chapter 3.

Exercise 4.5 (MaxWeight with a different weight function) Consider an $N \times N$ VOQ switch where the arrivals into queue (i,j) are Bernoulli with mean λ_{ij}. Assume that

$$\sum_i \lambda_{ik} < 1 \qquad \text{and} \qquad \sum_j \lambda_{lj} < 1, \qquad \forall k, l.$$

We have shown that the MaxWeight algorithm, which maximizes the sum of the product of the service rates and the queue lengths, subject to scheduling constraints, achieves 100% throughput in a high-speed switch. Instead, consider the following algorithm, where the schedule at time slot k is chosen to maximize

$$\sum_{ij} M_{ij} q_{ij}^2(k)$$

subject to scheduling constraints. In other words, we are now choosing the link weights to be the square of the queue lengths. Show that this algorithm achieves 100% throughput.

Hint: Consider the Lyapunov function $\sum_{i,j} q_{ij}^3$.

Exercise 4.6 (The capacity of crossbar switches) Consider an $N \times N$ crossbar switch. Prove that

$$\left\{ \lambda : \lambda \geq 0, \sum_{i=1}^N \lambda_{ij} \leq 1 \text{ for any } j \text{ and } \sum_{j=1}^N \lambda_{ij} \leq 1 \text{ for any } i \right\} = Co\left(\left\{ M^{(h)} \right\}_{h=1,\dots,H} \right).$$

Hint: Lemmas 4.2.2 and 4.2.3 will be helpful.

Exercise 4.7 (Stability with Markovian arrivals) Consider an $N \times N$ VOQ switch where the arrival process $a_{ij}(t)$ into queue (i,j) is described by the Markov chain in Figure 4.15. The source generating packets for queue (i,j) can be in one of two states: ON or OFF. In the ON state, it generates one packet per time slot; in the OFF state, it does not generate a packet. The source switches between the ON and OFF states according to a Markov chain. The arrival processes are independent across queues.

(1) Let $M(t)$ denote the matching selected by the MaxWeight scheduling algorithm at time t. Show that, for any $\tau \geq 1$,

$$\sum_{ij} M_{ij}(t+\tau) q_{ij}(t) \geq \sum_{ij} M_{ij}(t) q_{ij}(t) - 2N^2 \tau.$$

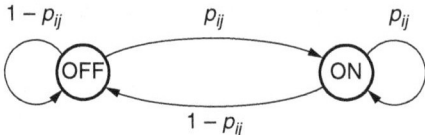

Figure 4.15 The Markov chain that describes the arrival process into queue (i,j).

(2) Show that the MaxWeight scheduling algorithm is throughput optimal for the above arrival process. Hint: Consider the Lyapunov function $V(t) = \sum_{ij} q_{ij}^2(t)$. Show that $E[V(t+m) - V(t)|q(t), a(t)] < -\delta$ when m is sufficiently large and $q(t) \notin \mathcal{S}$ for some finite set $\mathcal{S}$. You may need the following version of the law of large numbers for Markov chains:

$$\lim_{m \to \infty} \frac{1}{m} E\left[\sum_{\tau=0}^{m-1} a_{ij}(t+\tau) \middle| a_{ij}(t) \right] = \lambda_{ij}$$

for any t and $a(t)$.

Exercise 4.8 (The throughput of the pick-and-compare scheduling algorithm) Consider an $N \times N$ VOQ switch where the arrivals into queue (i,j) are Bernoulli with mean λ_{ij}. Prove that the pick-and-compare scheduling algorithm can support any arrival rate vector that satisfies

$$\sum_i \lambda_{ik} < 1 \quad \text{and} \quad \sum_j \lambda_{lj} < 1, \qquad \forall k, l.$$

Exercise 4.9 (The complexity of the PIM algorithm) Consider the PIM algorithm used in obtaining a maximal matching. Recall that each unmatched input sends a request message to each unmatched output. A request is said to be *resolved* if one of the following events occurs: (i) the request is granted, (ii) the input port that sends the request is matched, or (iii) the output port that receives the request is matched. Prove that the number of requests resolved at each iteration is at least 75% on average. Hint: To solve this problem, you can focus on the requests sent to a specific output port, say output port j. Suppose m requests are sent to output port j. The m input ports that sent the requests can be classified into classes. An input port is in class 1 if it receives a grant from an output port other than output port j; otherwise, the input port is in class 2. Assume there are k class-2 input ports. Then, with probability k/m, a class-2 input port is selected by output port j and all m requests are resolved; with probability $1 - k/m$, a class-1 input port is selected and $m - k$ requests are resolved (because class-1 input ports will be matched).

Exercise 4.10 (The throughput optimality of randomized load balancing) Consider the load-balanced switch discussed in Section 4.3.3. Instead of applying the N schedules in a round-robin fashion, each switch selects a schedule from the N schedules, uniformly at random, at each time slot. Prove that this randomized load-balanced switch is also throughput optimal.

4.6   Notes

Switch fabrics are discussed in detail in [59, 89]. Packet scheduling using the MaxWeight algorithm was introduced in [163] for wireless networks, which can be viewed as a more general model of packet switches. The case of packet switches was specifically considered in [112]. The throughput of maximal scheduling was studied in [31, 172]. The PIM algorithm was presented in [4], and an algorithm called iSLIP based on PIM was presented in [111]. The pick-and-compare algorithm was proposed in [162], and the load-balanced switch was introduced in [22].

Scheduling in wireless networks

So far we have looked at resource allocation algorithms in networks with wireline links. In this chapter, we consider networks with wireless components. The major difference between wireless and wireline networks is that in wireless networks links contend for a common resource, namely the wireless spectrum. As a result, we have to design Medium Access Control (MAC) algorithms to decide which links access the wireless medium at each time instant. As we will see, wireless MAC algorithms have features similar to scheduling algorithms for high-speed switches, which were studied in Chapter 4. However, there are some differences: wireless networks are subject to time-varying link quality due to channel fluctuations, also known as channel fading; and transmissions in wireless networks may interfere with each other, so transmissions have to be scheduled to avoid interference. In addition, some wireless networks do not have a central coordinator to perform scheduling, so scheduling decisions have to be taken independently by each link. In this chapter, we will address the following issues specific to wireless networks.

- *Does channel-state information play a critical role in scheduling in wireless networks?*
- *What is the capacity region of a cellular network, and what scheduling algorithm can be used to achieve the full capacity region?*
- *What is the capacity region of an ad hoc wireless network, what scheduling algorithm can be used to support the capacity region, and can the algorithm be implemented in a distributed fashion?*

5.1 Wireless communications

Before we discuss MAC algorithms, we first present a quick overview of wireless communications. Wireless communications use electromagnetic waves to carry information over the air. Electromagnetic wave propagation over the air suffers from several forms of degradation: *signal attenuation, multi-path propagation*, and *interference*.

Signal power attenuates as a function of distance during wireless transmission. If the transmit power is p and the distance between a transmitter and its receiver is d, the received power is $pd^{-\alpha}$, where α is called the path-loss exponent, ranging from 2 to 4 in most cases. This degradation in power as a function of distance is also called path loss.

Multi-path propagation refers to the fact that in wireless communications a signal may simultaneously travel over multiple paths (bouncing off nearby objects) to reach the receiver, as shown in Figure 5.1. Electromagnetic waves traveling over different paths may constructively or destructively interfere with each other depending on the location of the receiver; see Figure 5.2. Even small changes in the locations of the transmitter, receiver or nearby objects can shift the phase of the received signal over the various paths significantly, and thus change the interference from destructive to constructive, or vice versa. Therefore, the received power can vary significantly as a function of time.

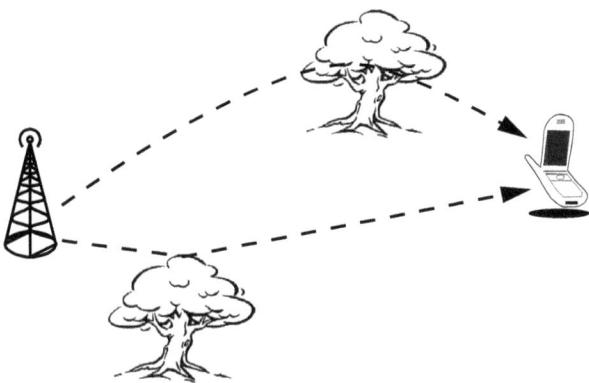

Figure 5.1 Multi-path propagation in wireless communication.

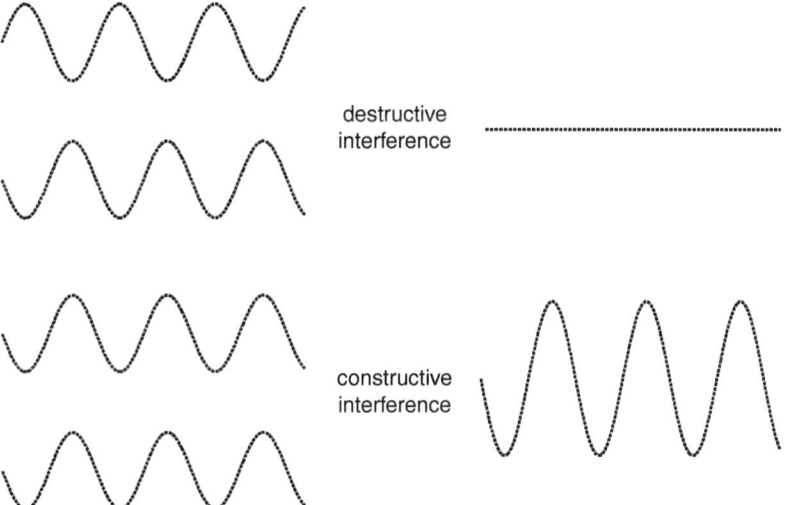

Figure 5.2 Destructive interference weakens the received signal, whereas constructive interference strengthens the signal.

This phenomenon is known as *fading*. Assuming the transmit power from radio i is p_i, the received power can be written as $p_i h_{ij}(t)$, where $h_{ij}(t)$ is called the channel gain from transmitter i to receiver j at time t. The parameter $h_{ij}(t)$ is the product of two terms: one is $d_{ij}^{-\alpha}$ due to path loss and the other is a random variable which reflects fading.

Given a single transmitter i and its receiver j, the rate at which data transmission occurs is some function of the Signal-to-Noise Ratio (SNR) given by

$$\text{SNR} = \frac{h_{ij}(t)p_{ij}}{\sigma_j^2},$$

where σ_j^2 is the thermal noise at the receiver. From a networking point of view, we will assume that time is slotted, i.e., the network operates in discrete time, and model transmission over a single wireless link (i.e., a transmitter–receiver pair) in one of the following ways.

- The transmission rate is a constant r packets/time slot: this is an abstraction of the situation where there is no fading, and hence the channel gain $h_{ij}(t)$ is constant (independent of t) and the transmitter uses a constant power level.
- The transmission rate can be chosen from a finite, discrete set $\{r_1, r_2, \ldots, r_K\}$: this is an abstraction of the case where the channel gain is constant as above but the transmitter uses a discrete number K of power levels and each transmit power level leads to a different rate.
- The channel is in one of a finite number of states $1, 2, \ldots, M$, and the channel gain in each of these states is given by $h_1, h_2, \ldots, h_M$, respectively. The probability that the channel gain is h_m is given by π_m. When the channel gain (also known as the channel state) is h_m, the possible transmission rates are $r_1(m), r_2(m), \ldots, r_K(m)$. This scenario corresponds to a time-varying fading channel and a transmitter that is capable of choosing from one of K power levels. However, since the channel gain is different in each state, the possible transmission rates in each channel state are different.

The first wireless network scenario (as opposed to a single wireless link as above) that we will encounter is one where a single transmitter (called a base station or access point) wants to transmit different packets to different receivers. In this case, for simplicity, we will assume that the transmitter can transmit to only a single receiver at a time. (There are communication schemes under which it is possible to transmit the same or different packets to multiple receivers at the same time, and such schemes can also be accommodated in the theory to be developed later. However, to keep the discussion simple, we will not consider these extensions.) At each time instant, each link's (transmitter–receiver pair) channel state may be different because fading is typically independent across receivers. The networking problem is to decide when the transmitter should transmit to a particular receiver. This decision is based on the joint channel states of all the links, and we will discuss algorithms to make the optimal transmission decision at each time instant. Once the transmitter makes a decision to transmit to a receiver, the network operates as a single wireless link and all the earlier comments regarding single-link transmission models apply.

The second wireless networking scenario that we will consider is one in which multiple transmitters want to transmit to multiple receivers. In the case where multiple transmitter–receiver pairs are active at the same time, the transmission rate to a single receiver is a function of the Signal-to-Interference and Noise Ratio (SINR) at the receiver:

$$\text{SINR} = \frac{\text{received signal power}}{\text{thermal noise at the receiver} + \text{interference power}}.$$

The SINR determines the maximum transmission rate over the link. We present an example to illustrate data rates in a simple wireless network with interference.

Example 5.1.1

Consider a four-node network as shown in Figure 5.3. Node 1 wants to transmit to node 2, and node 3 wants to transmit to node 4. The SINR at node 2 is given by

$$\frac{p_1 h_{12}(t)}{n_2 + p_3 h_{32}(t)},$$

and the SINR at node 4 is given by

$$\frac{p_3 h_{34}(t)}{n_4 + p_1 h_{14}(t)}.$$

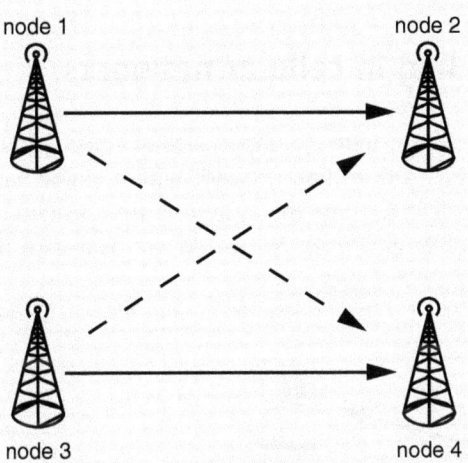

node 1   node 2

node 3   node 4

Figure 5.3 Four-node wireless network. Solid lines denote intended transmissions; dashed lines denote interference.

A well-known formula from information theory states that the corresponding transmission rates using simple signal transmission schemes are given by

$$r_{12}(t) = \frac{1}{2} \log_2 \left(1 + \frac{p_1 h_{12}(t)}{n_2 + p_3 h_{32}(t)} \right),$$

$$r_{34}(t) = \frac{1}{2} \log_2 \left(1 + \frac{p_3 h_{34}(t)}{n_4 + p_1 h_{14}(t)} \right),$$

where n_i is the power of the background noise experienced by receiver node i and r_{ij} is the transmission rate from node i to node j.

If we assume that the vector $(h_{12}, h_{14}, h_{34}, h_{32})$ can take on one of a finite set of possible values and that the transmit powers can take on only a finite, discrete set of values, this situation can be abstracted by the following model: the vector of channel gains is one of M states, with state m occurring with probability π_m. In each state, the rate vector (r_{12}, r_{34}) can take of one of K values.

In the simplest instance of this model, the channel gains are in only one possible state (i.e., the channel gains are invariant), and the three possible rate vectors are $(1, 0)$, $(0, 1)$, and $(0,0)$. In other words, link $(1,2)$ is activated, link $(3,4)$ is activated, or neither link is activated, and when activated a link can transmit one packet in a time slot.

The key problem in networks with multiple transmitter–receiver pairs is to choose the appropriate rate vector without using a central coordinator. Such networks are called ad hoc networks since these networks can operate in an "ad hoc" fashion without the need for a central infrastructure.   □

In the following sections, we will first consider the problem of a single transmitter transmitting to multiple receivers. This situation models a cellular network where a single base station transmits to multiple mobiles, or a wireless LAN in which an access point transmits to multiple mobiles. Then we will consider ad hoc wireless networks.

5.2   Channel-aware scheduling in cellular networks

First we use a simple example to demonstrate the importance of considering channel-state information in scheduling. Consider a downlink network with a single base station and two mobiles, as shown in Figure 5.4. Assume the system is time slotted. The base station can only transmit to one mobile at a time. So we do not deal with interference in this example.

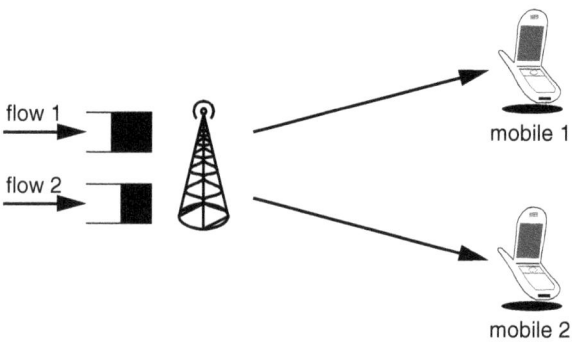

Figure 5.4 Downlink network with a single base station and two mobiles.

We further assume that the two channels are independent ON–OFF channels, and that each channel is ON with probability 1/2. When a channel is ON, one packet can be transmitted over the channel. This ON–OFF channel model is an abstraction of a situation where the quality of each channel is time varying due to fading. If we assume that the base station uses a fix transmit power, and that mobiles can successfully receive a packet transmission only when the Signal to Noise Ratio (SNR) is above a threshold, then the ON state corresponds to the state of the channel when the SNR is above the threshold.

Next we consider two scheduling policies, a channel-unaware policy and a channel-aware scheduling policy, to demonstrate the importance of including channel-state information in scheduling. The channel-unaware scheduling, which schedules the mobiles independent of their channel states, is now presented.

Channel-unaware scheduling

The base station selects mobile i with probability α_i and transmits a packet to mobile i.

Under this channel-unaware scheduling, the rate allocated to mobile i is

$$\alpha_i \Pr(\text{channel } i \text{ is ON}) = \frac{\alpha_i}{2} \text{ (packets/slot)}.$$

Thus, by varying α_1 and α_2, any service rate vector of the form $(\alpha_1/2, \alpha_2/2)$ can be achieved by the channel-unaware scheduling algorithm, where $\alpha_1 + \alpha_2 \leq 1$.

The channel-aware scheduling is presented below, where the base station schedules the mobiles based on their channel states.

Channel-aware scheduling

Assume that the base station knows the states of both channels through the feedback from the mobiles. If only one mobile's channel is ON, the base station schedules the mobile with the ON channel; otherwise, the base station schedules mobile i with probability α_i.

Under this scheduling algorithm, mobile 1 (2) is scheduled with probability 1 when its channel is ON, and mobile 2's (1's) channel is OFF; mobile 1 is scheduled with probability α_i when both channels are ON. Therefore, the rate allocated to mobile i ($i = 1, 2$) is

$$\Pr(\text{channel } i \text{ is ON, the other channel is OFF}) + \alpha_i \Pr(\text{both channels are ON})$$

$$= \frac{1}{4} + \alpha_i \frac{1}{4}$$

$$= \frac{(1 + \alpha_i)}{4}.$$

The set of rate allocations (μ_1, μ_2) such that

$$\mu_1 \leq \frac{(1 + \alpha_1)}{4} \quad \text{and} \quad \mu_2 \leq \frac{(1 + \alpha_2)}{4} \tag{5.1}$$

is feasible under the channel-aware scheduling.

By varying α_1 and α_2, we can plot the capacity regions of both scheduling algorithms, as shown in Figure 5.5. The capacity region of channel-aware scheduling is clearly larger than that of channel-unaware scheduling. This example demonstrates the importance of exploiting channel-state information in scheduling. To implement channel-aware scheduling, one needs to know the state of the channels. We will assume that such channel-state information can be obtained without much overhead.

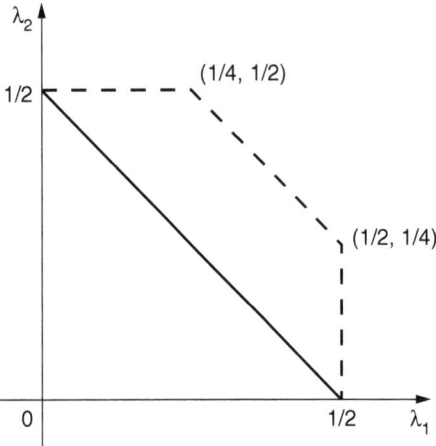

Figure 5.5 The region $Co\{(0,0),(0,1/2),(1/2,0)\}$ is the capacity region of the channel-unaware scheduling, and the region $Co\{(0,0),(0,1/2),(1/4,1/2),(1/2,1/4),(1/2,0)\}$ is the capacity region of the channel-aware scheduling. Recall that Co denotes the convex hull of a set.

5.3   The MaxWeight algorithm for the cellular downlink

Motivated by the simple example above, in this section we consider channel-aware scheduling in a general cellular downlink network with a single base station and N mobiles. We also make the following assumptions.

(i) Assume the network has M channel states, indexed by m, where the channel state refers to the states of all channels in the network.

(ii) The base station can transmit to one mobile at a time.

(iii) Define c_m to be the channel rate vector when the channel is in state m, where $c_{m,i}$ is the maximum number of packets that can be transmitted to mobile i, if it is scheduled. Note that this definition allows the base station to transmit fewer than $c_{m,i}$ packets to mobile i if required. Under this definition, note that c_m is a vector of non-negative integer values. Assume there exists $c_{max} < \infty$ such that $c_{m,i} \leq c_{max}$ for all i and m. Let $\mu_i(t)$ denote the transmission rate to mobile i at time t, so $\mu_i(t) = c_{m,i}$ if mobile i is scheduled in state m.

(iv) The channel state process is i.i.d. across time slots. The probability that the channel is in state m is denoted by π_m.

(v) Packet arrival processes are independent across users and time slots. Let $a_i(t)$ denote the number of packet arrivals for mobile i at time slot t, which is a random variable that takes positive integer values, with mean λ_i and a finite variance σ_i^2 for user i. We further assume that $\Pr(a_i(t) = 0) > 0$ for all i.

(vi) The base station maintains a separate queue for each mobile. Let $q_i(t)$ denote the number of packets buffered at the base station for mobile i at the beginning of time slot t, before arrivals occur.

The assumptions that the base station can transmit to one mobile at a time, i.i.d. channel fading, and i.i.d. arrival processes can be relaxed. We impose these assumptions to simplify the analysis.

Similar to switch scheduling, the arrival rate vector λ is said to be *supportable* if there exists a scheduling algorithm under which

$$\lim_{C \to \infty} \lim_{t \to \infty} \Pr\left(|q(t)| \geq C\right) = 0,$$

where $|q(t)|$ is the sum of the lengths of all the queues in the network at time t. Next we characterize the capacity region of the cellular downlink defined above, and then present a scheduling algorithm that supports any arrival rate vector that lies strictly in the capacity region.

Definition 5.3.1 (Capacity region C) An arrival rate vector λ is in C if there exists $\{\alpha_{m,i} \geq 0\}$ (interpret $\alpha_{m,i}$ as the probability user i is scheduled in channel state m) such that

$$\lambda_i \leq \sum_{m=1}^{M} \alpha_{m,i} c_{m,i} \pi_m, \quad \forall i \tag{5.2}$$

and

$$\sum_i \alpha_{m,i} \leq 1 \quad \forall m. \qquad \square$$

Due to assumption (iii) above, note that the set C satisfies the property that, if $\lambda \in C$, then $\lambda^{(i)}$, such that $\lambda_i^{(i)} = 0$ and $\lambda_j^{(i)} = \lambda_j$ for $j \neq i$, also belongs to C. A convex set with this property is called *coordinate convex*.

Example 5.3.1

Consider the network with a single base station and two mobiles, presented in Section 5.2. The network has four states: $(0,0)$ (both channels are OFF), $(0,1)$ (channel 1 is OFF and channel 2 is ON), $(1,0)$(channel 1 is ON and channel 2 is OFF), and $(1,1)$ (both channels are ON). Since a channel is equally likely to be ON or OFF, we have

$$\pi_{(0,0)} = \pi_{(0,1)} = \pi_{(1,0)} = \pi_{(1,1)} = 0.25.$$

The capacity region C is the set of λs such that

$$\lambda_1 \leq \alpha_{(0,0),1} \times 0 \times \pi_{(0,0)} + \alpha_{(0,1),1} \times 0 \times \pi_{(0,1)} + \alpha_{(1,0),1} \times 1$$

$$\times \pi_{(1,0)} + \alpha_{(1,1),1} \times 1 \times \pi_{(1,1)},$$

$$\lambda_2 \leq \alpha_{(0,0),2} \times 0 \times \pi_{(0,0)} + \alpha_{(0,1),2} \times 1 \times \pi_{(0,1)} + \alpha_{(1,0),2} \times 0$$

$$\times \pi_{(1,0)} + \alpha_{(1,1),2} \times 1 \times \pi_{(1,1)}.$$

Clearly, to maximize throughput, we should not schedule an OFF channel, so

$$\alpha_{(0,0),1} = \alpha_{(0,0),2} = \alpha_{(0,1),1} = \alpha_{(1,0),2} = 0,$$

$$\alpha_{(0,1),2} = \alpha_{(1,0),1} = 1.$$

Therefore, the capacity region can be written as

$$\lambda_1 \leq \pi_{(1,0)} + \alpha_{(1,1),1}\pi_{(1,1)} = \frac{1 + \alpha_{(1,1),1}}{4},$$

$$\lambda_2 \leq \pi_{(0,1)} + \alpha_{(1,1),2}\pi_{(1,1)} = \frac{1 + \alpha_{(1,1),2}}{4},$$

which is equivalent to (5.1). ☐

The following theorem shows that any $\lambda \notin \mathcal{C}$ is not supportable.

Theorem 5.3.1 No scheduling algorithm can support arrival rate vector $\lambda \notin \mathcal{C}$.

Proof Note that $\mathcal{C}$ is convex. According to the strict separation theorem (Result 2.1.4), if $\lambda \notin \mathcal{C}$, there exist $\beta \in \mathcal{R}^N$, $\beta \neq 0$, and $\delta > 0$ such that

$$\sum_i \beta_i \lambda_i \geq \sum_i \beta_i x_i + \delta \tag{5.3}$$

for any $x \in \mathcal{C}$.

Now suppose $\beta \geq 0$. Define $V(t) = \sum_i \beta_i q_i(t)$, and let $\mu_i(t)$ denote the amount of service allocated to user i in time slot t, i.e., if user i is scheduled in time slot t, then $\mu_i(t) = c_{m,i}$ if the channel is in state m at time t. If user i is not scheduled in time t, then $\mu_i(t) = 0$. Thus,

$$q_i(t+1) = (q_i(t) + a_i(t) - \mu_i(t))^+ \geq q_i(t) + a_i(t) - \mu_i(t),$$

which implies that

$$V(t+1) \geq \sum_{s=1}^{t} \sum_i \beta_i (a_i(s) - \mu_i(s)).$$

Denote by $m(t)$ the channel state at time t. According to the SLLN, with probability 1,

$$\lim_{t \to \infty} \frac{1}{t} \sum_{s=1}^{t} \sum_i \beta_i a_i(s) = \sum_i \beta_i \lambda_i,$$

$$\lim_{t \to \infty} \frac{1}{t} \sum_{s=1}^{t} \mathbb{I}_{m(s)=l} = \pi_l.$$

Due to inequality (5.3), we can further obtain that, with probability 1,

$$\lim_{t \to \infty} \frac{1}{t} \sum_{s=1}^{t} \sum_i \beta_i a_i(s) \geq \lim_{t \to \infty} \frac{1}{t} \sum_{s=1}^{t} \sum_i \beta_i \mu_i(s) + \delta,$$

which implies that $\lim_{t \to \infty} V(t) = \infty$ with probability 1. So λ is not supportable.

We now prove that there exists a $\beta \geq 0$ satisfying condition (5.3) to complete the theorem. Given any β that satisfies (5.3), we define $\tilde{\beta}$ to be $\tilde{\beta}_i = \beta_i$ if $\beta_i \geq 0$ and $\tilde{\beta}_i = 0$ otherwise. Since $\lambda \geq 0$, we first have

$$\sum_i \tilde{\beta}_i \lambda_i \geq \sum_i \beta_i \lambda_i. \tag{5.4}$$

Next, given any $x \in \mathcal{C}$, we define $\tilde{x}$ such that $\tilde{x}_i = x_i$ if $\beta_i \geq 0$ and $\tilde{x}_i = 0$ otherwise. So we have $\tilde{\beta}_i x_i = \beta_i \tilde{x}_i$, and

$$\sum_i \tilde{\beta}_i x_i = \sum_i \beta_i \tilde{x}_i \leq \sum_i \beta_i \lambda_i, \tag{5.5}$$

where the last inequality holds because, according to the definition of $\mathcal{C}$, $\tilde{x}$ also belongs to $\mathcal{C}$ given $x \in \mathcal{C}$. Combining inequalities (5.4) and (5.5), we conclude that $\tilde{\beta}$ is a non-negative vector that satisfies condition (5.3). $\qquad\square$

If the average arrival rate λ and channel-state distribution π are known, we can choose an α that satisfies (5.2), and schedule the mobiles according to the channel state and α. Knowing the arrival rate λ and the channel-state distribution π, however, is difficult in reality. The question, therefore, is *can we design a traffic-blind policy, such as MaxWeight for high-speed switches, for wireless networks?*

We note that scheduling in wireless networks shares some similarities with that in high-speed switches. In fact, if the channel state is fixed instead of time varying, a downlink network can be viewed as a switch with a single input port and N output ports. So, a natural question to ask is whether a variation of the MaxWeight algorithm can be used to achieve the maximum throughput in cellular networks. Next, we present the MaxWeight algorithm for wireless downlink networks, and prove its throughput optimality.

MaxWeight scheduling for downlink networks

At each time instant t, the base station transmits to mobile i such that

$$i \in \arg\max_j c_{m(t),j} q_j(t),$$

with rate $c_{m(t),i}$, breaking ties at random.

Example 5.3.2

Again, consider the network with a single base station and two mobiles, presented in Section 5.2, which has four states:

- when both channels are OFF, no mobile is scheduled;
- when channel 1 is OFF and channel 2 is ON, $c_1(t)q_1(t) = 0 \leq c_2(t)q_2(t) = q_2(t)$, so mobile 2 is scheduled;

- when channel 1 is ON and channel 2 is OFF, $c_1(t)q_1(t) = q_1(t) \geq c_2(t)q_2(t) = 0$, so mobile 1 is scheduled;
- when both channels are ON, $c_1(t)q_1(t) = q_1(t)$ and $c_2(t)q_2(t) = q_2(t)$, so the base station schedules the mobile with the longer queue length. ☐

In the following theorem, we show that MaxWeight scheduling can stabilize any λ that lies strictly inside set C.

Theorem 5.3.2 Given any arrival rate vector λ such that $(1+\epsilon)\lambda \in C$ for some $\epsilon > 0$, $q(t)$ is positive recurrent under the MaxWeight algorithm.

Proof It is easy to verify that $q(t)$ is an irreducible Markov chain under the assumption that $\Pr(a_i(t) = 0) > 0$ all i. We first prove the following fact, which is the key to this proof. Given $q(t) = q$, under MaxWeight scheduling,

$$ E\left[\sum_i q_i \mu_i(t)\right] \geq \sum_i q_i \gamma_i \quad \forall \, \gamma \in C. \tag{5.6} $$

According to the definition of C, for any $\gamma \in C$ there exists α such that $\gamma_i \leq \sum_{m=1}^{M} \alpha_{m,i} c_{m,i} \pi_m$ for all i. So,

$$ \sum_i q_i \gamma_i \leq \sum_i q_i \left(\sum_{m=1}^{M} \alpha_{m,i} c_{m,i} \pi_m\right) $$

$$ = \sum_{m=1}^{M} \pi_m \left(\sum_i q_i \alpha_{m,i} c_{m,i}\right) $$

$$ \leq \sum_{m=1}^{M} \pi_m \left(\max_i q_i c_{m,i}\right), $$

where the last inequality holds because $\sum_i \alpha_{m,i} \leq 1$ for all m. Given $q(t) = q$ and $c(t) = c_m$, MaxWeight scheduling selects a mobile with the largest $q_i c_{m,i}$, so

$$ E\left[\sum_i q_i \mu_i(t)\right] = \sum_{m=1}^{M} \pi_m \left(\max_i q_i c_{m,i}\right) \geq \sum_i q_i \gamma_i, $$

and (5.6) holds.

Define the Lyapunov function to be

$$ V(q(t)) = \sum_{i=1}^{N} q_i^2(t). $$

The conditional drift of the Lyapunov function is given by

$$E\left[V(q(t+1)) - V(q(t))|q(t) = q\right]$$

$$= E\left[\sum_{i=1}^{N}(q_i^2(t+1) - q_i^2(t))\middle|q(t) = q\right]$$

$$= E\left[\sum_{i=1}^{N}\left((q_i(t) + a_i(t) - \mu_i(t))^+\right)^2 - q_i^2(t)\middle|q(t) = q\right]$$

$$\leq E\left[\sum_{i=1}^{N}(q_i(t) + a_i(t) - \mu_i(t))^2 - q_i^2(t)\middle|q(t) = q\right]$$

$$= E\left[\sum_{i=1}^{N}(a_i(t) - \mu_i(t))^2 + 2q_i(t)(a_i(t) - \mu_i(t))\middle|q(t) = q\right] \tag{5.7}$$

$$= E\left[\sum_{i=1}^{N}(a_i(t) - \mu_i(t))^2\middle|q(t) = q\right] + 2\sum_{i} q_i \left(\lambda_i - E\left[\mu_i(t)|q(t) = q\right]\right). \tag{5.8}$$

Since the arrival processes are assumed to be independent of the queue state,

$$E\left[\sum_{i=1}^{N}(a_i(t) - \mu_i(t))^2\middle|q(t) = q\right]$$

$$= E\left[\sum_{i=1}^{N}a_i^2(t) - 2a_i(t)\mu_i(t) + \mu_i^2(t)\middle|q(t) = q\right]$$

$$\leq \sum_{i=1}^{N}\left(\sigma_i^2 + \lambda_i^2 + c_{\max}^2\right). \tag{5.9}$$

Since there exists $\epsilon > 0$ such that $(1 + \epsilon)\lambda \in \mathcal{C}$, based on the fact (5.6), we have

$$\sum_{i} q_i(1 + \epsilon)\lambda_i \leq \sum_{i} q_i E\left[\mu_i(t)|q(t) = q\right]. \tag{5.10}$$

Substituting inequalities (5.9) and (5.10) into (5.8), we obtain

$$E\left[V(q(t+1)) - V(q(t))|q(t) = q\right] \leq \sum_{i=1}^{N}\left(\sigma_i^2 + \lambda_i^2 + c_{\max}^2\right) - 2\epsilon \sum_{i=1}^{N}q_i\lambda_i.$$

So the Markov chain $q(t)$ is positive recurrent according to Theorem 3.3.8, and the theorem holds. $\square$

5.4   MaxWeight scheduling for ad hoc P2P wireless networks

Ad hoc wireless networks refer to wireless networks in which wireless devices communicate directly with each other instead of going through base stations or access points. Although wireless access networks, including cellular networks and Wireless Local Area Networks (WLANs), are most common today, ad hoc wireless networks have emerged for a wide range of applications, such as city-wide wireless mesh networks, wireless sensor networks, and vehicular networks.

In this and the following section, we will focus on the modeling and design of scheduling algorithms for ad hoc wireless networks. We will assume that the ad hoc network operates in a Peer-to-Peer (P2P) fashion, i.e., a source node directly transmits a packet to its destination without using multi-hop routing. Multi-hop wireless networks will be considered in Chapter 6.

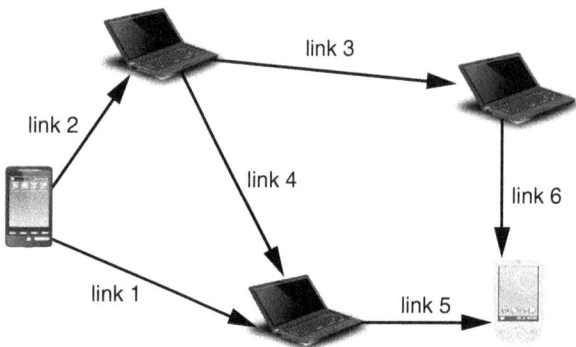

Figure 5.6 Ad hoc P2P wireless network.

Figure 5.6 shows an example of an ad hoc P2P network, where nodes are distributed in a two-dimensional space and form multiple source-destination pairs (called links). We will focus our attention on resolving interference in ad hoc wireless networks by assuming all channels are time-invariant channels, i.e., no fading. We further model the interference as a conflict graph in which each node represents a wireless link and an edge from node a to node b means link a interferes link b, so link a should keep silent when link b is active. For example, if only non-adjacent links can transmit simultaneously, the conflict graph associated with the network in Figure 5.6 is shown in Figure 5.7.

Note that a collection of nodes (which are wireless links in the real network) in the conflict graph can be scheduled simultaneously if they do not interfere with each other. Such a set of nodes is called a *schedule*, and because of the importance of this concept we present it as a definition.

Definition 5.4.1 (Schedule) A set of links that can be active simultaneously without interfering each other.   □

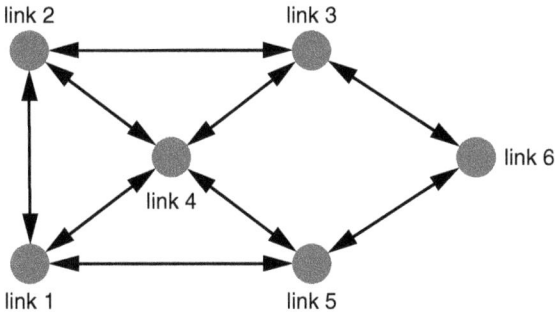

Figure 5.7 A conflict graph associated with the network in Figure 5.6.

Consider the conflict graph in Figure 5.7; the link set $\{1, 3\}$ is a schedule. The scheduling problem can be thought of as finding a good schedule in the conflict graph at each time instant.

Inspired by the MaxWeight algorithm for high-speed switches and for the cellular downlink, we are interested in the performance of MaxWeight scheduling in ad hoc P2P wireless networks. For simplicity, we assume Bernoulli packet arrivals to each link, and also assume that one packet can be transmitted in one time slot if a link is scheduled. Let $q_l(t)$ denote the number of packets queued at the transmitter of link l. We also assume the conflict graph has H distinct schedules. Let $M^{(1)}, M^{(2)}, \ldots, M^{(H)}$ represent the H schedules, where each M is a vector of size L, and L is the number of links in the network. According to this definition,

$$M_l^{(h)} = \begin{cases} 1, & \text{if link } l \text{ is in schedule } M^{(h)}, \\ 0, & \text{otherwise.} \end{cases}$$

The MaxWeight algorithm for ad hoc P2P wireless networks is presented below, where, in each time instant, a schedule with the largest sum-queue length is used.

MaxWeight scheduling for ad hoc P2P wireless networks

At each time instant t, the network schedules the links in schedule $M(t)$ such that

$$M(t) \in \arg \max_{M^{(h)}} \sum_l M_l^{(h)} q_l(t). \tag{5.11}$$

Break ties arbitrarily.

Example 5.4.1

Consider the conflict graph shown in Figure 5.8. MaxWeight scheduling will schedule $\{4, 6\}$, which has the maximum sum weight 21.   □

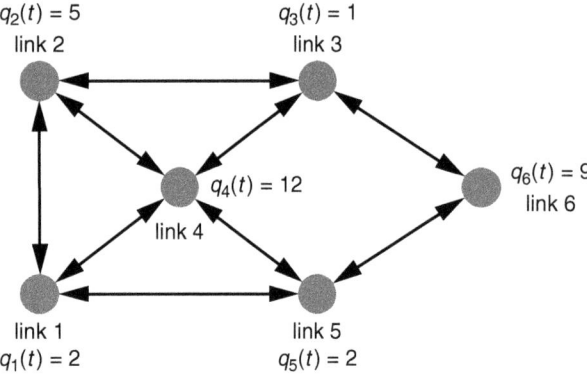

$q_2(t) = 5$
link 2

$q_3(t) = 1$
link 3

$q_4(t) = 12$

link 4

$q_6(t) = 9$
link 6

link 1
$q_1(t) = 2$

link 5
$q_5(t) = 2$

Figure 5.8 A conflict graph.

Our goal is to show that the MaxWeight scheduling algorithm is throughput optimal. To do this, we first have to understand the set of packet arrival rates that can be supported by the network. Define the set $\mathcal{C}$ to be

$$\mathcal{C} = \left\{ \lambda : \lambda_l \le \sum_{h=1}^{H} \alpha_h M_l^{(h)} \quad \forall l, \quad \text{for some } \alpha \ge 0 \text{ such that } \sum_h \alpha_h \le 1 \right\},$$

where α_h can be viewed as the probability of choosing schedule $M^{(h)}$. It should be intuitively obvious that any arrival rate outside $\mathcal{C}$ cannot be supported by the network if we interpret α_h to be the fraction of time that schedule $M^{(h)}$ is chosen. If the arrival rate vector lies outside $\mathcal{C}$, this means that there is no choice of fractions α_h such that, if schedule $M^{(h)}$ is chosen for a fraction α_h of the time, the service rates provided on all links are greater than their arrival rates. The following theorem formalizes this intuition.

Theorem 5.4.1 No scheduling algorithm can support arrival rate vector $\lambda \notin \mathcal{C}$.

Proof The proof is based on the strict separation theorem and SLLN, and is similar to the proof of Theorem 5.3.1. $\square$

We now prove that the MaxWeight algorithm is throughput optimal in ad hoc P2P wireless networks.

Theorem 5.4.2 Given an arrival rate vector λ such that $(1 + \epsilon)\lambda \in \mathcal{C}$ for some $\epsilon > 0$, $q(t)$ is positive recurrent under MaxWeight scheduling.

Proof According to the definition of $\mathcal{C}$, given any $\gamma \in \mathcal{C}$, there exists $\alpha \ge 0$ such that $\sum_h \alpha_h \le 1$ and

$$\gamma \le \sum_h \alpha_h M^{(h)}.$$

So, given $q(t) = q$,

$$\sum_l q_l \gamma_l \le \sum_l q_l \left(\sum_h \alpha_h M_l^{(h)} \right) = \sum_h \alpha_h \left(\sum_l q_l M_l^{(h)} \right)$$

$$\le \max_h \sum_l q_l M_l^{(h)} = \sum_l q_l M_l(t),$$

where the last equality follows from the definition of MaxWeight scheduling.

If there exists $\epsilon > 0$ such that $(1 + \epsilon)\lambda \in \mathcal{C}$, then, given $q(t) = q$,

$$\sum_l q_l (1 + \epsilon) \lambda_l \le \sum_l q_l M_l(t),$$

and

$$\sum_l q_l \lambda_l - \sum_l q_l M_l(t) \le -\epsilon \sum_l q_l \lambda_l.$$

Considering the Lyapunov function $V(q(t)) = \sum_l q_l^2(t)$ and following the argument given in the proof of Theorem 5.3.2, we can prove that $q(t)$ is positive recurrent. □

5.5   General MaxWeight algorithms

We have now seen that the MaxWeight algorithm is a powerful scheduling mechanism, which is throughput optimal in the contexts of switch scheduling, cellular downlink networks, and ad hoc P2P networks. The MaxWeight algorithm uses queue lengths as link weights so that if a flow does not receive enough service, its queue builds up, which forces the algorithm to allocate more resources to the flow. This interaction between queue lengths and scheduling guarantees the throughput optimality of resource allocation. The reader may wonder whether other choices of link weights can work as well. In this section, we will see that a large class of functions (of queue lengths) that are increasing and differentiable can be used as link weights while maintaining throughput optimality.

Denoting by $w_l(q_l)$ the weight associated with link l, we present the general MaxWeight algorithm, in the setting of ad hoc P2P networks without fading. The algorithm can be easily extended to networks with fading channels.

General MaxWeight scheduling for ad hoc P2P wireless networks

At time instant t, the network schedules the links in schedule $M(t)$ such that

$$M(t) \in \arg \max_{M^{(h)}} \sum_l w_l(q_l(t)) M_l^{(h)}.$$

If there is more than one such schedule, break ties arbitrarily.

Next, we introduce a class of weight functions, called valid weight functions, such that the general MaxWeight scheduling is throughput optimal if all link weights are valid weight functions.

Definition 5.5.1 (Valid weight functions for MaxWeight scheduling) We say a function $w(\cdot)$ is a *valid weight function* if the following conditions hold:

(i) function $w(x)$ is increasing and differentiable for $x \geq 0$, and $w(0) = 0$;
(ii) as $x \to \infty$, $w(x) \to \infty$;
(iii) given any $\delta > 0$, there exists a constant $B_\delta \geq 0$ such that, for any $x \geq 0$,

$$(1 - \delta)w(x + 1) - B_\delta \leq w(x) \leq (1 + \delta)w\left((x - 1)^+\right) + B_\delta. \qquad (5.12)$$

□

Note that the third condition simply states that $w(x)$ does not change significantly when x increases or decreases by 1. Specifically, when $w(x)$ is small, $|w(x) - w(x \pm 1)|$ is bounded by a constant, and, when $w(x)$ is large, $w(x + 1)/w(x)$ and $w(x)/w(x - 1)$ are bounded by constants.

The following theorem states that the general MaxWeight scheduling is throughput optimal if all weight functions are valid weight functions. To simplify notation and analysis, we assume arrivals are Bernoulli.

Theorem 5.5.1 Assume that the arrival processes to links in an ad hoc P2P network are Bernoulli processes. The general MaxWeight scheduling can support any λ such that $(1 + \epsilon)\lambda \in C$.

Proof Following the proof of Theorem 5.4.1, given $q(t) = q$, the general MaxWeight algorithm guarantees that

$$(1 + \epsilon) \sum_l w_l(q_l)\lambda_l \leq \sum_l w_l(q_l)M_l(t). \qquad (5.13)$$

Now we need to choose a proper Lyapunov function to show that $q(t)$ is positive recurrent. We consider the following Lyapunov function:

$$V(q(t)) = \sum_l \int_0^{q_l(t)} w_l(x)\mathrm{d}x.$$

The conditional drift of the Lyapunov function is given by

$$E\left[V(q(t + 1)) - V(q(t))|q(t) = q\right]$$

$$= E\left[\sum_l \left(\int_0^{q_l(t+1)} w_l(x)\mathrm{d}x - \int_0^{q_l(t)} w_l(x)\mathrm{d}x\right)\middle| q(t) = q\right]$$

$$= E\left[\sum_l w_l(\tilde{q}_l)(q_l(t + 1) - q_l(t))\middle| q(t) = q\right],$$

where the last equality follows from the mean-value theorem and

$$\tilde{q}_l \in [\min(q_l(t), q_l(t+1)), \max(q_l(t), q_l(t+1))].$$

Note that if $q_l(t) + a_l(t) > 0$, then $q_l(t+1) = q_l(t) + a_l(t) - M_l(t)$; otherwise, $q_l(t+1) = q_l(t) = 0$ and $\tilde{q}_l = 0$. Therefore,

$$E\left[V(q(t+1)) - V(q(t))|q(t) = q\right] \leq E\left[\sum_l w_l(\tilde{q}_l)(a_l(t) - M_l(t)) \middle| q(t) = q\right]. \quad (5.14)$$

Since $q_l(t)$ can at most increase or decrease by 1 in one time slot, $\tilde{q}_l \in [(q_l(t) - 1)^+, q_l(t) + 1]$. Since $w_l(x)$ is a valid weight function, it is increasing in x and satisfies condition (5.12). Therefore, given any $\delta > 0$, there exists $B_{l,\delta} > 0$ such that

$$w_l(\tilde{q}_l) \geq w_l((q_l(t) - 1)^+) \geq \frac{w_l(q_l(t)) - B_{l,\delta}}{1 + \delta},$$

$$w_l(\tilde{q}_l) \leq w_l(q_l(t) + 1) \leq \frac{w_l(q_l(t)) + B_{l,\delta}}{1 - \delta}.$$

Replacing $w_l(\tilde{q}_l)$ in inequality (5.14) using the above two inequalities, we obtain

$$E\left[V(q(t+1)) - V(q(t))|q(t) = q\right]$$

$$\leq \sum_l \frac{w_l(q_l) + B_{l,\delta}}{1 - \delta}\lambda_l - \sum_l \frac{w_l(q_l) - B_{l,\delta}}{1 + \delta}E[M_l(t)|q(t) = q]$$

$$\leq K + \frac{1}{1 - \delta}\sum_l w_l(q_l)\lambda_l - \frac{1}{1 + \delta}\sum_l w_l(q_l)E[M_l(t)|q(t) = q]$$

$$= K + \frac{1}{1 + \delta}\left(\frac{1 + \delta}{1 - \delta}\sum_l w_l(q_l)\lambda_l - \sum_l w_l(q_l)E[M_l(t)|q(t) = q]\right),$$

where

$$K = \sum_l \frac{B_{l,\delta}\lambda_l}{1 - \delta} + \sum_l \frac{B_{l,\delta}}{1 + \delta}.$$

Now choose a small enough δ such that $(1 + \delta)/(1 - \delta) \leq 1 + \epsilon/2$. From inequality (5.13) we have

$$E\left[V(q(t+1)) - V(q(t))|q(t) = q\right] \leq K - \frac{\epsilon}{2}\frac{1}{1 + \delta}\sum_l w_l(q_l)\lambda_l.$$

Since $w_l(q) \to \infty$ as $q \to \infty$, we can conclude that $q(t)$ is positive recurrent by invoking the Foster–Lyapunov theorem, which completes the proof. □

So, by selecting different weight functions, we can construct different throughput-optimal scheduling algorithms. While all of these algorithms are throughput optimal, they may differ in their delay performances. However, the delay performance of these algorithms is hard to analyze and will not be considered here. We end this section by giving two examples of valid weight functions.

Example 5.5.1

Weight function $w(q) = \kappa q^m$ for some $m > 0$ and $\kappa > 0$ is a valid weight function. Note that

$$\kappa q^m = \frac{1}{(1 + 1/q)^m} \kappa (q + 1)^m.$$

So, given $\delta > 0$, there exists q_δ such that, for any $q \geq q_\delta$,

$$\kappa q^m \geq (1 - \delta)\kappa(q + 1)^m.$$

So, for any $q \geq 0$,

$$\kappa q^m \geq (1 - \delta)\kappa(q + 1)^m - (1 - \delta)\kappa(q_\delta + 1)^m.$$

Similarly, when $q \geq 1$,

$$\kappa q^m = \frac{1}{(1 - 1/q)^m} \kappa (q - 1)^m.$$

So, given $\delta > 0$, there exists $\tilde{q}_\delta$ such that, for any $q \geq \tilde{q}_\delta$,

$$\kappa q^m \leq (1 + \delta)\kappa(q - 1)^m,$$

which implies that, for any $q \geq 0$,

$$\kappa q^m \leq (1 + \delta)\kappa((q - 1)^+)^m + \kappa \tilde{q}_\delta^m.$$

Choosing $B_\delta = \max\{(1 - \delta)\kappa(q_\delta + 1)^m, \kappa \tilde{q}_\delta^m\}$, we conclude that κq^m is a valid weight function for $m > 0$ and $\kappa > 0$. $\quad\square$

Example 5.5.2

Weight function $w(q) = \kappa \log(1 + q)$ for some $\kappa > 0$ is a valid weight function. We note that

$$\kappa \log(1 + q + 1) = \kappa \log(1 + q)\left(1 + \frac{1}{1 + q}\right)$$
$$= \kappa \log(1 + q) + \kappa \log\left(1 + \frac{1}{1 + q}\right)$$
$$\leq \kappa \log(1 + q) + \kappa \log 2.$$

If $q \geq 1$, we have

$$\kappa \log(1 + q - 1) = \kappa \log(1 + q) + \kappa \log\left(1 - \frac{1}{1 + q}\right)$$
$$\geq \kappa \log(1 + q) - \kappa \log 2.$$

If $0 \leq q < 1$, we have

$$\kappa \log \left(1 + (q-1)^+\right) = \kappa \log 1 = 0$$

$$\geq \kappa \log(1+q) - \kappa \log 2.$$

We therefore conclude that $\kappa \log(1 + q)$ is a valid weight function by choosing $B_\delta = \kappa \log 2$. □

5.6   Q-CSMA: a distributed algorithm for ad hoc P2P networks

The MaxWeight algorithm operates by finding a maximum weight schedule of the conflict graph. In wireless access networks, the base station (or access point) can collect channel- and queue-state information and solve the maximum weight independent set problem easily, as, in this case, it simply reduces to finding a single link with the largest weight. In ad hoc P2P wireless networks, due to the lack of a centralized infrastructure, nodes have to make their own decisions on turning ON or OFF based on local information, which requires us to seek distributed scheduling algorithms. In this section, we present the Queue-length-based CSMA/CA (Q-CSMA) algorithm for ad hoc P2P networks. Q-CSMA approximates MaxWeight scheduling in a distributed fashion, and can be shown to be throughput optimal.

5.6.1   The idea behind Q-CSMA

Assume that the conflict graph of an ad hoc network has H possible schedules. The key idea behind Q-CSMA is to select independent sets according to the following distribution:

$$\pi_h = \frac{1}{Z} \exp\left(\sum_l w_l M_l^{(h)}\right), \tag{5.15}$$

where w_l is the weight associated with link l and

$$Z = \sum_{h=1}^{H} \exp\left(\sum_l w_l M_l^{(h)}\right)$$

is the normalization factor. The reason we choose such a distribution is that the expected weight $E[\sum_l w_l M_l]$ under this distribution satisfies

$$E\left[\sum_l w_l M_l\right] \geq \max_h \sum_l w_l M_l^{(h)} - H, \tag{5.16}$$

which is close to the maximum weight when $\{w_l\}$ are sufficiently large. So, when link weights are fixed and sufficiently large, and if we generate a sufficient number of independent sets according to distribution (5.15), the algorithm performs like MaxWeight scheduling.

We next prove result (5.16). Define

$$\gamma_h = \sum_l w_l M_l^{(h)},$$

i.e., the sum weight associated with independent set $M^{(h)}$. Let $\mathcal{I}^*$ denote the set of maximum weight independent sets, i.e.,

$$
\mathcal{I}^* = \left\{ h : \sum_l w_l M_l^{(h)} = \max_k \sum_l w_l M_l^{(k)} \right\},
$$

and let h^* denote an element of $\mathcal{I}^*$. The expected weight by selecting independent sets according to distribution (5.15) can be written as

$$
E\left[\sum_l w_l M_l \right] = \sum_h \pi_h \gamma_h
$$

$$
= \gamma_{h^*} - \sum_{h \notin \mathcal{I}^*} \frac{e^{\gamma_h}}{Z}(\gamma_{h^*} - \gamma_h)
$$

$$
= \gamma_{h^*} - \frac{e^{\gamma_{h^*}}}{Z} \sum_{h \notin \mathcal{I}^*} \left(e^{\gamma_h - \gamma_{h^*}} \right)(\gamma_{h^*} - \gamma_h)
$$

$$
\geq_{(a)} \gamma_{h^*} - \sum_{h \notin \mathcal{I}^*} 1
$$

$$
= \gamma_{h^*} - (H - |\mathcal{I}^*|)
$$

$$
\geq \gamma_{h^*} - H,
$$

where inequality (a) holds because $xe^{-x} \leq 1$ and $e^{\gamma_{h^*}}/Z \leq 1$.

5.6.2   Q-CSMA

We now present a distributed algorithm that generates independent sets according to distribution (5.15). We assume that the w_l's are fixed and do not change with time. In reality, $w_l(q_l)$ will change, but if it changes very slowly, for example if $w_l(q_l)$ is chosen to be slightly smaller than $\log(1 + q_l)$, one can show that the stability results will not be affected, but we will not do this here. Here we will only consider the case of fixed weights and will describe a DTMC whose states are the independent sets, and show that the steady-state distribution of this DTMC has the desired form. We will then describe a distributed algorithm under which the MAC layer behaves like the DTMC.

Recall that a schedule is a set of links that are allowed to transmit at the same time. Further, define the transmission state of a link to be either 1 or 0, indicating whether the link is transmitting or not. We consider an algorithm that behaves as follows.

(i) The network first picks a set of links, say $\mathcal{D}$, that are allowed to change their transmission states. All other links are not allowed to change their transmission states. We assume that $\mathcal{D}$ itself is a valid schedule, although we do not use $\mathcal{D}$ for transmission. We call set $\mathcal{D}$ a *decision schedule*.

(ii) Links in $\mathcal{D}$ that have a neighbor in the conflict graph that was transmitting in the previous slot will not be allowed to change their transmission states.

(iii) Among the remaining links, each of them decides to turn ON with probability α_l, for link l, and turn OFF with probability $(1 - \alpha_l)$.

Note that, since $\mathcal{D}$ is a valid schedule and links that can interfere with links that were part of the schedule at the previous time instant were removed in step (ii), after step (iii) the set of ON links forms a valid schedule.

Let $\mathcal{S}(t)$ denote the schedule used in time slot t. Suppose $\mathcal{D}$ is chosen with some probability $p(\mathcal{D})$, independently at the beginning of each time slot. Then, $\mathcal{S}(t)$ is determined by $\mathcal{D}(t)$ and $\mathcal{S}(t-1)$ under the algorithm above, and $\mathcal{S}(t)$ is a DTMC. If we are lucky, we can verify that the desired distribution over schedules (see equation (5.15)) satisfies local balance equations, thus establishing that (5.15) is the steady-state distribution of the DTMC $\mathcal{S}(t)$.

We now derive conditions on $p(\mathcal{D})$ and α_l to ensure that the DTMC satisfies local balance with the desired steady-state distribution. For this purpose, let x be a schedule (independent set) and let $x + m_1 - m_2$ be another schedule, where m_1 are links that are not in x, and m_2 are links that are in x. Let us consider the transition from x to $x + m_1 - m_2$ and vice versa, as shown in Figure 5.9.

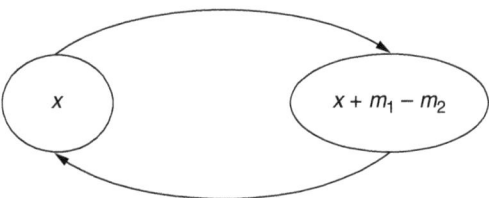

Figure 5.9 The transition from x to $x + m_1 - m_2$ and vice versa.

A transition from x to $x + m_1 - m_2$ means the network will turn ON those links in m_1 and turn OFF those links in m_2. Similarly, the transition from $x + m_1 - m_2$ to x means the network will turn OFF those links in m_1 and turn ON links in m_2. For the transition from x to $x+m_1-m_2$ to occur, the decision schedule $\mathcal{D}$ must contain both m_1 and m_2, and similarly when in state $x + m_1 - m_2$. Pictorially, $\mathcal{D}$ will look like something like Figure 5.10.

Let γ_x be the sum weight associated with schedule x, i.e., $\gamma_x = \sum_{l \in x} w_l$. According to Theorem 3.3.6, the DTMC has (5.15) as the stationary distribution if $\pi_x P_{x,y} = \pi_y P_{y,x}$,

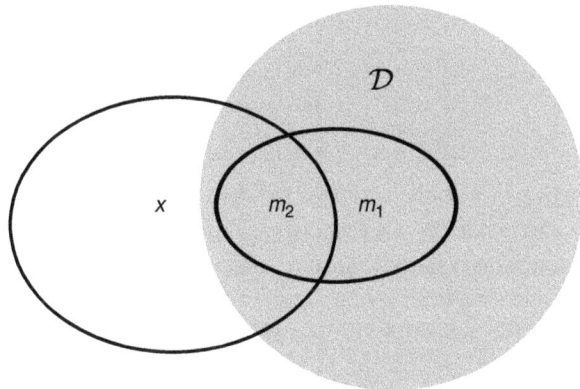

Figure 5.10 The gray area is $\mathcal{D}$ and $m_1 \cup m_2 \subseteq \mathcal{D}$. Other links can also belong to $\mathcal{D}$.

where x and y are two schedules. In other words, the DTMC has a stationary distribution (5.15) when the following equation holds:

$$\frac{e^{\gamma_x}}{Z} P_{x,x+m_1-m_2} = \frac{e^{\gamma_x+\gamma_{m_1}-\gamma_{m_2}}}{Z} P_{x+m_1-m_2,x},$$

which is equivalent to

$$\frac{e^{\gamma_x}}{Z} \sum_{\mathcal{D}:m_1 \cup m_2 \subseteq \mathcal{D}} \Pr(\mathcal{D}\setminus(m_1 \cup m_2) \text{ do not change state})$$

$$\times \Pr(m_1 \text{ turns ON}) \times \Pr(m_2 \text{ turns OFF}) \times p(\mathcal{D})$$

$$= \frac{e^{\gamma_x+\gamma_{m_1}-\gamma_{m_2}}}{Z} \sum_{\mathcal{D}:m_1 \cup m_2 \subseteq \mathcal{D}} \Pr(\mathcal{D}\setminus(m_1 \cup m_2) \text{ do not change state})$$

$$\times \Pr(m_1 \text{ turns OFF}) \times \Pr(m_2 \text{ turns ON}) \times p(\mathcal{D}).$$

A sufficient condition for the above equality to hold is

$$e^{\gamma_{m_2}} \Pr(m_1 \text{ turns ON}) \times \Pr(m_2 \text{ turns OFF})$$

$$= e^{\gamma_{m_1}} \Pr(m_1 \text{ turns OFF}) \times \Pr(m_2 \text{ turns ON}). \qquad (5.17)$$

Note that $\Pr(\mathcal{D}\setminus(m_1 \cup m_2) \text{ do not change state})$ is a complicated expression, but it is the same for each $\mathcal{D}$ on the right-hand side and the left-hand side. It even depends on x, but this fact is irrelevant to us.

A sufficient condition for (5.17) to hold is, for any subset of links m,

$$e^{\gamma_m} \Pr(m \text{ turns OFF}) = \Pr(m \text{ turns ON}),$$

which is equivalent to

$$\prod_{l \in m} e^{w_l}(1 - \alpha_l) = \prod_{l \in m} \alpha_l, \qquad (5.18)$$

where α_l is the probability that link l turns ON if it is part of the decision schedule and is allowed to change its transmission state. Matching the terms for the same l, we have that a sufficient condition for (5.18) to hold is

$$e^{w_l}(1 - \alpha_l) = \alpha_l,$$

which requires

$$\alpha_l = \frac{e^{w_l}}{1 + e^{w_l}}.$$

We note that α_l is a function of w_l only, and does not depend on the weights of other links, which means that link l can make ON–OFF decisions based on local information.

In summary, if link l knows it is in the decision schedule and none of its neighbors is in the schedule used in the previous time instant, then it can turn ON–OFF probabilistically according to its weight w_l. The schedule $\mathcal{S}(t)$ is a DTMC, and its steady-state distribution is (5.15).

We assume that each link can sense whether other links in its neighborhood were transmitting in the previous time slot using carrier sensing, so both steps (ii) and (iii) can be carried out locally. So, to make the algorithm fully distributed, we need a distributed mechanism to pick the decision set $\mathcal{D}$. Further, we need to ensure that the DTMC is aperiodic and irreducible. The following protocol is one way to generate $\mathcal{D}$ in a distributed fashion.

(a) At the beginning of each time slot, each link transmits a "control" message with probability β, independent of all other links. The parameter β can be any arbitrary fixed value in $(0, 1)$.

(b) If a control message collides with a neighbor's control message, the link does not become a part of $\mathcal{D}$. Otherwise, it becomes a part of decision set $\mathcal{D}$.

Clearly, $\mathcal{D}$ is an independent set because two conflicting links cannot both be present in $\mathcal{D}$. Further, from any state x, the above process guarantees that the transition probability from x to the schedule in which all links are OFF is strictly positive, so the Markov chain is irreducible and aperiodic.

The Q-CSMA algorithm is presented in Algorithm 3.

Algorithm 3 Q-CSMA

1: At time slot t, set $w_l = \log(1 + q_l(t))$.
2: At the beginning of each time slot, each link transmits a "control" message with probability β, independent of all other links. The parameter β can be any arbitrary fixed value in $(0, 1)$.
3: **if** a control message collides with a neighbor's control message, **then**
4:   the link does not become a part of $\mathcal{D}$.
5: **else**
6:   the link becomes a part of decision set $\mathcal{D}$.
7: **end if**
8: **for** $l \in \mathcal{D}$ **do**
9:   **if** link l has a neighbor in the conflict graph which was transmitting in the previous time slot, **then**
10:     link l does not change its transmission state.
11:   **else**
12:     link l turns on with probability $e^{w_l}/(1 + e^{w_l})$, and turns off with probability $1/(1 + e^{w_l})$.
13:   **end if**
14: **end for**

Example 5.6.1

Consider a three-link ad hoc network and its conflict graph, as shown in Figure 5.11. Assume $x(t - 1) = \{1, 0, 0\}$, i.e., link 1 is ON and links 2 and 3 are OFF. In the following, we illustrate the Q-CSMA at time t for two specific decision schedules $\mathcal{D}_1 = \{1, 3\}$ and $\mathcal{D}_2 = \{2\}$.

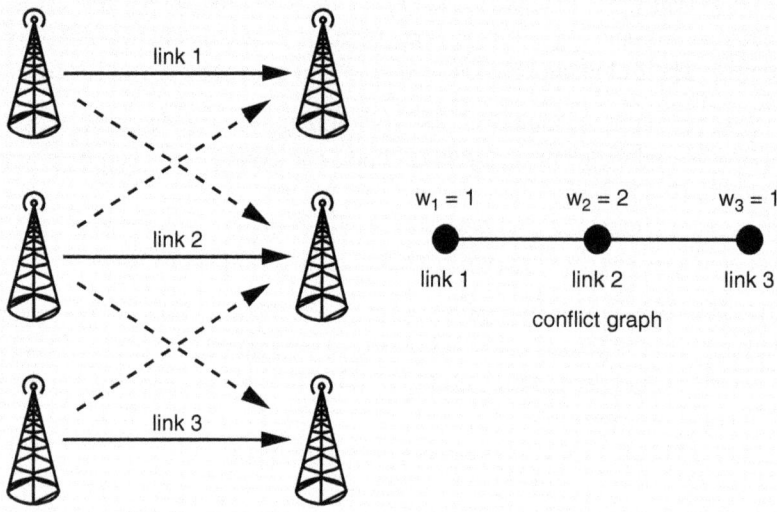

Figure 5.11 Three-link example illustrating the Q-CSMA.

- If the decision schedule is $\{1, 3\}$, link 1 will continue to transmit with probability $e^{w_1}/(1 + e^{w_1}) = e/(1 + e)$, and turns OFF with probability $1/(1 + e)$, and link 3 turns ON with probability $e^{w_3}/(1 + e^{w_3}) = e/(1 + e)$ and turns OFF with probability $1/(1 + e)$.
- If the decision schedule is $\{2\}$, since link 1 was transmitting in the previous time slot, link 2 does not change its transmission state and will remain OFF.   □

We finally comment that we have made a number of assumptions in describing Algorithm 3.

- Each link can sense whether other links in its neighborhood were transmitting in the previous time slot. This is accomplished by a protocol called "carrier sensing." Each link in our model senses the presence or absence of a "carrier" by detecting the transmission energy in its neighborhood. Hence, the algorithm is Carrier Sensing Multiple Access.
- We also assume that when two control messages collide, the collision can be detected.
- We assume that the process of selecting $\mathcal{D}$ takes negligible time. This is a reasonable assumption if packet sizes are much longer than the time to transmit an intent message and detect collisions.
- We assume time is slotted.

Some of these assumptions will not hold in practice. At the end of the chapter, we provide references which address these issues.

5.7   Summary

- **The MaxWeight algorithm for the cellular downlink** Recall that $c_{m,i}$ is the maximum number of packets that can be transmitted to mobile i when the channel is in state m and mobile i is scheduled. At each time t, the MaxWeight algorithm chooses to transmit to a mobile i which solves

$$\max_{j} c_{m(t),j} q_j(t).$$

The MaxWeight algorithm is throughput optimal.

- **The MaxWeight algorithm for ad hoc P2P wireless networks** Recall that a schedule is a set of links that can be active simultaneously without interfering with each other. The MaxWeight algorithm selects a schedule $M(t)$ such that

$$M(t) \in \arg\max_{M^{(h)}} \sum_{l} M_l^{(h)} q_l(t).$$

The algorithm is throughput optimal.

- **General MaxWeight scheduling** Recall that $w_l(q_l)$ is the weight associated with link l. General MaxWeight scheduling schedules the links in schedule $M(t)$ such that

$$M(t) \in \arg\max_{M^{(h)}} \sum_{l} w_l(q_l(t)) M_l^{(h)}.$$

The algorithm is throughput optimal if the weight functions are valid weight functions (see Definition 5.5.1).

- **Q-CSMA** Q-CSMA is a distributed approximation of the MaxWeight algorithm for ad hoc P2P wireless networks. In Q-CSMA, a link makes transmission decisions based on the states of other links in its neighborhood in the previous time slot. The schedules under Q-CSMA forms a DTMC, and the stationary distribution of the DTMC satisfies

$$\pi_h = \frac{1}{Z} \exp\left(\sum_l w_l M_l^{(h)}\right), \tag{5.19}$$

where w_l is the weight associated with link l and

$$Z = \sum_{h=1}^{H} \exp\left(\sum_l w_l M_l^{(h)}\right)$$

is the normalization factor.

5.8   Exercises

Exercise 5.1 (The capacity region of a cellular network) Consider a cellular wireless network consisting of a base station and two receivers, mobile 1 and mobile 2. The network can be in one of two equally likely channel states: $c_1 = (1, 2)$ and $c_2 = (3, 1)$. Assume the channel state process is i.i.d. across time and that the base station can transmit to only one mobile in each time slot. Draw the capacity region of this wireless network.

Exercise 5.2 (The throughput under a power constraint) Consider a single wireless link, which can be in one of two states indexed by $m = 0, 1$. The transmitter has two power levels, indexed by $j = 0, 1$, such that $p_0 = 1$ and $p_1 = 5$. The maximum numbers of packets that can be transmitted over the link under different channel states and transmit powers are summarized in Table 5.1. Assume that two channel states occur equally likely, i.e., $\pi_0 = \pi_1 = 0.5$.

Table 5.1 The maximum numbers of packets that can be transmitted over the link under different channel states and transmit powers

	$m = 0$ (bad)	$m = 1$ (good)
$p_0 = 1$	1	2
$p_1 = 5$	3	5

(1) Let $\alpha_{m,j}$ denote the fraction of time that the transmitter transmits at power level j when the channel is in state m. Given $\alpha_{00} + 5\alpha_{01} = x$ for $0 \le x \le 5$, i.e., the average transmit power when the channel is in state 0 is x, calculate the maximum achievable throughput when the channel is in state 0, i.e.,

$$\max \alpha_{00} + 3\alpha_{01} \quad \text{subject to} \quad \alpha_{00} + 5\alpha_{01} = x.$$

(2) Given $\alpha_{10} + 5\alpha_{11} = y$ for $0 \le y \le 5$, i.e., the average transmit power when the channel is in state 1 is y, calculate the maximum achievable throughput when the channel is in state 1, i.e.,

$$\max 2\alpha_{10} + 5\alpha_{11} \quad \text{subject to} \quad \alpha_{10} + 5\alpha_{11} = y.$$

(3) Suppose that we impose a constraint that the average power used by the transmitter must be no more than 3, where the average is taken over the channel-state distribution and the probability with which the transmitter chooses a certain power level in each state. What is the maximum throughput that the link can achieve?

Exercise 5.3 (Joint scheduling and power control) In Exercise 5.2, we saw that, when there is an average power constraint, the transmission power level in each channel state has to be carefully chosen to maximize throughput. In this exercise, we extend the model to an uplink network, i.e., one in which multiple mobiles transmit to a single base station receiver. We will show that the MaxWeight scheduling algorithm can be used to perform throughput-optimal scheduling and power control without knowing the channel or packet arrival statistics.

Consider an uplink network with N mobiles. The packet arrival process to mobile i is an ON–OFF process, i.e., the source is in one of two states, ON or OFF, with a probability of being in the ON state of β_i. Thus, the probability of being in the OFF state is $1 - \beta_i$. The source generates $A_{\max}$ packets when ON and zero packets when OFF. So the average packet arrival rate at mobile i is $\lambda_i = A_{\max}\beta_i$ packets per time slot. Assume that each mobile can transmit at one of J power levels, p_1, p_2 through p_J, and that the channel state of the network (i.e, the joint channel states of all the mobiles) can be in one of M states. Assume that, for the transmission of a mobile to be successful in a time slot, none of the other mobiles must be transmitting in the time slot. Let $c_{m,i,j}$ denote the maximum number of packets that can be transmitted from mobile i at power level p_j when the network is in state m. Assume that the average power (averaged over time) used by mobile i is constrained to be less than or equal to $P_{\max,i}$.

(1) Let $\alpha_{m,i,j}$ denote the fraction of time mobile i transmits at power level j when the network is in state m. List the necessary conditions for λ to be supportable by the network.

(2) To guarantee that the average power constraints are not violated, each mobile maintains a virtual queue, as shown in Figure 5.12. The arrival to virtual queue i in each time slot is the power level used by mobile i, while the virtual queue is served at rate $P_{max,i}$ in each time slot. Thus, in Figure 5.12, $b_i(t)$ is equal to the transmit power used by mobile i at time t, so $b_i(t) = p_j$ if mobile i transmits at power level j and $b_i(t) = 0$ otherwise. So the virtual queue will be stable if and only if the average used power is less than the maximum allowable average power. Let $d_i(t)$ denote the length of this virtual queue at time slot t, so

$$d_i(t + 1) = \left(d_i(t) + b_i(t) - P_{\max,i}\right)^+.$$

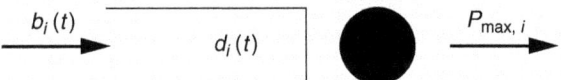

$b_i(t)$ $d_i(t)$ $P_{\max, i}$

Figure 5.12 The virtual queue for power consumption.

Let $q_i(t)$ denote the queue length at mobile i and $m(t)$ denote the state of the network at time t. Consider a joint scheduling and power control algorithm that schedules mobile $i^*(t)$ with power level $j^*(t)$ at time t such that

$$(i^*(t), j^*(t)) \in \arg \max_{(i,j)} q_i(t)c_{m(t),i,j} - d_i(t)p_j.$$

Show that the algorithm is throughput optimal in the following sense: the packet queues and the virtual queues are all stable when the packet arrival rates to the mobile lie within the capacity region (the set of supportable arrival rates) of the network. Assume that the available power levels and the power constraints $P_{max,i}$ at each link are all positive integers, so that $(q(t), d(t))$ is a Markov chain.

Exercise 5.4 (Scheduling real-time traffic) Consider a downlink network, with one base station transmitting to two mobiles. Assume that the channel states from the base station to the two mobiles are constant, and assume that each link (from the base station to each mobile) can transmit one packet per time slot when scheduled. Arrivals to the base station destined for mobile i arrive according to a Bernoulli process with parameter λ_i, and the arrival processes destined for the two mobiles are independent of each other. Suppose that the packets belong to real-time flows, and that each packet has to delivered to its destination in the same time slot that it arrived; otherwise, the packet is dropped.

(1) Let s_i denote the average throughput (average number of packets successfully delivered to the destination per time slot) of mobile i. Characterize the set of $s := (s_1, s_2)$ that can be supported by the network for a given $\lambda := (\lambda_1, \lambda_2)$.

(2) Now assume that mobile i has a minimum requirement on the average throughput, denoted by $s_{\min,i}$. To guarantee this minimum average throughput, each mobile maintains a virtual queue, as in Figure 5.13, where $\mu_i(t) = 1$ if a packet is delivered to mobile i during time slot t and $\mu_i(t) = 0$ otherwise, and

$$d_i(t + 1) = \left(d_i(t) + s_{\min,i} - \mu_i(t)\right)^+.$$

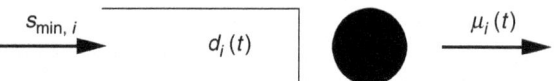

Figure 5.13 The virtual queue for real-time communication.

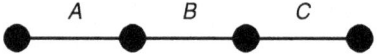

Figure 5.14 Wireless network for Exercise 5.7.

Note that the virtual queue is stable if and only if the average rate at which packets are delivered to mobile i is greater than the minimum requirement $s_{min,i}$.

Now consider the following algorithm that schedules mobile $i^*(t)$ at time t such that

$$i^*(t) \in \arg\max_i a_i(t)q_i(t),$$

where $a_i(t)$ is the number of packets arriving at the base station destined for mobile i in time slot t. The scheduling algorithm simply states that if there is one mobile with a backlogged packet, that mobile is scheduled; if both mobiles have packets destined for them, schedule the mobile which has a larger virtual queue length. Thus, $\mu_{i^*}(t) = a_{i^*}(t)$ and $\mu_i(t) = 0$ if $i \neq i^*(t)$. Prove that the algorithm is throughput optimal in the sense that it will achieve minimum rates $s_{min,i}$ if they are feasible.

Exercise 5.5 (Scheduling real-time traffic over ON–OFF channels) Consider the same network as in Exercise 5.4, but assume ON–OFF channels such that the link between the base station and mobile i is ON with probability γ_i. If a channel is ON, one packet may be transmitted over the channel; otherwise, no packet can be transmitted over the channel. Answer questions (1) and (2) in Exercise 5.4.

Exercise 5.6 (The capacity region of ad hoc P2P wireless networks) Consider the model of an ad hoc P2P wireless network studied in Section 5.4. Use the strong law of large numbers (SLLN) to prove that, if $\lambda \notin C$, no scheduling algorithm can support arrival rate vector λ. Hint: The following fact will be helpful: based on the strict separation theorem, if $\lambda \notin C$, there exist $\beta \geq 0$, $\beta \neq 0$, and $\delta > 0$ such that

$$\sum_l \beta_l \lambda_l \geq \sum_l \beta_l M_l^{(h)} + \delta$$

for any schedule $M^{(h)}$.

Exercise 5.7 (Scheduling in an ad hoc P2P wireless network) Consider the three-link wireless network in Figure 5.14, with links A, B, and C. The interference constraints are such that links A and C can be active simultaneously or link B can be active. In other words, the possible schedules are $\{A, C\}$ and $\{B\}$. When a link is scheduled, it can serve one packet. Let the arrivals to the links be independent Bernoulli processes with means λ_A, λ_B, λ_C, respectively.

(1) What is the capacity region of the network?

(2) Assuming arrivals occur first, then departures, recall that

$$q_l(k+1) = [q_l(k) + a_l(k) - I_l(k)]^+,$$

where $q_l(k)$ is the queue length of link l at the beginning of time slot k, $a_l(k)$ is the number of arrivals to link l in time slot k, and $I_l(k)$ is an indicator function indicating whether link l is scheduled in time slot k. Consider a scheduling algorithm in which scheduling decisions are based on observing $\tilde{q}_l(k) = q_l(k) + a_l(k)$. Further, suppose that the scheduling decisions are made according to the following rule: schedule link B if $\tilde{q}_B(k) > 0$; otherwise, schedule links A and C. Is this scheduling rule throughput optimal?

Exercise 5.8 (Randomized scheduling algorithm for ad hoc P2P wireless networks) In ad hoc P2P wireless networks, recall that the throughput-optimal scheduling algorithm requires the network to solve a MaxWeight independent set problem with queue lengths as weights. In this exercise, we show that an approximation to the MaxWeight independent set scheduling algorithm is also throughput optimal. Towards this end, suppose that the scheduling algorithm is a randomized algorithm which picks a schedule in each time slot according to some probabilistic rule. Further, suppose that the probabilistic rule satisfies the following property: given any $\delta > 0$ and $\epsilon > 0$, there exists a finite set $\mathcal{B}_{\delta,\epsilon}$ such that, with probability greater than or equal to $(1 - \delta)$, the randomized algorithm produces a schedule whose weight is greater than or equal to $(1-\epsilon)$ times the weight of the MaxWeight schedule whenever the vector of queue lengths lie in $\mathcal{B}_{\delta,\epsilon}^c$. Show that such a randomized algorithm is also throughput optimal.

Exercise 5.9 (MaxWeight with delayed queue length information) In this exercise, we consider the case where instantaneous queue lengths of each link are not available. Consider an ad hoc P2P wireless network, and assume that a central scheduler is available and that it only has access only to the delayed queue length information. In particular, at time t, the scheduler has access only to $q_l(t - D_l)$ for link l, where D_l is a delay in propagating the queue length from link l to the scheduler. Assume that $D_l \le D_{\max}$ for all l, and that packet arrivals to each link at each time slot are independent Bernoulli random variables with mean λ_l at link l.

(1) Prove that $y(t) = (q(t - D_{\max}), \ldots, q(t))$ is a Markov chain. Note that, in this case, $q(t)$ is *not* a Markov chain.

(2) Define the state space of $y(t)$ to

$$\mathcal{S} = \{y : y \text{ is reachable from } 0 \text{ under the MaxWeight scheduling algorithm}\}.$$

Show that $y(t)$ with state space $\mathcal{S}$ is an irreducible Markov chain.

(3) Prove that the MaxWeight scheduling algorithm using delayed queue lengths is throughput optimal. Hint: Consider the Lyapunov function $V(t) = \sum_l q_l^2(t)$. First show that there exists a constant A such that

$$E[V(t+1) - V(t)|y(t)] \le A + 2\sum_l \lambda_l q_l(t - D_l) - \sum_l E[q_l(t - D_l)M_l(t)|y(t)].$$

Exercise 5.10 (Maximal scheduling algorithm) Consider a simple interference model for ad hoc P2P wireless networks, where nodes cannot simultaneously transmit and receive. Assume that there are no other interference constraints. This would be the case in a network where each link transmits over a different frequency, for example. A schedule is called a *maximal schedule* if no link can be added into the schedule without violating the interference constraint. We then consider the following maximal scheduling algorithm for ad hoc P2P wireless networks.

Maximal scheduling for ad hoc P2P wireless networks

At each time instant t, the scheduler first removes all links with zero queue lengths from the conflict graph, and then finds a maximal schedule $M(t)$ in the remaining conflict graph. One packet is transmitted for each link in $M(t)$.

---

(1) Given a link l, let E_l be the set of all links (including l itself) that can interfere with it. Thus, if s and t are the two end points of link l, then E_l is the set of all links in which either s is a transmitter or a receiver, or t is a transmitter or a receiver. Here, by a link, we mean a transmit–receive pair of nodes. Show that, if $q_l(t) > 0$, the following inequality must hold under maximal scheduling:

$$1 \leq \sum_{k \in E_l} M_k(t) \leq 2. \tag{5.20}$$

(2) Prove that the maximal scheduling algorithm can support any λ such that $2(1+\epsilon)\lambda \in \mathcal{C}$ for some $\epsilon > 0$. Hint: See the proof of Theorem 4.3.1 for switch scheduling.

Exercise 5.11 (Q-CSMA example) Consider Example 5.6.1, and assume $x(t-1) = (0,0,1)$. Describe the actions of Q-CSMA at time t for the following two cases: (*i*) the decision schedule is $\{1,3\}$, and (*ii*) the decision schedule is $\{2\}$.

Exercise 5.12 (Stationary distribution over schedules under Q-CSMA) Consider the conflict graph shown in Figure 5.11. Assume that, at each time slot, the decision set is selected from $\{1,3\}$ or $\{2\}$ with equal probability. Write down the transition probability matrix of the Markov chain (x_1, x_2, x_3), where x_l is the transmission state of link l with $x_l = 0$ if the link is OFF and $x_l = 1$ if the link is ON. Compute the stationary distribution of x and $E[\sum_l w_l x_l]$.

5.9   Notes

Packet scheduling using the MaxWeight algorithm was introduced in [163] and generalized in [40, 98, 121, 130]. A continuous-time Markov chain model of the CSMA algorithm was introduced and studied in [14], and further analyzed in [100, 169]. In [64, 105], it was suggested that, by using weights as functions of queue lengths, one can achieve throughput optimality. Further, a network utility maximization interpretation was provided for the CSMA algorithm. The model presented in this chapter is a discrete-time version of these continuous-time models

[133, 134]. When the weights are chosen to be equal to some function of queue length, the analysis presented in this chapter is valid only if it is assumed that the weights are assumed to change very slowly compared to the time required for the CSMA Markov chain to reach steady state. This assumption has been justified under certain mild global knowledge assumptions in [142, 149] for the continuous-time CSMA Markov chain, and in [45] for the discrete-time CSMA Markov chain presented in this chapter. Q-CSMA models with collisions have been analyzed in [65, 84, 85]. An experimental study of Q-CSMA is presented in [128].

Virtual queues to enforce average power constraints were presented in [131, 161]. Virtual queues for enforcing minimum guaranteed throughput for delay-constrained traffic were presented in [56, 57]. An optimization-based derivation of these virtual queues and a Lyapunov function-based stability analysis were provided in [62, 63].

6 Back to network utility maximization

In Chapter 5, we looked at scheduling in wireless networks, where the focus was to design a scheduling algorithm that can support any traffic strictly within the capacity region. Implicitly, we assumed that data flows were regulated by congestion control so the incoming traffic was always within the capacity region. However, in practice, the behavior of congestion control algorithms themselves may be affected by scheduling algorithms. For example, a dual congestion control algorithm which reacts to delays, so the source rates would be regulated based on queueing delays, which could be quite different under different scheduling algorithms. Therefore, to achieve fair resource allocation in wireless networks, we need to revisit the network utility maximization formulation introduced in Chapter 2. In Chapter 2, we assumed that routes from sources to their destinations are given. In this chapter, we will study a more general model in which routes are not given and are part of the resource allocation decision process. We will consider the network utility maximization problem for wireless networks, and answer the following questions.

- *How should the Lagrangian duality to derive joint congestion control, routing, and scheduling algorithms that maximize the sum network utility be used?*
- *How should the performance of the joint algorithms using discrete-time Markov chains be analyzed?*
- *What is the difference between the network utility maximization formulation for the Internet and that for wireless networks, and what is the rationale behind the difference?*
- *What is the relationship between proposed models/algorithms and practical MAC/physical layer protocols?*

6.1 Joint formulation of the transport, network, and MAC problems

Consider a wireless network with N nodes. We are going to take into account both interference and channel fading in the same model. The transmission rates between node pairs are determined by the channel state in the network, denoted by m. Recall that the channel state is a variable which describes the channel conditions in the network. For example, if the channel conditions are good, the realized data rates will be high; otherwise, the rates will be low.

In addition to the channel state, interference determines the data rates at which different node pairs can communicate at the same time. Let r denote the data-rate vector, where r_{ij} is the data rate from node i to node j. The vector r is assumed to take on values in a discrete set $\mathcal{R}_m$ when the channel is in state m. The set $\mathcal{R}_m$ captures interference effects and power constraints. We further assume that, given any $r \in \mathcal{R}_m$, $r^{(i)}$, such that $r_i^{(i)} = 0$ and $r_j^{(i)} = r_j$ for $j \neq i$, also belongs to $\mathcal{R}_m$. Under this reasonable assumption, $Co(\mathcal{R}_m)$ is coordinate convex. Example 6.1.1 illustrates the idea.

node 1   node 2

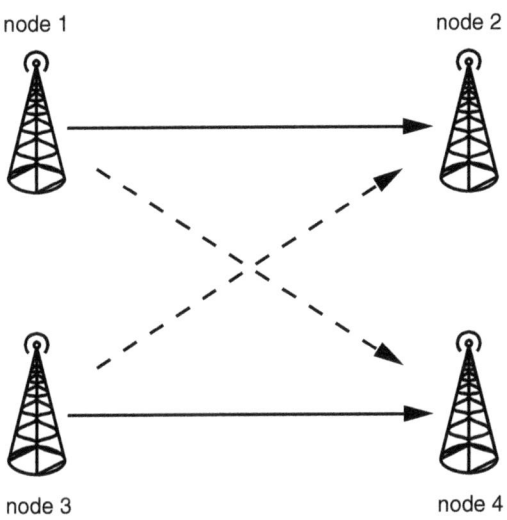

node 3   node 4

Figure 6.1 Four-node wireless network. Solid lines represent intended transmissions; dashed lines represent interference.

Example 6.1.1

Consider a four-node network, as shown in Figure 6.1. Node 1 wants to transmit to node 2, and node 3 wants to transmit to node 4. For simplicity, assume each node has three different power levels: $\{0, p_{1a}, p_{1b}\}$ are the power levels available at node 1 and $\{0, p_{3a}, p_{3b}\}$ are the power levels available at node 3. Further suppose that h_{ij} is the channel gain from node i to node j, i.e., if node i transmits at power P_i, the received power at node j is $P_i h_{ij}$. The maximum transmission rates are functions of SINRs, i.e.,

$$r_{12} = f(P_1, P_3, h_{12}, h_{32}, n_2),$$
$$r_{34} = f(P_3, P_1, h_{34}, h_{14}, n_4),$$

where n_i is the variance of the Gaussian background noise experienced by node i and f is some function. By varying the power levels P_1 and P_3 over $\{0, p_{1a}, p_{1b}\}$ and $\{0, p_{3a}, p_{3b}\}$, respectively, we obtain different values of (r_{12}, r_{34}), which is what we denote by the set $\mathcal{R}_m$. In this example, the channel state can be thought of as depicting the channel gains $\{h_{ij}\}$. We assume that, if $r \in \mathcal{R}_m$, the same vector with one element replaced by 0 also belongs to $\mathcal{R}_m$. In other words, a user has the choice to transmit at rate 0.

Assuming simple coding and modulation schemes, the transmission rates can be written as

$$r_{12} = \frac{1}{2} \log_2 \left(1 + \frac{P_1 h_{12}}{n_2 + P_3 h_{32}} \right),$$

$$r_{34} = \frac{1}{2} \log_2 \left(1 + \frac{P_3 h_{34}}{n_4 + P_1 h_{14}} \right).$$

Assuming the channel gains, power levels, and variances of the noises are as shown in Table 6.1, the set $\mathcal{R}_m$ is illustrated in Figure 6.2. Note that a vector in $Co(\mathcal{R}_m)$ is a linear combination of vectors in $\mathcal{R}_m$, so any rate vector in $Co(\mathcal{R}_m)$ can be achieved by time sharing. The capacity region given channel state m, therefore, is $Co(\mathcal{R}_m)$. Given the parameters in Table 6.1, the capacity region is shown in Figure 6.3.   □

Table 6.1 Channel gains, power levels, and variances of noises

Channel gain	$h_{12} = h_{34} = 1$ and $h_{14} = h_{32} = 0.1$
Power levels	$P_1 \in \{0, 10, 15\}$ and $P_3 \in \{0, 5, 10\}$
Variances of noises	$n_2 = n_4 = 1$

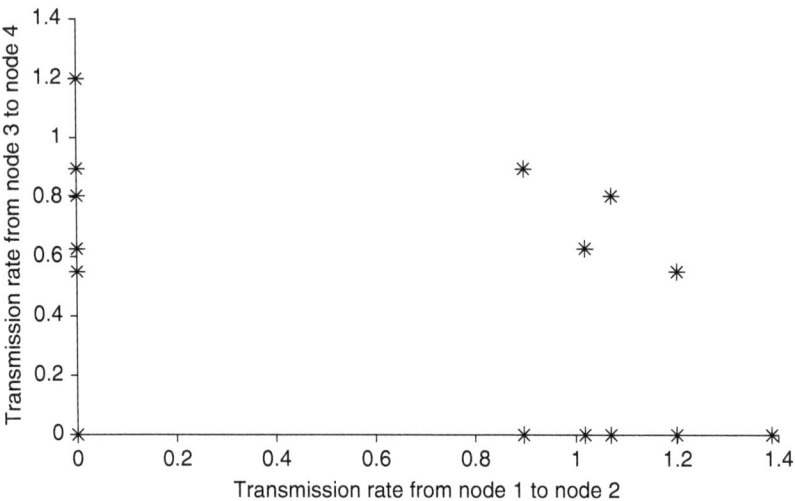

Figure 6.2 The set $\mathcal{R}_m$.

In general, the channel state can vary over time. In our discrete-time model, we assume that π_m is the fraction of time that the channel is in state m. Assume there are M possible states, then

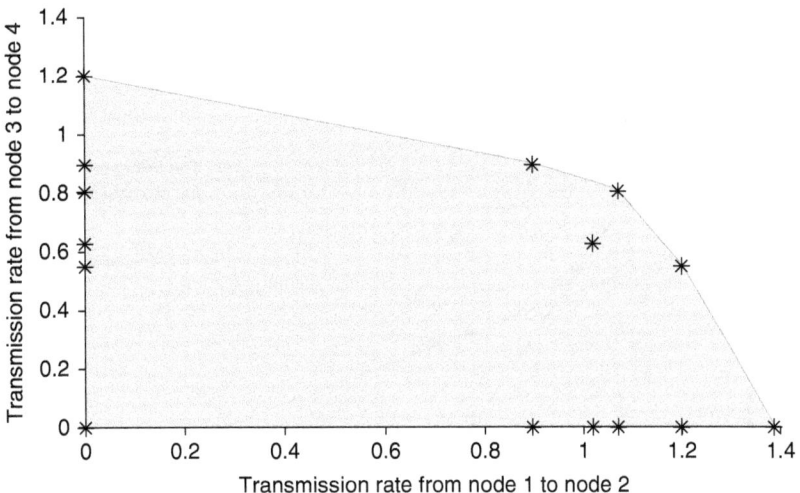

Figure 6.3 The set $Co(\mathcal{R}_m)$.

$$\sum_{1 \leq m \leq M} \pi_m = 1.$$

Now let r be an element of the set $\mathcal{R}_m$, and let $\alpha_{m,r}$ be the fraction of time that the rate vector r is used by the network when the channel state is m. Then the data rate realized from node i to node j is given by

$$\mu_{ij} = \sum_m \pi_m \sum_{r \in \mathcal{R}_m} \alpha_{m,r} r_{ij}.$$

The set of achievable rates (achieved by varying α) is called the link capacity region, and is denoted by $\mathcal{C}$.

We note that, for given m,

$$Co(\mathcal{R}_m) = \left\{ \mu : \mu = \sum_{r \in \mathcal{R}_m} \alpha_{m,r} r \text{ such that } \alpha_m \geq 0 \text{ and } \sum_{r \in \mathcal{R}_m} \alpha_{m,r} = 1 \right\}.$$

Therefore, an equivalent way to express $\mathcal{C}$ is as follows:

$$\mathcal{C} = \left\{ \mu : \mu = \sum_m \pi_m r_m, \qquad r_m \in Co(\mathcal{R}_m) \right\}.$$

Example 6.1.2

Consider a network with two links. Assume the network has two states: state 1 and state 2. Each state occurs with probability 1/2. The link capacity region is plotted in Figure 6.4. □

We now introduce the network utility maximization problem for general wireless networks, where multi-hop flows are allowed. Let $\mathcal{F}$ be the set of flows in the network. A flow

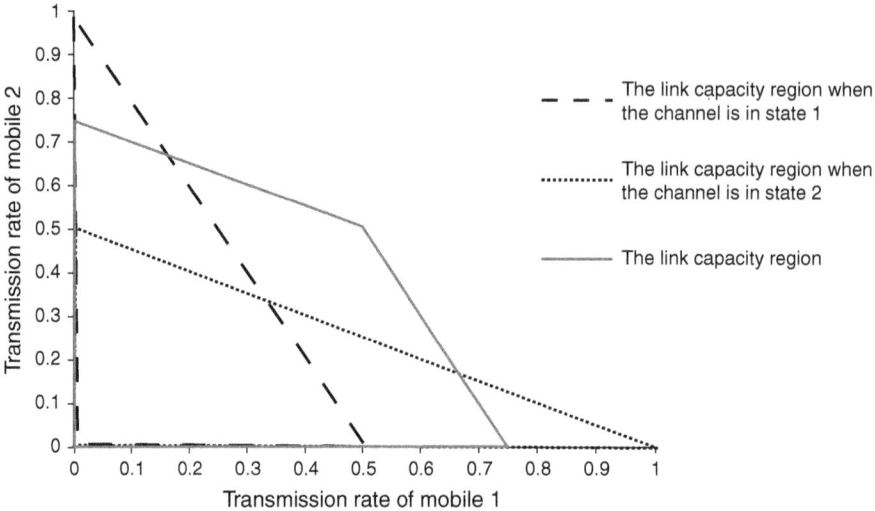

Figure 6.4 Link capacity region with fading channels.

f is associated with a source node $s(f)$ and a destination node $d(f)$. We assume that a flow can split its packets along multiple routes. In our model, the routes are not pre-specified, in general. The algorithm that we design will automatically find appropriate routes through the network.

Let x_f be the rate at which source $s(f)$ generates data. Let $\mu_{ij}^{(d)}$ be the rate at which the data to destination d are transmitted over link (i, j). For $\{x_f\}$ to be accommodated in the network, $\{\mu_{ij}^{(d)}\}$ must satisfy the following equation for each node i and each destination d:

$$\sum_f x_f \mathbb{I}_{s(f)=i,d(f)=d} + \sum_{j:(j,i)\in\mathcal{L}} \mu_{ji}^{(d)} \leq \sum_{k:(i,k)\in\mathcal{L}} \mu_{ik}^{(d)}, \tag{6.1}$$

where $\mathcal{L}$ is the set of links. This constraint says that, at each node i, the total incoming data rate for destination d (the left-hand side of the constraint) must be less than or equal to the total outgoing rate allocated to destination d (the right-hand side of the constraint).

Associate a concave utility function $U_f(x_f)$ with flow f. Our resource allocation problem is to find the appropriate α to solve the following network utility maximization problem:

$$\max_{x,\alpha\geq 0} \sum_f U_f(x_f), \tag{6.2}$$

subject to:

$$\sum_f x_f \mathbb{I}_{s(f)=i,d(f)=d} + \sum_j \mu_{ji}^{(d)} \leq \sum_k \mu_{ik}^{(d)} \qquad \forall i \quad \forall d, \tag{6.3}$$

$$\sum_d \mu_{ij}^{(d)} = \sum_m \pi_m \sum_{r\in\mathcal{R}_m} \alpha_{m,r} r_{ij} \qquad \forall (i,j) \in \mathcal{L}, \tag{6.4}$$

$$\sum_{r\in\mathcal{R}_m} \alpha_{m,r} = 1 \qquad \forall m. \tag{6.5}$$

By abusing notation for convenience, we will denote constraints (6.4) and (6.5) by

$$\mu \in \mathcal{C}.$$

To solve the problem, append (6.3) to the objective using Lagrange multipliers to obtain

$$\max_{x,\mu \geq 0} \sum_f U_f(x_f) - \sum_{i,d} \phi_{id} \left(\sum_f x_f \mathbb{I}_{s(f)=i,d(f)=d} + \sum_j \mu_{ji}^{(d)} - \sum_k \mu_{ik}^{(d)} \right),$$

subject to

$$\mu \in \mathcal{C}.$$

Note that, for fixed ϕ_{id}, the maximization over x and μ can be separated. The optimization problem can be decomposed into the following two optimization problems:

$$\textbf{sub-problem 1:} \quad \sum_f \max_{x_f \geq 0} \left(U_f(x_f) - \phi_{s(f)d(f)} x_f \right) \tag{6.6}$$

and

$$\textbf{sub-problem 2:} \quad \max_{\mu \in \mathcal{C}} \sum_{i,d} \phi_{id} \left(\sum_k \mu_{ik}^{(d)} - \sum_j \mu_{ji}^{(d)} \right). \tag{6.7}$$

If the Lagrange multipliers ϕ are known, sub-problem 1 (6.6) is equivalent to

$$\max_{x_f \geq 0} U_f(x_f) - \phi_{s(f)d(f)} x_f \qquad \forall f. \tag{6.8}$$

Thus, each user can determine its transmission rate based on the Lagrange multiplier associated with its ingress node and its destination.

To understand (6.7), note that

$$\sum_{i,d} \phi_{id} \sum_k \mu_{ik}^{(d)} = \sum_{i,k} \sum_d \mu_{ik}^{(d)} \phi_{id} = \sum_{i,j} \sum_d \mu_{ij}^{(d)} \phi_{id}$$

and

$$\sum_{i,d} \phi_{id} \sum_j \mu_{ji}^{(d)} = \sum_{i,j} \sum_d \mu_{ji}^{(d)} \phi_{id} = \sum_{i,j} \sum_d \mu_{ij}^{(d)} \phi_{jd},$$

where the second equality is obtained by switching the indices i and j. Therefore, (6.7) can be written as

$$\max_{\mu \in \mathcal{C}} \sum_{i,j} \sum_d \mu_{ij}^{(d)} (\phi_{id} - \phi_{jd}) = \max_{\mu \in \mathcal{C}} \sum_{ij} \max_d (\phi_{id} - \phi_{jd}) \left(\sum_d \mu_{ij}^{(d)} \right).$$

Substituting for $\mu_{ij}^{(d)}$ from (6.4) yields

$$\max_\alpha \sum_{i,j} \max_d (\phi_{id} - \phi_{jd}) \left(\sum_m \pi_m \sum_{r \in \mathcal{R}_m} \alpha_{m,r} r_{ij} \right).$$

Using the linearity of the above expression, it can be simplified to (this is left as Exercise 6.4 for the reader):

$$\max_{r \in \mathcal{R}_m} \sum_{i,j} r_{ij} \max_d (\phi_{id} - \phi_{jd}). \tag{6.9}$$

Algorithm (6.9) is the so-called *backpressure algorithm*. It is a MaxWeight algorithm, where the weight of each link is

$$\max_d(\phi_{id} - \phi_{jd}),$$

and $(\phi_{id} - \phi_{jd})$ is called the *backpressure* of destination d on link (i, j). Note that if

$$\max_d(\phi_{id} - \phi_{jd}) < 0,$$

the MaxWeight algorithm will choose $r_{ij} = 0$.

If the Lagrange multipliers $\{\phi_{id}\}$ were known, the congestion control algorithm (6.8) and the MaxWeight scheduling algorithm (6.9) solve the resource allocation problem. So the algorithm is fully specified if there is a way to estimate the Lagrange multipliers.

As in the dual algorithm for the Internet, a natural algorithm to estimate ϕ_{id} is the following:

$$\phi_{id}(t+1) = \left(\phi_{id}(t) + \epsilon\left(\sum_f x_f(t)\mathbb{I}_{s(f)=i,d(f)=d} + \sum_j \mu_{ji}^{(d)}(t) - \sum_k \mu_{ik}^{(d)}(t)\right)\right)^+, \tag{6.10}$$

where $\mu_{ji}^{(d)}(t)$ and $x_f(t)$ are the solutions of (6.6) and (6.7) with ϕ_{id} replaced by the estimate $\phi_{id}(t)$. Note that (6.10) is a difference equation counterpart of the differential equation in the dual algorithm for the Internet, and ϵ is a step-size parameter.

Computing $\mu_{ji}^{(d)}(t)$ requires the knowledge of π_m, which may be unknown. Suppose that the channel state at time t is $m(t)$, and that $r(t)$ is the solution to (6.9), then we can try to use the following difference equation to estimate the Lagrange multipliers:

$$\phi_{id}(t+1) = \left[\phi_{id}(t) + \epsilon\left(\sum_f x_f(t)\mathbb{I}_{s(f)=i,d(f)=d} + \sum_j r_{ji}^{(d)}(t) - \sum_k r_{ik}^{(d)}(t)\right)\right]^+, \tag{6.11}$$

where

$$r_{ij}^{(d)}(t) = \begin{cases} r_{ij}(t), & \text{if } d = d^*, \\ 0, & \text{otherwise,} \end{cases} \tag{6.12}$$

and d^* is a solution to

$$\max_d(\phi_{id}(t) - \phi_{jd}(t)).$$

If more than one d achieves the maximum, an arbitrary such d is chosen to be d^*.

The behavior of ϕ_{id}/ϵ in equation (6.11) looks almost like the queue length at node i for destination d. The only difference from the real queue dynamics is that it may not be possible to transfer $r_{ji}^{(d)}(t)$ packets from q_{jd} to q_{id} as there may not be enough packets in q_{jd}.[1] Thus, the true queue dynamics would satisfy

$$q_{id}(t+1) \leq \left[q_{id}(t) + \left(\sum_f x_f(t)\mathbb{I}_{s(f)=i,d(f)=d} + \sum_j r_{ji}^{(d)}(t) - \sum_k r_{ik}^{(d)}(t)\right)\right]^+. \tag{6.13}$$

---

1 Abusing the terminology, we use q_{jd} to denote the queue at node i for destination d and also to denote the corresponding queue length.

The exact dynamics would depend on the policy used to allocate packets along multiple links at each node when there are not enough packets to transfer, as dictated by the MaxWeight algorithm. Using the queue length q_{id} to approximate ϕ_{id} turns out to be good enough.

Based on the above derivation, we obtain a joint congestion control, routing, and scheduling algorithm, where each source adjusts its transmission rate based on the optimization problem (6.8), and the network determines routing and scheduling by solving optimization problem (6.9). The Lagrange multipliers in (6.8) and (6.9) are replaced by per-destination queues maintained at each node. A detailed description is presented in Algorithm 4.

Algorithm 4 Joint congestion control, routing, and scheduling algorithm

---

1: **Congestion control**: In each time slot $t \geq 0$, each source f computes its transmission rate $x_f(t)$ by solving the following optimization problem:

$$\max_{0 \leq x_f \leq x_{max}} U_f(x_f) - \epsilon q_{s(f)d(f)}(t)x_f, \qquad (6.14)$$

and then generates $a_f(t)$ packets, where $a_f(t)$ is a random variable taking integer values and with mean $x_f(t)$ (e.g., $a_f(t)$ can be a Poisson random variable with parameter $x_f(t)$).

2: **Routing and scheduling**: In each time slot $t \geq 0$, the network schedules a link-rate vector $r(t)$ such that

$$r(t) \in \arg\max_{r \in \mathcal{R}_{m(t)}} \sum_{i,j} w_{ij}(t)r_{ij},$$

where $w_{ij}(t) = \max_d \left(q_{id}(t) - q_{jd}(t) \right)$. Once the link rates are selected, the network has to decide which packets will be routed from node i to node j, for each i and j. Packets belonging to destination d^* are selected for transmission over link (i, j) if

$$d^*_{ij}(t) \in \arg\max_d (q_{id}(t) - q_{jd}(t)).$$

Ties are broken randomly.

3: **Data transmission**: Node i transmits $\mu_{ij}(t)$ packets with destination d^*_{ij} to node j, which will be stored in queue $q_{jd^*_{ij}}$, where $\mu_{ij}(t) = \min\{r_{ij}(t), q_{id^*_{ij}(t)}(t)\}$.

4: **Queue dynamics**: The queue q_{id} evolves as follows:

$$q_{id}(t+1) = q_{id}(t) + a_f(t)\mathbb{I}_{s(f)=i, d(f)=d}$$

$$+ \sum_j \mu_{ji}^{(d)}(t) - \sum_k \mu_{ik}^{(d)}(t).$$

Note that in (6.14) we added a constraint $x_f \leq x_{max}$ so that the source will not transmit with infinite speed when $q_{s(f),d(f)}(t) = 0$. This constraint does not change the optimal solution when $x_{max} > x_f^*$, which can be guaranteed by choosing a sufficiently large x_{max}, e.g., $x_{max} \geq r_{max}$, where r_{max} is the maximum link capacity in the network.

Example 6.1.3

Consider a network with three nodes, as shown in Figure 6.5, where we have two flows: one flow from node 1 to node 4, and another flow from node 1 to node 2. In this network, all nodes maintain a queue for node 4 because node 4 is reachable from all other nodes in the network. Only nodes 1 and 2 maintain queues for node 2 because node 2 cannot be reached from other nodes. We further note that $q_{dd} \equiv 0$, so $q_{22} = q_{44} = 0$.

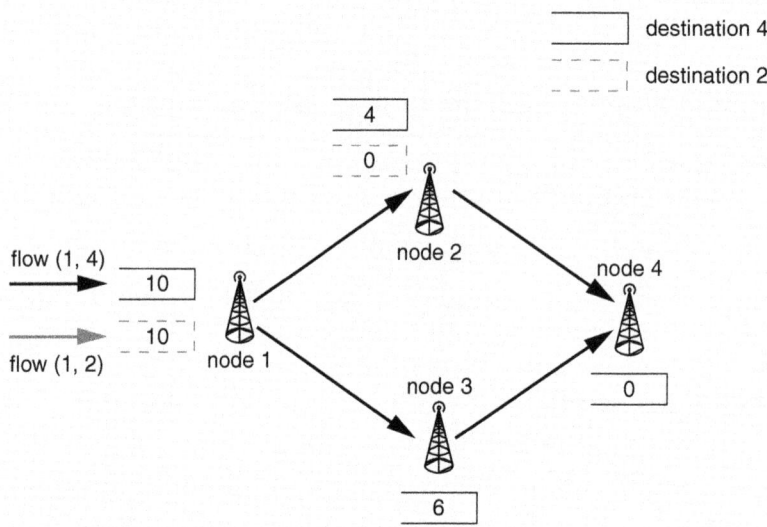

Figure 6.5 Four-node wireless network.

Assume that the utility function of flow $(1,4)$ is $2\log x_{(1,4)}$ and that the utility function of flow $(1,2)$ is $\log x_{(1,2)}$. Further, assume that $\epsilon = 0.1$, $x_{max} = 10$, and $\mathcal{R}_{m(t)} = \{(r_{1,2}, r_{1,3}, r_{2,4}, r_{3,4}) : (4,0,0,5), (0,2,3,0)\}$. Then Algorithm 4 works as follows.

- Given the utility function, we can explicitly solve (6.14). We have

$$x_{(1,4)}(t) = \min\left\{\frac{2}{\epsilon q_{14}(t)}, x_{max}\right\} = 2,$$

$$x_{(1,2)}(t) = \min\left\{\frac{1}{\epsilon q_{12}(t)}, x_{max}\right\} = 1.$$

- At link $(1,2)$, the backpressure towards destination 4 is 6 and the backpressure towards destination 2 is 10, so link $(1,2)$ will transmit packets to destination 2, and the link weight $w_{12} = 10$. All other links serve packets to destination 4, and the link weights are $w_{13} = 4$, $w_{34} = 6$, and $w_{24} = 4$.
- Given the link weights, it is not hard to verify that the link-rate vector $\{4,0,0,5\}$ results in a larger $\sum_{i,j} w_{ij} r_{ij}$ than $\{0,2,3,0\}$. Therefore, node 1 transmits four packets with destination 2 to node 2, and node 3 transmits five packets with destination 4 to node 4. □

To understand the performance of the joint algorithm, we will assume that $\{a_f(t)\}$ are independent across time and that the channel-state process is i.i.d. across time, with π_m the probability of the channel being in state m. Under these conditions, $q(t)$ is a DTMC. We further assume that the arrival processes and channel-state process are such that the DTMC is irreducible. We then have the following theorem, which states that, under the joint algorithm, the expected sum-queue length is bounded and the resource allocation maximizes the network utility when $\epsilon \to 0$.

Theorem 6.1.1 The irreducible DTMC $q(t)$ is positive recurrent under the joint algorithm and

$$\lim_{t \to \infty} E\left[\sum_{i,d} q_{id}(t)\right] \leq \frac{B_1}{\epsilon} \tag{6.15}$$

for some $B_1 < \infty$.
  Further, let

$$y_f = \lim_{t \to \infty} E\left[x_f(t)\right].$$

Then

$$\sum_f U_f(x_f^*) - B_2\epsilon \leq \sum_f U_f(y_f) \tag{6.16}$$

for some $B_2 < \infty$, where $\{x_f^*\}$ is the solution to (6.2) subject to (6.3), (6.4), and (6.5). □

Note that the theorem suggests a tradeoff: if our realized utility $\sum_f U_f(y_f)$ is close to the optimal utility $\sum_f U_f(x_f^*)$, the upper bound on the sum of the queue lengths is large, $O(1/\epsilon)$. While the estimate of the sum-queue lengths is only an upper bound, this behavior is to be expected for the following reason: if we get close to the optimal solution, it means that some of the constraints (6.3) must be close to tight. But these constraints simply state that the arrival rate is no more than the departure rate at a queue. Recalling our preceding discussion of discrete-time queueing systems, when the arrival rate is close to the service rate, the expected queue length in the steady state can be large. While this is a useful intuition, we note that the discrete-time queueing models do not directly apply here because the arrival processes in the current model are controlled as a function of queue lengths whereas the queueing theory models assume that the arrival process is independent of the queue length.
  We will not prove Theorem 6.1.1. We will discuss some special cases and provide a proof for one of these special cases.

6.2   Stability and convergence: a cellular network example

We consider a cellular downlink network, as shown in Figure 6.6, where packets are injected by the sources into the base station, and the base station needs to deliver the packets to the mobiles. As each flow goes over only one link (from the base station to a mobile), the notation can be simplified as follows:

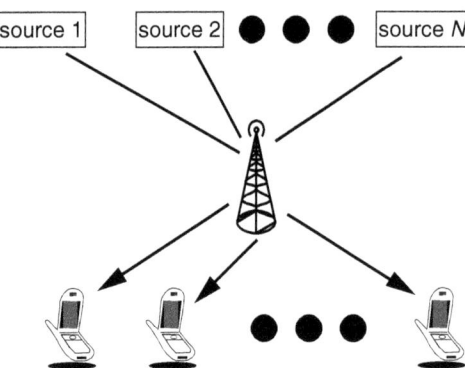

Figure 6.6 Cellular network with N mobiles.

- x_i is the average rate at which the source sending content to mobile i injects data into the base station;
- μ_i is the average rate at which the base station transmits to mobile i.

The rest of the variables are as before.

The NUM problem in this case is

$$\max_{x,\alpha,\mu \geq 0} \sum_{i=1}^{N} U_i(x_i),$$

$$x_i \leq \mu_i,$$

subject to

$$\mu_i = \sum_m \pi_m \left(\sum_{r \in \mathcal{R}_m} \alpha_{m,r} r_i \right).$$

Using the decomposition as before, we arrive at Algorithm 5.

We will give a proof of Theorem 6.1.1 for this special case. It is easy to verify that $q(t)$ is an irreducible Markov chain under the assumption that $\Pr(a_i(t) = 0) > \delta$ for some $\delta > 0$ all i and t.

Proof In this proof, we will first show that $q(t)$ is positive recurrent, and then prove that the resource allocation under the joint algorithm is close to the optimal resource allocation based on the fact that $q(t)$ is positive recurrent. Consider the Lyapunov function

$$V(q(t)) = \frac{1}{2} \sum_i q_i^2(t).$$

Algorithm 5 Joint congestion control and scheduling for cellular downlink networks

---

1: **Congestion control:** Source i injects $a_i(t)$ packets into the base station such that $E[a_i(t)] = x_i(t)$, and

$$x_i(t) \in \arg\max_{0 \le x_i \le x_{\max}} U_i(x_i) - \epsilon q_i(t) x_i.$$

2: **MaxWeight scheduling:** The transmission rate vector $r(t)$ is chosen to be a solution of the following optimization problem:

$$\max_{r \in \mathcal{R}_{m(t)}} \sum_{i=1}^{N} r_i q_i(t).$$

Ties are broken at random.

3: **The evolution of the queues:** The queue maintained for mobile i evolves as follows:

$$q_i(t+1) = (q_i(t) + a_i(t) - r_i(t))^+.$$

The drift of the Lyapunov function, given the queue and channel states, is

$$E\left[V(q(t+1)) - V(q(t))|q(t) = q\right]$$

$$= \frac{1}{2} E\left[\sum_i q_i^2(t+1) - q_i^2(t)\,\middle|\, q(t) = q\right]$$

$$= \frac{1}{2} E\left[\sum_i ((q_i(t) + a_i(t) - r_i(t))^+)^2 - q_i^2(t)\,\middle|\, q(t) = q\right]$$

$$\le \frac{1}{2} E\left[\sum_i (q_i(t) + a_i(t) - r_i(t))^2 - q_i^2(t)\,\middle|\, q(t) = q\right]$$

$$= \frac{1}{2} E\left[\sum_i (a_i(t) - r_i(t))^2 + 2q_i(t)(a_i(t) - r_i(t))\,\middle|\, q(t) = q\right]$$

$$\le K + \sum_i q_i x_i(t) - \sum_i q_i E[r_i(t)|q(t) = q],$$

where we assume $E[a_i^2(t)] \le \sigma_{\max}^2$ and $K = N\left(\sigma_{\max}^2 + r_{\max}^2\right)/2$. Adding and subtracting $(1/\epsilon) \sum_i U_i(x_i)$ on the right-hand side of the above inequality results in

$$E\left[V(q(t+1)) - V(q(t))|q(t) = q\right]$$

$$\le K + \sum_i q_i x_i(t) - \frac{1}{\epsilon} \sum_i U_i(x_i(t)) + \frac{1}{\epsilon} \sum_i U_i(x_i(t)) - \sum_i q_i \bar{r}_i,$$

where $\bar{r}_i = E[r_i(t)|q(t) = q]$.

Since $x_i(t)$ maximizes $U_i(x) - \epsilon q_i x$ for $0 \le x \le x_{\max}$, and $x_i^* \le x_{\max}$ (recall that x^* is the optimal resource allocation), we have

$$\frac{1}{\epsilon}U_i(x_i^*(1-\delta)) - q_i x_i^*(1-\delta) \le \frac{1}{\epsilon}U_i(x_i(t)) - q_i x_i(t),$$

where $0 \le \delta < 1$. So we can obtain

$$E\left[V(q(t+1)) - V(q(t))|q(t) = q\right]$$

$$\le K + \sum_i q_i x_i^*(1-\delta) - \sum_i \frac{1}{\epsilon}U_i(x_i^*(1-\delta)) + \frac{1}{\epsilon}\sum_i U_i(x_i(t)) - \sum_i q_i \bar{r}_i$$

$$= K + \sum_i q_i(x_i^* - \bar{r}_i) - \delta \sum_i q_i x_i^* + \frac{1}{\epsilon}\sum_i (U_i(x_i(t)) - U_i(x_i^*(1-\delta)))$$

$$\le \tilde{K} + \sum_i q_i(x_i^* - \bar{r}_i) - \delta \sum_i q_i x_i^*, \tag{6.17}$$

where $\tilde{K} = K + (1/\epsilon)\sum_i (U_i(x_{\max}) - U_i(x_i^*(1-\delta)))$. It is not difficult to show that, under MaxWeight scheduling, $r(t)$ solves

$$\max_{r \in \mathcal{R}_{m(t)}} \sum_i r_{m(t),i} q_i,$$

and $\bar{r} = E[r(t)|q(t) = q]$ is a maximizer to the following problem:

$$\max_{r \in \mathcal{C}} \sum_i r_i q_i.$$

So $\sum_i q_i x_i^* \le \sum_i q_i \bar{r}_i$, and

$$E\left[V(q(t+1)) - V(q(t))|q(t) = q\right] \le \tilde{K} - \delta \sum_i q_i x_i^*. \tag{6.18}$$

As δ can be chosen to be any value between 0 and 1, $q(t)$ is positive recurrent according to the Foster–Lyapunov theorem if $x_i^* > 0$ for all i.

Thus, we do not even need the strict concavity of $U_i(x_i)$. We only need concavity and the requirement that there exists an optimal solution x^* such that $x_i^* > 0$ for all i, which we assume to be true.

Since $q(t)$ is positive recurrent, a stationary distribution for $q(t)$ exists, and

$$E\left[V(q(t+1)) - V(q(t))\right] = 0$$

in the steady state. So, assuming $q(t)$ is in the steady state, we take expectations on both sides of inequality (6.18) to obtain

$$0 \le \tilde{K} - \delta \sum_i E[q_i(t)] x_i^*,$$

which implies that

$$E\left[\sum_i q_i(t)\right] \leq \frac{\tilde{K}}{\delta \min_i x_i^*}$$

when $q(t)$ is in the steady state.

Since $\sum_i q_i x_i^* \leq \sum_i q_i \bar{r}_i$, by choosing $\delta = 0$, we can further obtain from (6.17) that

$$E\left[\sum_i U_i(x_i^*) - U_i(x_i(t))\right] \leq \epsilon K,$$

or

$$E\left[\sum_i U_i(x_i^*)\right] \leq E\left[\sum_i U_i(x_i(t))\right] + \epsilon K,$$

when $q(t)$ is in the steady state.

We remark that that the upper bound on the sum-queue length is of the order of $O(1/\epsilon)$, whereas the difference between the network utility under the joint algorithm and the optimal network utility is of the order of $O(\epsilon)$. $\qquad\square$

6.3 Ad hoc P2P wireless networks

We now consider network utility maximization in an ad hoc P2P network. For simplicity, we assume that the channel state does not vary, so we omit the subscript m. Further, instead of using a double subscript (i, j) to denote a link, we use l to denote a link. We let x_l denote the rate at which the source using link l transmits data, and μ_l denote the rate allocated to link l by the network. The necessary conditions for x_l to be supportable include

$$x_l \leq \mu_l \quad \forall l, \tag{6.19}$$

$$\mu_l = \sum_{r \in \mathcal{C}} \alpha_r r_l \quad \forall l, \tag{6.20}$$

$$\sum_{r \in \mathcal{R}} \alpha_r = 1. \tag{6.21}$$

The NUM problem is reduced to

$$\max_{x,\mu,\alpha \geq 0} \sum_l U_l(x_l)$$

subject to the constraints in (6.19)–(6.21).

The joint congestion control and scheduling algorithm is presented in Algorithm 6. The proof is similar to the cellular case, and left as Exercise 6.6.

Often, the interference constraints in the ad hoc network can be represented by a conflict graph, as we explained in Chapter 5, which is a special case of the model we described here. To see this, we will consider a simple example.

Algorithm 6 Joint congestion control and scheduling for ad hoc P2P networks

---

1: **Congestion control**: The transmitter of link l computes the rate

$$x_l(t) = \arg \max_{0 \le x_l \le x_{max}} U_l(x_l) - \epsilon q_l(t)x_l,$$

and injects $a_l(t)$ packets into its buffer, where $a_l(t)$ is an integer and $E[a_l(t)] = x_l(t)$.

2: **MaxWeight scheduling**: The network selects the link-rate vector $r(t)$ such that

$$r(t) \in \arg \max_{r \in \mathcal{R}} \sum_l q_l(t)r_l.$$

Ties are broken at random.

3: **Queue evoluation**: The queue maintained at the transmitter of link l evolves as follows:

$$q_l(t+1) = (q_l(t) + a_l(t) - r_l(t))^+.$$

Example 6.3.1

Consider the conflict graph shown in Figure 6.7. In this example, link 1 and link 4 can be scheduled simultaneously, or link 2 and link 3 can be scheduled simultaneously. Assume that when link l is scheduled, it can transmit at a maximum rate of c_l, which we will assume is an integer. Then, the possible rate vectors under the two schedules are given by

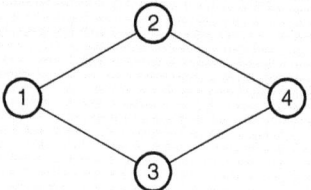

Figure 6.7 Example of a conflict graph, in which a node represents a wireless link and a link represents a conflict of two links.

$$a = \begin{pmatrix} c_1 \\ 0 \\ 0 \\ c_4 \end{pmatrix} \quad \text{and} \quad b = \begin{pmatrix} 0 \\ c_2 \\ c_3 \\ 0 \end{pmatrix}.$$

The convex hull of these two rate vectors and vector 0 is given by

$$Co(\{a, b, 0\}) = \{r : r = \alpha a + \beta b \text{ for some } \alpha, \beta \ge 0 \text{ and } \alpha + \beta \le 1\},$$

which is also the capacity region obtained based on the conflict graph and convex hull model described in Chapter 5.

Instead, one can also enumerate all possible rate vectors as follows:

$$\mathcal{A} = \left\{ \begin{pmatrix} r_1 \\ 0 \\ 0 \\ r_4 \end{pmatrix} : r_1 \leq c_1, r_4 \leq c_4, r_1, r_4 \geq 0, \text{integers} \right\}$$

and

$$\mathcal{B} = \left\{ \begin{pmatrix} 0 \\ r_2 \\ r_3 \\ 0 \end{pmatrix} : r_2 \leq c_2, r_3 \leq c_3, r_2, r_3 \geq 0, \text{integers} \right\}.$$

Letting $\hat{\mathcal{C}} = \mathcal{A} \cup \mathcal{B}$, $\mathcal{C}$ can be equivalently defined as

$$\mathcal{C} = \{r : \exists \gamma_i \in \hat{\mathcal{C}}, \alpha_i \geq 0, \sum_i \alpha_i = 1 \text{ such that } r = \sum_i \alpha_i \gamma_i\}.$$

This is the definition that we have used in this chapter. It is not difficult to see that these definitions are equivalent. □

6.4   Internet versus wireless formulations: an example

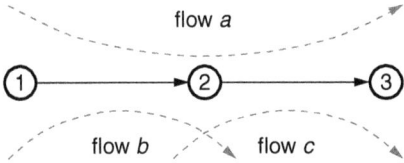

Figure 6.8 Simple line network with two links and three flows.

Consider a simple two-link, three-flow network, as shown in Figure 6.8. Assume that the two links can operate simultaneously. For example, it could be a wireline network or two wireless links operating in different frequency bands so that they do not interfere with each other. Let both links have capacity of 1.

Then, in the network utility maximization formulations in this chapter, we set up the following resource allocation problem:

formulation 1: $\max_{x,\mu \geq 0} \sum_{f=a,b,c} U_f(x_f),$

$$x_a \leq \mu_{12}^{(3)},$$

$$x_b \leq \mu_{12}^{(2)},$$

$$x_c + \mu_{12}^{(3)} \leq \mu_{23}^{(3)},$$

$$\mu_{12}^{(2)} + \mu_{12}^{(3)} \leq 1,$$

$$\mu_{23}^{(3)} \leq 1,$$

where the x's are the flow rates and the μ's are the data rates allocated to per-destination queues at each link. As we have seen, this results in congestion control, which acts on the ingress queues at the sources, along with a back-pressure algorithm, as shown in Figure 6.9. In this case, the back-pressure algorithm is not used to schedule links as two links can be ON simultaneously. However, it determines which queue will be served on each link at each time instant.

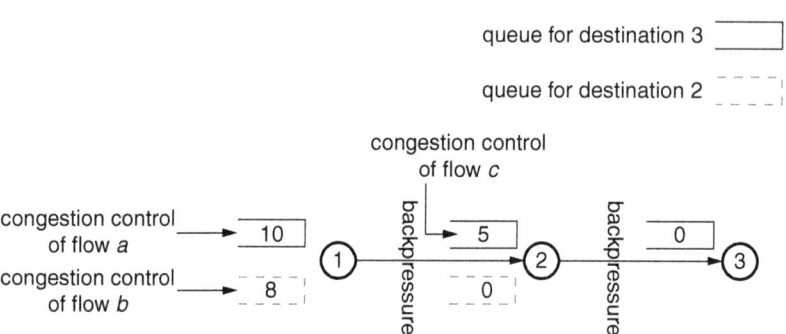

Figure 6.9 NUM architecture of formulation 1.

An alternative formulation was presented in Chapter 2:

formulation 2: $\max_{x \geq 0} U_a(x_a) + U_b(x_b) + U_c(x_c),$

$$x_a + x_b \leq 1,$$

$$x_a + x_c \leq 1.$$

This results in a congestion control algorithm which simply uses the price feedback from the network. There is no backpressure algorithm. In other words, we did not need to maintain separate queues for different destinations, and all packets can be stored in a FIFO queue, as shown in Figure 6.10. Next, we discuss the difference between these two formulations.

In formulation 1, we correctly model the fact that a flow's packets do not arrive at each of the links in its path instantaneously. Flow a's packets have to arrive at node 1, then at node 2, and finally at node 3 to depart the network. So the departures from the first link are the arrivals to the second link. Hence, the constraints are

$$x_a \leq \mu_{12}^{(3)} \qquad \text{and} \qquad x_b + \mu_{12}^{(3)} \leq \mu_{23}^{(3)}.$$

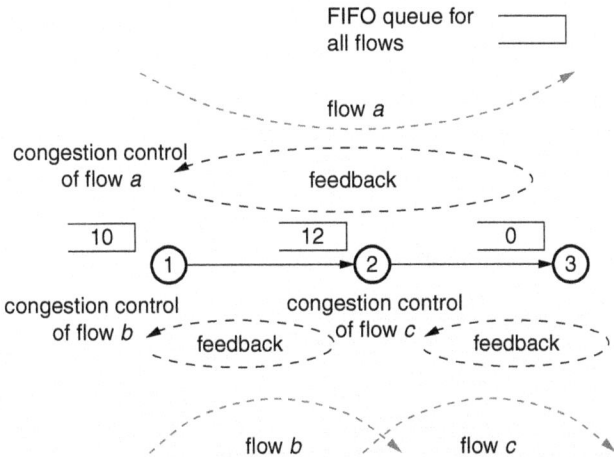

Figure 6.10 NUM architecture of formulation 2.

On the other hand, in formulation 2, we assume that packets from flow a arrive at all the links on their paths simultaneously. This leads to a simpler algorithm.

While formulation 1 is exact, we believe that formulation 2 is appropriate for the wired Internet. The reason is that the Internet, except at access points, operates at less than full capacity. Hence, queues tend to be small or at least can be made small with a well-designed algorithm. Consequently, the delays are small, and it seems reasonable to assume that packets arrive instantaneously at all links in their paths.

On the other hand, in wireless networks, such an assumption may not be reasonable. Due to interference constraints (for instance, if the two links in this example cannot transmit simultaneously), their queues will accumulate at one link when the other is being scheduled. So formulation 1 may be more reasonable.

In conclusion, formulation 1 is more reasonable in wireless networks with interference, while formulation 2 seems to be appropriate for wireline networks and leads to a simpler algorithm.

6.5   Summary

- **Network utility maximization for wireless networks** Algorithm 4 is a joint congestion control, routing, and scheduling algorithm that solves the network utility maximization $(\sum_f U_f(x_f))$ for wireless networks.
- **Stability and fairness** Under the joint algorithm, the realized utility is lower bounded by the optimal utility minus a constant, i.e., $\sum_f U_f(x_f^*) - B_2\epsilon \leq \sum_f U_f(y_f)$, where x^* is the optimal solution to the network utility maximization problem, y is the mean of the rate vector obtained under the joint algorithm, B_2 is a constant, and ϵ is a tuning parameter used in the congestion control. The mean of the sum-queue length under the joint algorithm is bounded by B_1/ϵ, where B_1 is also a constant.

6.6   Exercises

Exercise 6.1 (The capacity region) Consider the network in Figure 6.1, and assume all channels are AWGN channels. Plot the network capacity region given by the parameters in Table 6.2.

Table 6.2 Channel gains, power levels, and variances of noises

Channel gain	$h_{12} = h_{34} = 1$ and $h_{14} = h_{32} = 0.1$
Power levels	$p_1 \in \{0, 5, 10\}$ and $p_3 \in \{0, 1, 5\}$
Variances of noises	$n_2 = n_4 = 1$

Exercise 6.2 (The link-rate region) Consider the line network in Figure 6.11, where both links $(1, 2)$ and $(2, 3)$ are ON–OFF channels such that the probability of being ON is 0.5. One packet can be transmitted when a link is ON. Plot the link-rate region of this network.

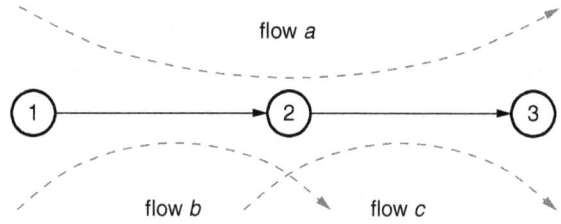

Figure 6.11 Line network.

Exercise 6.3 (An example of the joint congestion, routing, and scheduling algorithm) Consider the line network in Figure 6.11. Assume the utility functions associated with the three flows are $U_i(x_i) = -w_i/x_i$, where $i = a, b, c$ and w_i is a constant. Write down the explicit forms of the network utility maximization problem and Algorithm 4.

Exercise 6.4 (Backpressure routing) Recall that sub-problem 2 (6.7) is equivalent to the following problem:

$$\max_{\alpha} \sum_{i,j} \max_{d} (\phi_{id} - \phi_{jd}) \left(\sum_{m} \pi_m \sum_{r \in \mathcal{R}_m} \alpha_{m,r} r_{ij} \right).$$

Using the linearity of the above expression, prove that the optimization problem defined by the above equation can be further simplified to

$$\max_{r \in \mathcal{R}_m} \sum_{i,j} r_{ij} \max_{d} (\phi_{id} - \phi_{jd}).$$

Exercise 6.5 (Minimizing the transmit power) Consider the same network model as in Exercise 5.3, but assume that there is no average power constraint at each mobile. Instead, the base station needs to minimize the total average transmit power under the constraint that the packet queues in the network are stable. In other words, the base station needs to solve the following problem:

$$\min_{\alpha} \sum_m \sum_i \sum_j \pi_m \alpha_{m,i,j} p_j, \tag{6.22}$$

subject to the constraint that the network is stable. Consider a joint scheduling and power control algorithm that schedules mobile $i^*(t)$ at transmit power level $j^*(t)$ in time slot t such that

$$(i^*(t), j^*(t)) \in \arg\max_{(i,j)} q_i(t) c_{m(t),i,j} - Rp_j.$$

Let α^* denote the optimal solution to (6.22). Show that

$$E\left[\sum_i b_i(\infty)\right] = \sum_m \sum_i \sum_j \pi_m \alpha^*_{m,i,j} p_j + O\left(\frac{1}{R}\right),$$

i.e., the joint scheduling and power control algorithm approximately minimizes the total average transmit power when R is sufficiently large.

Exercise 6.6 (The optimality of Algorithm 6) Consider the joint congestion control and scheduling algorithm (Algorithm 6) for ad hoc P2P wireless networks, and assume that the arrival processes are Bernoulli (which implicitly assumes that the x_l's are always between 0 and 1).

(1) Consider the Lyapunov function $V(t) = \sum_l q_l^2(t)$ and prove that $q(t)$ is positive recurrent.

(2) Assume $q(t)$ is in steady state. Prove that

$$E\left[\sum_l U_l(x_l^*)\right] \le E\left[\sum_l U_l(x_l(t))\right] + \epsilon K$$

for some positive constant K, where ϵ is the tuning parameter in the congestion controller and x^* is the optimal solution to the network utility maximization problem.

Exercise 6.7 (The optimality of Algorithm 4) Consider the line network in Figure 6.11. Let $q(t)$ denote the queue lengths at time t. Further, assume that $x_{\max} < 1$ in Algorithm 4 and that $a_f(t)$ is a Bernoulli random variable with parameter $x_f(t)$.

(1) Prove that $q(t)$ is irreducible. Hint: Prove that any state q is reachable from state 0, and that state 0 is reachable from any state q.

(2) Prove that $q(t)$ is a positive recurrent Markov chain under Algorithm 4.

(3) Assume the network to be in steady state. Prove that the following bound holds for some $K > 0$ under Algorithm 4:

$$U_a(x_a^*) + U_b(x_b^*) + U_c(x_c^*) \le E[U_a(x_a(t)) + U_b(x_b(t)) + U_c(x_c(t))] + \epsilon K,$$

where x^* is the optimal solution to the network utility maximization problem.

Exercise 6.8 (Internet versus wireless formulations) Consider the network in Figure 6.12. Assume there is a flow from node 1 to node 3 and another flow from node 1 to node 4. Write down the queue architectures and the joint congestion control, routing, and scheduling algorithms under formulations 1 and 2 presented in Section 6.4. For formulation 2, assume the route for flow 1 is $1 \to 2 \to 3$ and the route for flow 2 is $1 \to 4$; derive the dual solution.

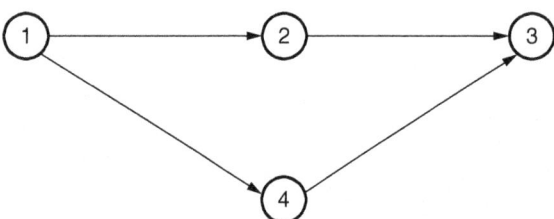

Figure 6.12 Simple network.

Exercise 6.9 (Slotted Aloha) Consider n radios transmitting to each other. The interference constraints are such that only one radio can transmit at a time. We assume that a transmitter can detect whether its transmission was successful or if it resulted in a collision. If, after a collision, a retransmission is attempted immediately, a collision will occur if all transmitters follow the same protocol. The Aloha protocol, introduced in the 1970s, solves this problem by requiring each transmitter to attempt a transmission with probability p when it has a packet to transmit, independent of past history of collisions or successful transmissions.

(1) Assume time is slotted, i.e., the network operates in discrete time. Also assume that all nodes have an infinite backlog of packets. What is the average throughput of a given node under Aloha? Assume $n \geq 2$ throughout.

(2) Find the optimal p to maximize the above throughput.

(3) Show that the total throughput of the network approaches $1/e$ as $n \to \infty$.

Note: The original Aloha protocol does not assume that time is slotted. The version of Aloha described here is called slotted Aloha to indicate time is slotted. The Aloha protocol has been very influential in the design of communication network protocols. For example, the backoff mechanism in 802.11 (discussed in Section 7.4) is similar in spirit to Aloha: the idea is to randomize transmissions so that repeated collisions are avoided. However, the RTS/CTS[2] protocol goes one step further and attempts to come up with a collision-free schedule.

Exercise 6.10 (The infinite population model of slotted Aloha) Consider the following model of packet arrivals, called the infinite population model of slotted Aloha: at the beginning of time slot t, let $N(t)$ be the number of users in the system. Each user has exactly one packet and departs the system when its packet has been transmitted successfully. Let p be

2 RTS and CTS are defined in Chapter 7.

the attempt probability (see Exercise 6.9). Let $S(t)$ be the number of successful transmissions in time slot t. Note that $S(t) \in \{0, 1\}$. At the end of the time slot, a new user joins the system with probability λ. Each new user also brings only one packet with it. Thus, $N(t)$ evolves according to

$$N(t + 1) = N(t) - S(t) + A(t),$$

where $A(t)$ is a Bernoulli random variable with mean λ.

(1) For any fixed p, show that the system is unstable, i.e., the Markov chain $N(t)$ is not positive recurrent.

(2) If $p(t) = 1/N(t)$, show that the Markov chain $N(t)$ is positive recurrent if $\lambda < 1/e$.

Exercise 6.11 (A network extension of Aloha) We now consider a network extension of Aloha, where all transmissions do not interfere with each other. Let $\mathcal{O}_i$ be the number of neighbors of node i that are within receiving range of transmissions from node i, and let $\mathcal{I}_i$ be the set of nodes whose transmissions can be heard by node i. We assume that a node cannot transmit and receive simultaneously, and that a node can transmit to only one other node at a time. Let p_{ij} be the probability that node i transmits to node j. The probability of successful transmission from node i to node j is given by

$$x_{ij} = p_{ij}(1 - P_j) \prod_{l \in \mathcal{I}_j \setminus \{i\}} (1 - P_l),$$

where P_j, the probability that node j transmits, is given by

$$P_j = \sum_{k \in \mathcal{O}_j} p_{jk}.$$

The goal in this exercise is to select $\{p_{ij}\}$ to maximize

$$\sum_{i,j:i \neq j} \log x_{ij}, \tag{6.23}$$

i.e., to achieve the proportional fairness in this network Aloha setting. The constraints are

$$\sum_{k \in \mathcal{O}_j} p_{jk} \leq 1 \quad \forall j,$$

$$p_{jk} \geq 0 \quad \forall k, j.$$

Show that the optimal $\{p_{ij}\}$'s are given by

$$p_{ij}^* = \frac{1}{|\mathcal{I}_i| + \sum_{k \in \mathcal{O}_i} |\mathcal{I}_k|}.$$

Hint: It is helpful to show first that the following elements of the vector x are functions of p_{ij}: x_{ij}, x_{ki} for $k \in \mathcal{I}_i$, and x_{kh} for k and h such that $h \in \mathcal{O}_i \cup \mathcal{O}_k$ and $k \neq i$.

6.7   Notes

Utility maximization which considers resource allocation at multiple layers simultaneously was introduced in [37, 38, 102, 131, 132, 160]. A tutorial focusing on the convex optimization aspects of the problem can be found in [103], and a comprehensive treatment of the stochastic analysis can be found in [44]. The stability analysis and performance bounds presented in this chapter are similar to the treatment in [44, 132]. Practical implementations of backpressure algorithms can be found in [154, 170, 171]. The Aloha protocol was introduced in [1]. The infinite population model of slotted Aloha has been studied in [41, 69, 88]. The network extension of Aloha was first studied in [67].

7 Network protocols

In this chapter, we explore the relationship between the algorithms discussed in the previous chapters and the protocols used in the Internet and wireless networks today. We will first discuss protocols used in the wired Internet, and then discuss protocols used in wireless networks.

It is important to note that Internet congestion control protocols were *not* designed using the optimization formulation of the resource allocation problem discussed earlier. The predominant concern while designing these protocols was to minimize the risk of congestion collapse, i.e., large-scale buffer overflows, and hence they tended to be rather conservative in their behaviors. Even though the current Internet protocols were not designed with clearly defined fairness and stability ideas in mind, they bear a strong resemblance to the ideas of fair resource allocation that we have discussed so far. In fact, the utility maximization methods presented earlier provide a solid framework that could be for understanding the operation of these congestion control algorithms. Further, going forward, the utility maximization approach seems like a natural candidate framework that could be used to modify existing protocols to adapt to the evolution of the Internet as it continues to grow faster.

In our earlier discussion on network utility maximization for the Internet, we assumed that routes are fixed. In this chapter, we discuss how the Internet chooses a route for each source-destination pair. Further, we also discuss the addressing protocol used in the Internet which allows one to implement routing protocols.

Finally, we will discuss protocols for wireless networks. In particular, we will discuss protocols used for fair resource allocation in cellular networks, and discuss protocols used for medium access control in wireless networks. Throughout, we will point out similarities and differences between the mathematical models in the previous chapters and the practical protocols implemented today.

The key questions that will be answered in this chapter include the following.

- *What are the principles behind the design of the most widely used congestion control protocols in the Internet today?*
- *What is the relationship between today's congestion control protocols and primal and dual algorithms derived in earlier chapters?*
- *How does the network control access to users in a wireless medium, in both cellular networks and in WiFi networks?*
- *What is the relationship between wireless protocols and the mathematical models of wireless networks developed in earlier chapters?*

7.1   Adaptive window flow control and TCP protocols

Congestion control algorithms deployed in the Internet use a concept called *window flow control*. Each user maintains a number, called a *window size*, which is the number of unacknowledged packets that are allowed to be sent into the network. Any new packet can be sent only when an acknowledgement for one of the previous sent packets is received by the sender, as shown in Figure 7.1.

Suppose that the link speeds in the network are so large that the amount of time it takes to process a packet at a link is much smaller than the Round-Trip Time (RTT), which is the amount of time that elapses between the time that a packet is released by the source and the time when it receives the acknowledgement (ack) for the packet from the destination. Then, roughly speaking, the source will send W packets into the network, receive acknowledgements for all of them after one RTT, then send W more packets into the network, and so on. Thus, the source sends W packets once every RTT, which means that the average transmission rate x of the source can approximated by the formula $x = W/T$, where T is the RTT.

A significant problem with the simple window flow control scheme as described above is that the optimal choice of W depends on the number of other users in the network: clearly, the transmission rate of a user should be small if it shares a common link with many other users, but it can be large (but smaller than the capacity of its route) if there are few other users sharing the links on its route. Thus, what is required is an algorithm to adjust adaptively the window size in response to congestion information. This task is performed by the congestion control part of the *Transmission Control Protocol* (TCP) in today's Internet. The general idea is as follows: a sender increases its window size if it determines that there is no congestion, and decreases the window size if it detects congestion. Packet losses or excessive delays in the route from the source to the destination are used as indicators of congestion. Excessive delays in receiving an ack indicate large delays, while missing acks signal lost packets. With some exceptions, congestion detection

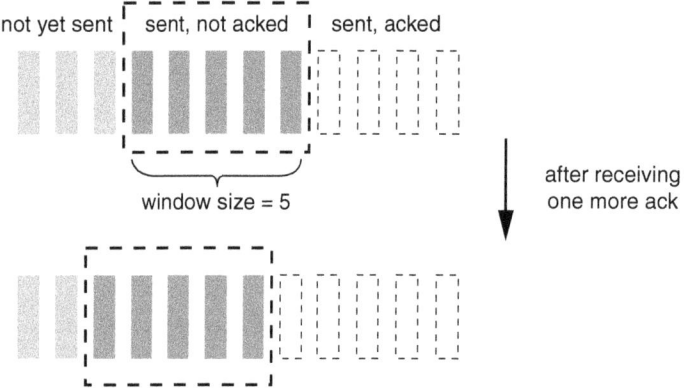

Figure 7.1 Window flow control. The window size is set to be 5, so at most five unacknowledged packets are allowed. After an additional acknowledgement (ack) is received, one more packet can be sent out.

is performed by the source at time instants when it receives an ack for a packet. This means that the congestion control algorithm makes decision at ack reception events, and does not explicitly rely on any clocks for timing information. In other words, TCP is a *self-clocking* protocol. Different versions of TCP use different algorithms to detect congestion and to increase/decrease the window sizes. We discuss two versions of TCP in the following sections.

7.1.1   TCP-Reno: a loss-based algorithm

The most common versions of TCP used in the Internet today use packet loss as the indicator of congestion. When a packet loss is detected, the window size is reduced, and when packets are being successfully received by the destination, the window size is increased slowly. There are many different versions of TCP which implement such a loss-based congestion control mechanism. The most common ones today are TCP-Reno and TCP-NewReno, both of which are variants of the first congestion control algorithm used in the Internet called TCP-Tahoe. These versions differ in the protocol details, and we will only present a simplified model of the congestion control algorithm in TCP-Reno, which captures the basic ideas behind these protocols.

First, we will describe how TCP-Reno detects lost packets. Upon reception of each packet, the destination sends an ack back to the source. The ack contains the sequence number of the next expected packet. For example, if the destination has already received packets 1–4 and receives packet 5 next, the ack indicates that the destination expects to receive packet 6. Now suppose that packet 7 is received next; the destination will again ask for packet 6 since it is the lowest-numbered packet that has not yet been received by the destination. If the source receives four acks asking for packet 6 (which is equivalent to saying that three duplicate acks, or dupacks, are received for packet 6), it assumes that packet 6 has been lost and cuts down the window size in a manner to be described shortly. However, sometimes all packets in a window are lost, in which case the destination cannot send any further acks since it receives no further packets. In such a case, TCP detects packet loss by using a pre-determined timer: if no acks are received over a time duration determined by the timer, TCP assumes that a loss event has occurred.

Given the above mechanism for packet-loss detection, TCP-Reno's window adaptation algorithm can be divided into two distinct phases, depending upon whether the current window size W is below or above a threshold, called *ssthresh*.

- **Slow-start phase** When the window size is below ssthresh, it is increased by one for every ack received. Once the window size reaches ssthresh, TCP-Reno switches to the congestion-avoidance phase.
- **Congestion-avoidance phase** In this phase, the window size is increased much more slowly than the slow-start phase, when there is no congestion indication. More precisely, W is set to $W + 1/W$ for every non-duplicate ack received. When packet loss is detected by the reception of three dupacks, the window size is halved, i.e., $W \leftarrow W/2$, and the algorithm continues in the congestion-avoidance phase. However, if packet loss is detected by timer expiry, ssthresh is set to $W/2$, the window size W is then set to 1, and TCP-Reno goes back to the slow-start phase. Thus, timer expiry is treated as a more serious loss event than three dupacks.

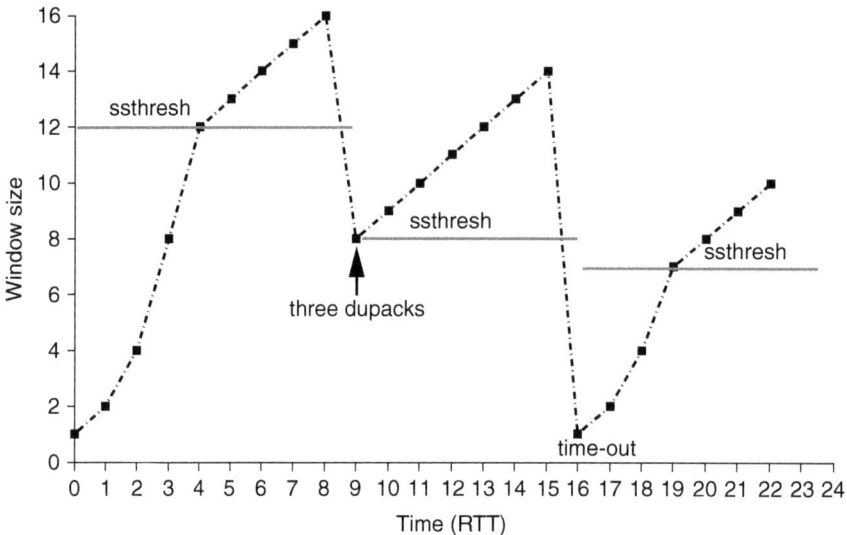

Figure 7.2 Operation of TCP-Reno. The window size exponentially increases during the slow-start phase until reaching ssthresh (= 12). Then TCP-Reno enters the congestion-avoidance phase and the window size increases linearly. When a loss is detected by receiving three dupacks at the 9th RTT, the ssthresh is set to be 8, and the window size is set to 8 and increases linearly. When a time-out occurs at the 16th RTT, the ssthresh is set to be 7. The window size is set to be 1, and TCP-Reno enters the slow-start phase.

We illustrate the operation of TCP-Reno in Figure 7.2. A few comments are in order.

- The initial value of ssthresh is chosen differently by different versions of TCP; before a TCP connection is set up, the source and destination exchange packets to let each other know that a TCP file transfer is going to be initiated. During this process, the sender may be able to determine the maximum rate at which the receiver can receive data and may choose ssthresh so that the transmission rate determined by the window size is below the reception capacity of the destination.
- Recall that the window size is increased by $1/W$ for each non-dupack in the congestion-avoidance phase. Roughly speaking, this means that the source has to receive W acks to increase the window size by 1. Again, roughly speaking, it takes one RTT to receive a whole window's worth of acks. Thus, the window size is approximately increased by 1 once every RTT. This means that the transmission rate roughly increases linearly as a function of time (measured in units of RTT) as long as no loss is detected. Upon loss detection through dupacks, the rate is decreased by one-half. Since the rate is increased additively and decreased multiplicatively, this type of rate adaptation mechanism is called an *additive increase–multiplicative decrease*, or AIMD, algorithm.
- The term "slow-start phase" may appear to be incorrect since the window size is increased more rapidly in this phase when compared to the congestion-avoidance phase. However, the term "slow start" refers to the fact that window size (and the rate at which it is increased) is small during this phase.

We will now develop a differential equation model to describe the dynamics of TCP Reno's congestion control algorithm. The slow-start phase of a flow is relatively insignificant if the flow consists of a large number of packets. So we will consider only the congestion-avoidance phase. Consider a flow r, and let $W_r(t)$ denote its window size and T_r its RTT. The RTT is a combination of propagation delay and queueing delay, but we ignore the fluctuations in queueing delay and assume that the RTT is a constant. In Section 2.4, we used the notation $q_r(t)$ to denote the price at route r. TCP uses packet loss probability as the price of a route. So we use the same notation $q_r(t)$ to denote the rate at which packets sent from the source at time $t - T_r$ are lost, which is the congestion feedback information received by the source at time t. Recalling that

$$x_r(t) = \frac{W_r(t)}{T_r},$$

we can model the congestion-avoidance phase of TCP-Reno as

$$\dot{W}_r(t) = \frac{x_r(t - T_r)(1 - q_r(t))}{W_r(t)} - \beta x_r(t - T_r)q_r(t)W_r(t). \tag{7.1}$$

The above equation can be derived as follows.

- The rate at which the source obtains ack is $x_r(t - T_r)(1 - q_r(t))$. Because each ack leads to an increase by $1/W_r(t)$, the rate at which the window size increases is given by the first term on the right-hand side.
- The rate at which packets are lost is given by $x_r(t - T_r)q_r(t)$. Such events would cause the window size to be decreased by a factor that we call β. This is the second term on the right-hand side. Considering the fact that there is a halving of window size due to loss of packets, β would naturally be taken to be $1/2$. However, studies show that a more precise value of β when making a continuous-time approximation of the TCP behavior is close to $2/3$.

To compare the TCP formulation above to the resource allocation framework, we write $W_r(t)$ in terms of $x_r(t)$, i.e. ($W_r(t) = x_r(t)T_r$), which yields

$$\dot{x}_r = \frac{x_r(t - T_r)(1 - q_r(t))}{T_r^2 x_r(t)} - \beta x_r(t - T_r)q_r(t)x_r(t). \tag{7.2}$$

The equilibrium value of x_r is found by setting $\dot{x}_r = 0$, and is seen to be

$$\hat{x}_r = \sqrt{\frac{1 - \hat{q}_r}{\beta \hat{q}_r}} \frac{1}{T_r},$$

where $\hat{q}_r$ is the equilibrium loss probability. For small values of $\hat{q}_r$ (which is what one desires in the Internet),

$$\hat{x}_r \propto 1/(T_r \sqrt{\hat{q}_r});$$

i.e., the equilibrium rate of TCP rate is inversely proportional to the RTT and the square-root of the loss probability. This result is well known and widely used in the performance analysis of TCP-Reno.

Relationship with primal congestion control algorithm

Consider the controller (7.2) again. Suppose that there were no feedback delay, but the equation is otherwise unchanged. So now T_r^2 that appears in (7.2) is just some constant.

Also, let $q_r(t)$ be small, i.e., the probability of losing a packet is not too large. Then the controller reduces to

$$\dot{x}_r = \frac{1}{T_r^2} - \beta x_r^2 q_r$$

$$= \beta x_r^2 \left(\frac{1}{\beta T_r^2 x_r^2} - q_r \right).$$

Comparing the above equation with the primal congestion controller (2.23), we find that the utility function of source r satisfies

$$U_r'(x_r) = \frac{1}{\beta T_r^2 x_r^2}.$$

We can find the source utility (up to an additive constant) by integrating the above equation, which yields

$$U_r(x_r) = -\frac{1}{\beta T_r^2 x_r}.$$

Thus, TCP-Reno can be approximately viewed as a control algorithm that attempts to achieve weighted minimum potential delay fairness.

If we do not assume that q_r is small, the delay-free differential equation is given by

$$\dot{x}_r = \frac{1 - q_r}{T_r^2} - \beta x_r^2 q_r$$

$$= (\beta x_r^2 + 1/T_r^2) \left(\frac{1}{\beta x_r^2 + \frac{1}{T_r^2}} \frac{1}{T_r^2} - q_r \right).$$

Thus,

$$U_r'(x_r) = \frac{1}{T_r^2} \frac{1}{\beta x_r^2 + \frac{1}{T_r^2}} \qquad \Rightarrow \qquad U_r(x_r) = \frac{1}{T_r \sqrt{\beta}} \tan^{-1} \left(\sqrt{\beta}\, T_r x_r \right),$$

where the utility function is determined up to an additive constant.

7.1.2 TCP-Reno with feedback delay

Our analysis of the stability and convergence of congestion control algorithms did not consider delays in the feedback path. However, feedback delays can be quite important in determining the stability of an algorithm. To understand the impact of feedback delays, it is useful to consider the time scales involved in a typical long-distance file transfer across the continental USA. The maximum RTT in the continental USA is roughly the amount of time that a packet may take to travel from the east to the west coast plus the time that it takes for an ack to return to the east coast. Approximating the one-way distance by 3000 miles, ignoring queueing and packet processing delays in the network, the RTT is the amount of time that it takes for light to travel 6000 miles. This is approximately 30 ms. On the other hand, link speeds in the Internet can be very high, from several Gbps (gigabits per second) to Tbps (terabits per second). Assuming a link operates at 1Tbps, and considering that a typical IP packet is 1500 bytes or 12 000 bits long, the amount of time it takes for a packet to be processed at a link is $0.12\,\mu s$. Thus, packet level phenomena occur at much faster time scales compared to the RTT, and that is why a delay differential equation model to study stability is important, even though an RTT of $32\,ms$ may appear to be very small based on our daily experience.

To analyze the stability of the congestion control algorithm, consider a simple model in which there is only source and one link in the network. Dropping the subscript r, (7.2) can be rewritten as

$$\dot{x} = x(t - T) \left(\frac{(1 - q(t))}{T^2 x(t)} - \frac{q(t)x(t - T)}{2} \right), \tag{7.3}$$

where we have taken β to be $1/2$. If we can express $q(t)$ as a function of $x(t)$, the above equation is a delay differential equation for $x(t)$. Recall that $q(t)$ is the loss rate experienced by packets that were sent at time $t - T$ from the source. Thus,

$$q(t) = f(x(t - T)),$$

where $f(y)$ is a function that determines the loss rate as a function of the arrival rate y. In Exercise 7.1, the interested reader is asked to linearize (7.3) and derive conditions under which the linear delay differential equation is asymptotically stable, based on Theorem 2.6.1.

7.1.3   TCP-Vegas: a delay-based algorithm

We now consider another variation of TCP called TCP-Vegas. TCP-Vegas uses queueing delay, instead of packet loss, to infer congestion in the network. To understand the idea behind TCP-Vegas, we first note that RTT is the sum of propagation delay and queueing delay. If one can estimate the propagation delay, congestion control can be implemented as follows: if the RTT is significantly larger than the propagation delay, decrease the transmission rate; otherwise, increase the transmission rate. We next describe the dynamics of TCP-Vegas' window evolution more precisely.

TCP-Vegas uses a very simple scheme to estimate the propagation delay T_p: the current estimate of T_p is the smallest RTT seen by a packet so far. The idea here is that occasionally packets will see empty queues, and such packets will experience only propagation delay and no queueing delay. Thus, the smallest RTT seen so far is a reasonable estimate of T_p. If there is no queueing delay, the expected throughput would be approximately given by

$$e = \frac{W}{T_p},$$

where W is the window size. TCP-Vegas compares this expected throughput to the current estimate of the actual throughput a. The actual throughput is estimated as follows: the source estimates the current RTT by sending a marked packet and recording the amount of time it takes to receive the ack back for the packet. This RTT will be the sum of T_p and the queueing delay T_q. The source also records the number of acks S it receives during this RTT. Then, a is estimated as

$$a = \frac{S}{T_p + T_q}.$$

When an ack is received for a marked packet at time t, window adaptation is performed based on the current estimates of the expected and actual throughputs, $a(t)$ and $e(t)$, respectively. TCP-Vegas uses two constant parameters, α and β, with $\alpha \le \beta$, and proceeds as follows:

- if $\alpha \le (e(t) - a(t)) \le \beta$, the window size is left unchanged as the expected and actual throughputs are reasonably close;
- if $(e(t) - a(t)) > \beta$, decrease the window size by 1 for the next RTT because the actual throughput is too small compared to the expected throughput due to the fact that the queueing delay is too large;
- if $(e(t) - a(t)) < \alpha$, increase the window size by 1 for the next RTT.

We note that TCP-Vegas also uses a slow-start phase, similar to that used by TCP-Reno. The behavior of TCP-Vegas under ideal conditions is shown in Figure 7.3.

Next, we interpret TCP-Vegas as a resource allocation algorithm in the utility maximization framework. We assume $\alpha = \beta$ and that the propagation delay is estimated accurately; i.e., for source r, $T_{pr}(t) \equiv T_{pr}$ for all t. At equilibrium, the estimated throughput is the same as the actual throughput, with the window size and the number of acks received in an RTT being the same. If we denote the equilibrium window size and queueing delay of source r by $\hat{W}_r$ and $\hat{T}_{qr}$, respectively (and, by assumption, the propagation delay T_{pr} is correctly estimated), we have

$$\frac{\hat{W}_r}{T_{pr}} - \frac{\hat{W}_r}{T_{pr} + \hat{T}_{qr}} = \alpha. \tag{7.4}$$

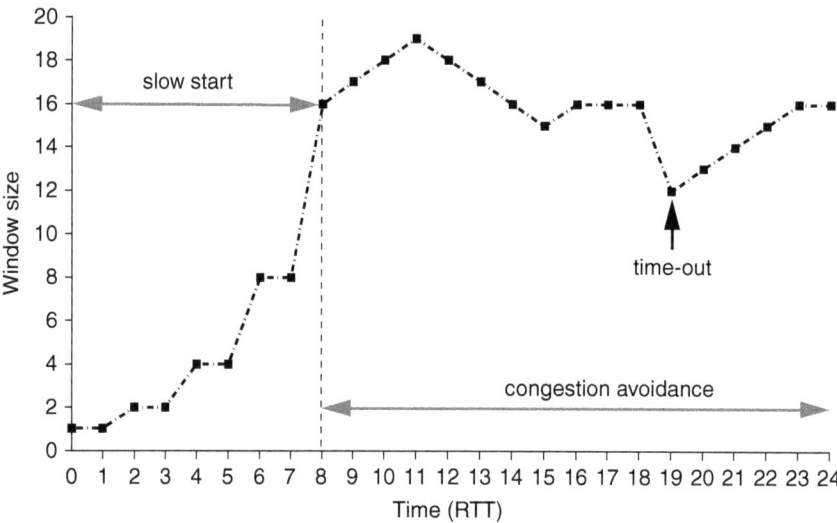

Figure 7.3 Operation of TCP-Vegas. It preserves the slow-start phase, but doubles the window size every other RTT for an accurate comparison between the actual throughput and the expected throughput. After switching to congestion-avoidance phase, the window size increases and decreases linearly. Ideally, the algorithm should converge to a stable window size (e.g., during the 17th and 18th RTTs).

At equilibrium, the transmission rate $\hat{x}_r$ is approximately given by

$$\hat{x}_r = \hat{W}_r/(T_{pr} + \hat{T}_{qr}),$$

which means that (7.4) can be simplified to

$$\frac{\alpha T_{pr}}{\hat{x}_r} = \hat{T}_{qr}. \qquad (7.5)$$

Now that we know what the equilibrium transmission rate looks like, let us study what the equilibrium queueing delay $\hat{T}_{qr}$ would look like. If the queue length at equilibrium at link l is denoted by $\hat{b}_l$, the equilibrium queueing delay at link l is $\hat{b}_l/c_l$ (where c_l is the capacity of link l). So we have

$$\hat{T}_{qr} = \sum_{l:l\in r} \frac{\hat{b}_l}{c_l}. \qquad (7.6)$$

Also, if

$$\sum_{k:l\in k} \hat{x}_k < c_l,$$

there is no queueing delay, i.e., $\hat{b}_l = 0$ in this case since we are using a fluid model. Note that, because the aggregate equilibrium transmission rate of all flows using link l cannot exceed the link capacity, we cannot possibly have

$$\sum_{k:l\in k} \hat{x}_k > c_l.$$

Thus, if $\hat{b}_l \neq 0$, it means that

$$\sum_{k:l\in k} \hat{x}_k = c_l.$$

Thus, we have, from the above and (7.5), the following equilibrium conditions:

$$\frac{\alpha T_{pr}}{\hat{x}_r} = \sum_{l:l\in r} \frac{\hat{b}_l}{c_l},$$

with

$$\frac{\hat{b}_l}{c_l}\left(\sum_{k:l\in k} \hat{x}_k - c_l\right) = 0 \qquad \forall l.$$

These are the KKT conditions for the utility maximization problem

$$\max_{\{x_r\}} \alpha T_{pr} \log x_r, \qquad (7.7)$$

subject to

$$\sum_{r:l\in r} x_r \leq c_l, \qquad \forall l,$$

$$x_r \geq 0, \qquad \forall r,$$

with $\hat{b}_l/c_l$ as the Lagrange multipliers. Thus, assuming it converges to some equilibrium rates, TCP-Vegas is weighted proportionally fair. If we let each flow have a different value of α, i.e., we associate α_r with route r, the equilibrium rates will maximize $\sum_r \alpha_r T_{pr} \log x_r$. Recall that we have assumed that $\alpha_r = \beta_r$ in this analysis.

Relationship to dual algorithms and extensions

We now discuss the relationship between TCP-Vegas and dual algorithms. From (2.36) and (2.37), the dual congestion control algorithm with utility function $w_r \log(x_r)$ for user r is given by

$$x_r = \frac{w_r}{q_r} \qquad (7.8)$$

and

$$\dot{p}_l = h_l (y_l - c_l)_{p_l}^+. \qquad (7.9)$$

To avoid confusion, we note that w_r is the weight assigned to source r and is unrelated to the window size W_r. Letting b_l denote the queue length at link l,

$$\dot{b}_l = (y_l - c_l)_{b_l}^+.$$

So,

$$\frac{\dot{b}_l}{c_l} = \frac{1}{c_l}(y_l - c_l)_{b_l}^+,$$

where b_l / c_l is the queueing delay experienced by packets using link l. Therefore, if we choose $h_l = 1/c_l$, the price function of a link becomes the queueing delay experienced by packets using that link, which, when added to a constant propagation delay, is the feedback that is used in the TCP-Vegas algorithm.

Next, let us consider the source rates achieved under TCP-Vegas so that we can compare them to the dynamics of the dual algorithm. From the algorithm description (with $\alpha = \beta$), TCP-Vegas updates its window size W_r based on whether

$$\frac{W_r}{T_{pr}} - \frac{W_r}{T_{pr} + T_{qr}} < \alpha \qquad \text{or} \qquad \frac{W_r}{T_{pr}} - \frac{W_r}{T_{pr} + T_{qr}} > \alpha. \qquad (7.10)$$

Using the approximation $x_r = \frac{W_r}{T_{pr} + T_{qr}}$, we can rewrite the conditions as

$$x_r T_{qr} < \alpha T_{pr} \qquad \text{or} \qquad x_r T_{qr} > \alpha T_{pr}. \qquad (7.11)$$

As in (7.6), we also have

$$T_{qr} = \sum_{l:l \in r} \frac{b_l}{c_l} = \sum_{l:l \in r} p_l, \qquad (7.12)$$

where b_l is the queue length at link l, and we have used p_l to denote the queueing delay at link l, which acts as the price function for TCP-Vegas. Combining the above expressions, the condition for increase/decrease becomes

$$x_r \sum_{l:l \in r} p_l < \alpha T_{pr} \qquad \text{or} \qquad x_r \sum_{l:l \in r} p_l > \alpha T_{pr}. \qquad (7.13)$$

Hence, the window control algorithm can be written as

$$W_r \leftarrow \left[W_r + \frac{1}{T_{pr} + T_{qr}} \operatorname{sgn}\left(\alpha T_{pr} - x_r T_{qr}\right) \right]_{W_r}^+, \qquad (7.14)$$

where $\mathrm{sgn}(z) = -1$ if $z < 0$, $\mathrm{sgn}(z) = 1$ if $z > 0$, and $\mathrm{sgn}(z) = 0$ if $z = 0$. Thus, we can now write down the differential equations describing TCP-Vegas as follows:

$$\dot{p}_l = \frac{1}{c_l}(y_l - c_l)^+_{p_l}, \tag{7.15}$$

$$\dot{W}_r = \left[\frac{1}{T_{pr} + T_{qr}} \, \mathrm{sgn}\left(\alpha T_{pr} - x_r \, T_{qr} \right) \right]^+_{W_r}, \tag{7.16}$$

$$x_r = \frac{W_r}{T_{pr} + T_{qr}}, \tag{7.17}$$

with $T_{qr} = \sum_{l:l \in r} p_l$. The above is not the same as the dual algorithm that we derived in Section 2.5. However, the price update dynamics are the same as those of the dual algorithm. Further, at the source, by attempting to increase or decrease the rate based on whether x_r is less than or greater than $\alpha T_{pr}/T_{qr}$, it is clear the source attempts to drive the system towards

$$x_r = \frac{\alpha T_{pr}}{T_{qr}} = \frac{\alpha T_{pr}}{\sum_{l \in r} p_l},$$

which is the desired source behavior for a dual congestion controller (as in (7.8)). Thus, the equilibrium point of TCP-Vegas is the same as that of the dual algorithm, and one can interpret TCP-Vegas as an algorithm that approximates the dual congestion control algorithm.

A modification of TCP-Vegas called FAST-TCP has been suggested for very high-speed networks. In FAST-TCP, the window size is increased or decreased depending upon how far the window size is from a desired equilibrium point. The fluid model describing the protocol is given by

$$\dot{p}_l = \frac{1}{c_l}(y_l - c_l)^+_{p_l}, \tag{7.18}$$

$$\dot{W}_r = \gamma_r \left(\alpha_r - x_r \, T_{qr} \right)^+_{W_r}, \tag{7.19}$$

$$x_r = \frac{W_r}{T_{pr} + T_{qr}}, \tag{7.20}$$

where α_r determines the desired equilibrium point and γ_r is a scaling constant. Replacing the sgn function in TCP-Vegas with the difference between the current operating point and the desired equilibrium allows FAST-TCP to approach the desired equilibrium point rapidly.

7.2 Routing algorithms: Dijkstra and Bellman–Ford algorithms

In Section 7.1 and Chapter 2, we assumed that routes from sources to their destinations are given. In the Internet, these routes are determined by routing algorithms, whose goal is to find "good" paths (routes) for the traffic flows in the network.

Consider a network represented by a directed graph $\mathcal{G} = (\mathcal{N}, \mathcal{L})$, where $\mathcal{N}$ is the set of nodes and $\mathcal{L}$ is the set of links. Let (i, j) denote the link from node i to node j and c_{ij}

the cost associated with link (i,j); we assume $c_{ij} > 0$. Note that the algorithm respects direction, the cost of the link from i to j can be different from the cost of the link from j to i. A common objective of routing algorithms is to find the route with the minimum cost from a source to its destination, where the cost of a route is simply the sum of the costs of the links in the route. For example, if $c_{ij} = 1$ for all links, the routing algorithm simply finds the shortest path from source to destination. Other choices of c_{ij} may reflect some preference among links; for example, if links with higher capacity are preferred, their costs may be chosen to be small to reflect this preference. In this section, we introduce two widely used minimum cost routing algorithms: Dijkstra's algorithm and the Bellman–Ford algorithm.

7.2.1   Dijkstra's algorithm: link-state routing

Dijkstra's algorithm is a link-state routing algorithm, where each node knows the states of all links in the network, i.e., the complete network topology and the costs of all links. Based on global link states, a node iteratively computes the minimum cost paths from itself to all other nodes in the network.

In Dijkstra's algorithm, each node (say node u) maintains a node set $\mathcal{K}$, which represents the set of nodes to which u knows the minimum cost paths, and two other pieces of information – ω_{ui} and p_{ui} – for every node (say node i) in the network, where ω_{ui} is the current minimum cost from node u to node i, and p_{ui} is the previous hop to node i on the current minimum cost path from u to i.

At each iteration, the algorithm selects the node with the smallest ω_{ui} among those not in set $\mathcal{K}$, and adds the selected node to the set. Then, the algorithm updates ω_{ui} and p_{ui} for

---

Algorithm 7 Dijkstra's algorithm at node u

1: Input: c_{ij} for all $(i, j) \in \mathcal{L}$.

2: Set $\mathcal{K} = \{u\}$,

3: **for** $i \in \mathcal{N}$ **do**

4:     Set $\omega_{ui} = c_{ui}$ if $(u, i) \in \mathcal{L}$, i.e., node i is a neighbor of node u, and $\omega_{ui} = \infty$ otherwise.

5:     Set $p_{ui} = u$ if $(u, i) \in \mathcal{L}$, i.e., node i is a neighbor of node u, and $p_{ui} = -1$ otherwise, where -1 indicates that the previous hop is unknown.

6: **end for**

7: **while** $\mathcal{K} \neq \mathcal{N}$ **do**

8:     Find node i^* such that $i^* \in \text{argmin}_{i \notin \mathcal{K}} \omega_{ui}$. Ties are broken arbitrarily.

9:     Set $\mathcal{K} = \mathcal{K} \cup \{i^*\}$.

10:     **for** $i \notin \mathcal{K}$ **do**

11:         **if** $\omega_{ui} > \omega_{ui^*} + c_{i^*i}$ **then**

12:             Set $\omega_{ui} = \omega_{ui^*} + c_{i^*i}$ and $p_{ui} = i^*$.

13:         **end if**

14:     **end for**

15: **end while**

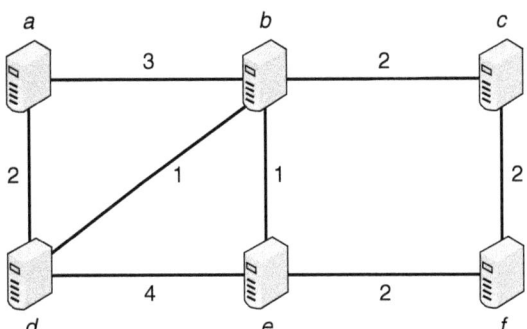

Figure 7.4 A simple network used for illustrating Dijkstra's algorithm.

the rest of the nodes that are not in $\mathcal{K}$ by checking whether going through the selected node reduces their path costs. A detailed description of the algorithm is presented in Algorithm 7.

Example 7.2.1

We illustrate Dijkstra's algorithm for the simple network shown in Figure 7.4. All links of this network are bi-directional links, i.e., $c_{ij} = c_{ji}$ for all i and j. The numbers in the figure are the link costs. Table 7.1 summarizes the result of Dijkstra's algorithm (at node a) at the end of each iteration.

Next, we elaborate on the computation process at the second iteration.

(1) After the first iteration, we have $\mathcal{K} = \{a, d\}$, $(\omega_{ab}, p_{ab}) = (3, a)$, $(\omega_{ac}, p_{ac}) = (\infty, -1)$, $(\omega_{ae}, p_{ae}) = (6, d)$, and $(\omega_{af}, p_{af}) = (\infty, -1)$.

(2) Line 8 of Algorithm 7: the algorithm first selects the node with the smallest (ω_{ai}) among possible node i not in $\mathcal{K}$, i.e., among nodes b, c, e, and f, which is node b.

(3) Line 9 of Algorithm 7: the algorithm adds node b to set $\mathcal{K}$, so $\mathcal{K}$ becomes $\{a, d, b\}$.

(4) Lines 11–12 of Algorithm 7: for the rest of nodes not in $\mathcal{K}$, i.e., nodes c, e, and f, the algorithm checks whether the path costs to these nodes can be reduced by going through node b.

- The path cost to node c via node b is 5, which is smaller than the current cost $\omega_{ac} = \infty$, so ω_{ac} is updated to 5 and p_{ac} is updated to b.
- The path cost to node e via node b is 4, which is smaller than the current cost $\omega_{ae} = 6$, so ω_{ae} is updated to 4 and p_{ae} is updated to b.
- Node f is not an immediate neighbor of any node in $\mathcal{K}$, so node f is still viewed as unreachable from node a. The cost and previous hop remain unchanged.

Once the algorithm has terminated, we can find the path from a to every other node by backtracking through the final values of the p_{ai}'s. For example, to find the path from node a to node f, we note that $p_{af} = e$, then $p_{ae} = b$, and $p_{ab} = a$. Thus, the minimum cost path from a to f is given by $a \rightarrow b \rightarrow e \rightarrow f$. The cost of this path is given by the final value of ω_{af}, which is equal to 6. □

Table 7.1 The per-iteration results of Dijkstra's algorithm at node a for the network in Figure 7.4

Iteration	$\mathcal{K}$	(ω_{ab}, p_{ab})	(ω_{ac}, p_{ac})	(ω_{ad}, p_{ad})	(ω_{ae}, p_{ae})	(ω_{af}, p_{af})
0	$\{a\}$	$(3, a)$	$(\infty, -1)$	$(2, a)$	$(\infty, -1)$	$(\infty, -1)$
1	$\{a, d\}$	$(3, a)$	$(\infty, -1)$	$\cdot$	$(6, d)$	$(\infty, -1)$
2	$\{a, d, b\}$	$\cdot$	$(5, b)$	$\cdot$	$(4, b)$	$(\infty, -1)$
3	$\{a, d, b, e\}$	$\cdot$	$(5, b)$	$\cdot$	$\cdot$	$(6, e)$
4	$\{a, d, b, e, c\}$	$\cdot$	$\cdot$	$\cdot$	$\cdot$	$(6, e)$
5	$\{a, d, b, e, c, f\}$	$\cdot$	$\cdot$	$\cdot$	$\cdot$	$\cdot$

Key properties of Dijkstra's algorithm

At any iteration, the set $\mathcal{K}$ under Dijkstra's algorithm is the set of nodes to which the minimum cost paths have already been determined. Denoting by ω_{ui}^* the minimum cost from node u to node i, similarly p_{ui}^*, we summarize the key properties of Dijkstra's algorithm in the following theorem.

Theorem 7.2.1 Dijkstra's algorithm described in Algorithm 7 satisfies the following properties after the kth iteration:

- Property 1: Set $\mathcal{K}$ contains $k + 1$ nodes, i.e., $|\mathcal{K}| = k + 1$.
- Property 2: For each $i \in \mathcal{K}$, $\omega_{ui} = \omega_{ui}^*$, i.e., the minimum cost from node u to node i is obtained.
- Property 3: For any two nodes i and j, such that $i \in \mathcal{K}$ and $j \notin \mathcal{K}$, $\omega_{ui}^* \leq \omega_{uj}^*$, i.e., the minimum cost to a node outside of set $\mathcal{K}$ is higher than the minimum cost to a node in set $\mathcal{K}$.

Proof We prove the theorem by induction. In the proof, we will consider the cost ω_{ui} at different iterations. For ease of understanding, we use superscript k to indicate the iteration, e.g., we denote by ω_{ui}^k the value of ω_{ui} after the kth iteration.

At iteration 0, it is trivial to prove the properties because $\mathcal{K}^0 = \{u\}$ and $\omega_{uu}^0 = \omega_{uu}^* = 0$.

Now suppose that, after the $(k-1)$th iteration, the three properties stated in the theorem hold. We then consider the computation at the kth iteration.

As one and at most one node is added into $\mathcal{K}$ during each iteration, property 1 holds after the kth iteration.

Now consider the set of nodes $\mathcal{S}$ such that

$$\mathcal{S} = \left\{ j : j \notin \mathcal{K}^{k-1} \text{ and } \omega_{uj}^* = \min_{l:l \notin \mathcal{K}^{k-1}} \omega_{ul}^* \right\}. \tag{7.21}$$

This is the set of nodes that have the smallest true minimum cost among those not in $\mathcal{K}^{k-1}$, e.g., in the second iteration of the example, $\mathcal{S} = \{b\}$.

Now, for any node $j \in \mathcal{S}$, the previous hop to node j on a minimum cost path, i.e., p_{uj}^*, must be a node in $\mathcal{K}^{k-1}$. (For example, in the second iteration of the example,

$p_{ab}^* = a \in \{a, b\}$.) Otherwise, the true minimum cost to node p_{uj}^* is smaller than that to node $j (\omega_{uj}^* = \omega_{up_{uj}^*}^* + c_{p_{uj}^* j} > \omega_{up_{uj}^*}^*)$, which contradicts the fact that node j is in set $\mathcal{S}$.

Therefore, for any node $j \in \mathcal{S}$, we have

$$\omega_{uj}^* \leq \omega_{uj}^{k-1} \leq \min_{i \in \mathcal{K}^{k-1}} \left(\omega_{ui}^* + c_{ij} \right) \leq \omega_{up_{uj}^*} + c_{p_{uj}^* j} = \omega_{uj}^*,$$

where the second inequality holds due to the computation in lines 11–13 of Dijkstra's algorithm and the property that $\omega_{ui} = \omega_{ui}^*$ when node i is added into set $\mathcal{K}^{k-1}$. We then conclude, for any $j \in \mathcal{S}$ and $l \notin \mathcal{K}^{k-1} \cup \mathcal{S}$,

$$\omega_{uj}^{k-1} = \omega_{uj}^* < \omega_{ul}^* \leq \omega_{ul}^{k-1}. \tag{7.22}$$

In other words, after the $(k-1)$th iteration, the current minimum costs to those nodes in set $\mathcal{S}$ are the true minimum costs, which are smaller than the current minimum costs to those nodes that are not in $\mathcal{S}$. For instance, after the first iteration of the example,

$$\omega_{ab}^1 = 3 = \omega_{ab}^* < \omega_{ae}^1 = 6 < \omega_{ac}^1 = \omega_{af}^1 = \infty.$$

Hence, at the kth iteration, the algorithm selects a node in set $\mathcal{S}$. So properties 2 and 3 hold according to inequality (7.22), and the theorem follows from the induction principle. $\qquad \square$

The following proposition is an immediate result of Theorem 7.2.1.

Proposition 7.2.2 In a network with N nodes, Dijkstra's algorithm terminates after $N-1$ iterations. Consider node u and denote by n_k the node that is added into set $\mathcal{K}$ at the kth iteration of Dijkstra's algorithm at node u. Then

$$\omega_{un_0}^* \leq \omega_{un_1}^* \leq \cdots \leq \omega_{un_{N-1}}^*,$$

where $n_0 = u$. $\qquad \square$

We remark that Dijkstra's algorithm requires each node to know the states of all links. If there is a link failure, and some nodes are unaware of the failure, it could lead to loops, as shown in Exercise 7.6.

7.2.2   Bellman–Ford algorithm: distance-vector routing

The advantage of Dijkstra's algorithm is that it allows a node to compute minimum cost paths to every other node in the network. However, the algorithm requires global knowledge of network topology and link costs. The Bellman–Ford algorithm takes a different approach, in which nodes only communicate with immediate neighbors and iteratively compute the minimum cost to a destination and the next hop to the destination. The Bellman–Ford algorithm can be implemented in a distributed and asynchronous fashion. But, instead of knowing the complete minimum cost path to a destination, a node only knows the next hop to the destination via a minimum cost path.

Recall that ω_{ui}^* denotes the minimum cost from node u to node i. The Bellman–Ford algorithm is based on the following recursive equation, called the Bellman–Ford equation:

$$\omega_{ui}^* = \min_{j:(u,j)\in\mathcal{L}} \left(c_{uj} + \omega_{ji}^* \right). \tag{7.23}$$

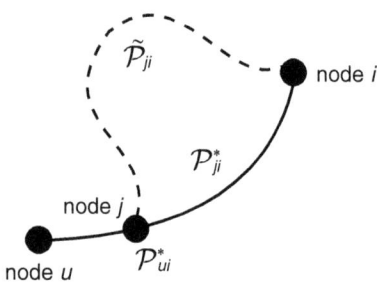

Figure 7.5 The solid line from node u to node i is a minimum cost path. The Bellman–Ford equation states that the solid line from node j to node i must be a minimum cost path from node j to node i. Suppose this is not true and there exists another path, say the dashed line, with a lower cost. Then the path that consists of the solid line from u to j and the dashed line from node j to node i has a lower cost than the solid line path from u to i, which contradicts the fact that the solid line from node u to node i is a minimum cost path.

The Bellman–Ford equation states that if a minimum cost path from node u to node i passes through node j, who is an immediate neighbor of node u, the part of the path from node j to node i is a minimum cost path from node j to node i. A pictorial explanation is given in Figure 7.5.

Therefore, a node can determine the minimum cost and the neighbor on a minimum cost path by finding the smallest $c_{uj} + \omega_{ji}^*$, the sum of link cost c_{uj} and the minimum cost from node j to node i. For example, consider nodes d and c in Figure 7.4.

Based on the Bellman–Ford equation, the Bellman–Ford algorithm requires each node (say node u) to maintain two pieces of information about every other node (say node i): (ω_{ui}, n_{ui}), where ω_{ui} is the current estimate of the minimum cost from node u to node i, and n_{ui} is the next hop on the current minimum cost path to destination i. The vector of these two pieces of information maintained by node i is called the *distance vector* at node i. Node u then periodically queries the distance vectors from its neighbors, and then updates ω_{ui} and n_{ui} based on the received information. So the Bellman–Ford algorithm is called a distance-vector routing. A detailed description of the algorithm is presented in Algorithm 8.

We note that Algorithm 8 is a synchronized Bellman–Ford algorithm where nodes first exchange their information (line 9) and then update minimum costs and next hops based on received information. The Bellman–Ford algorithm, however, can be implemented asynchronously, where nodes update their neighbors asynchronously and each node immediately updates its minimum cost and next-hop information upon receiving an update. If there is a change, the node notifies all neighbors immediately.

Example 7.2.2

We again use the network in Figure 7.4 to illustrate the computation process of the Bellman–Ford algorithm. For simplicity, we assume that nodes are interested in knowing minimum cost paths to node c only. Each node only maintains $\omega_{\cdot c}$ and $n_{\cdot c}$. The result of each iteration of the Bellman–Ford algorithm is summarized in Table 7.2.

We further elaborate on the computation performed by node e at the second iteration.

(1) After the first iteration, $\omega_{ec} = -\infty$ and $n_{ec} = -1$.
(2) Lines 8–10 of Algorithm 8: node e receives $(\omega_{bc}, \omega_{dc}, \omega_{fc}) = (2, \infty, 2)$.
(3) Line 13 of Algorithm 8: node e compares $\omega_{ec} = \infty$ and

$$\min_{j \in \{b,d,f\}} (c_{ej} + \omega_{jc}) = \min\{3, \infty, 4\} = 3.$$

(4) Line 14 of Algorithm 8: because $\infty > 3$, node e sets $\omega_{ec} = 3$ and $n_{ec} = b$. □

Table 7.2 The computation process of the Bellman–Ford algorithm for the network in Figure 7.4

Iteration	(ω_{ac}, n_{ac})	(ω_{bc}, n_{bc})	(ω_{cc}, n_{cc})	(ω_{dc}, n_{dc})	(ω_{ec}, n_{ec})	(ω_{fc}, n_{fc})
0	$(\infty, -1)$	$(\infty, -1)$	$(0, c)$	$(\infty, -1)$	$(\infty, -1)$	$(\infty, -1)$
1	$(\infty, -1)$	$(2, c)$	$(0, c)$	$(\infty, -1)$	$(\infty, -1)$	$(2, c)$
2	$(5, b)$	$(2, c)$	$(0, c)$	$(3, b)$	$(3, b)$	$(2, c)$

Algorithm 8 Bellman–Ford algorithm

```
 1: for u ∈ 𝒩 do
 2:    for i ∈ 𝒩 \ {u} do
 3:       Set ω_ui = c_ui if (u, i) ∈ ℒ, i.e., node i is a neighbor of node u, and
          ω_ui = ∞ otherwise.
 4:       Set n_ui = i if (u, i) ∈ ℒ, i.e., node i is a neighbor of node u, and
          n_ui = −1 otherwise, where −1 indicates the next hop is unknown.
 5:    end for
 6: end for
 7: while t ≥ 0 do
 8:    for u ∈ 𝒩 do
 9:       Node u sends out its ω_u = [ω_uj]_{j∈𝒩} to all its neighbors if ω_u was
          updated during iteration t − 1.
10:    end for
11:    for u ∈ 𝒩 do
12:       for i ∈ 𝒩 \ {u} do
13:          if ω_ui ≠ min_{j:(u, j)∈ℒ} (c_uj + ω_ji) then
14:             ω_ui = min_{j:(u, j)∈ℒ} (c_uj + ω_ji)
15:             n_ui ∈ argmin_{j:(u, j)∈ℒ} (c_uj + ω_ji)
16:          end if
17:       end for
18:    end for
19: end while
```

Key properties of the Bellman–Ford algorithm

The following theorem summarizes the key properties of the Bellman–Ford algorithm.

Theorem 7.2.3 Assume that a minimum cost path from node u to node i consists of k hops. Node u obtains the true minimum cost and the next hop after the kth iteration, i.e., $\omega_{ui} = \omega_{ui}^*$ after the kth iteration. If the network has N nodes, the Bellman–Ford algorithm terminates after at most $N - 1$ iterations.

Proof We prove this theorem by induction. If a minimum cost path from node u to node i has only one hop, i.e., the direct link from u to i is the minimum cost path, $\omega_{ui}^* = c_{ui}$. Clearly, after one iteration, $\omega_{ui} = c_{ui} = \omega_{ui}^*$, which does not change after that.

Now suppose that after the $(k-1)$th iteration, node u obtains the minimum cost to node n if a minimum cost path from node u to node n has $k - 1$ hops.

We then consider node i, to which a minimum cost path from node u has k hops. We denote by j^* the first hop of the path. Note that j^* must be an immediate neighbor of node u. From the Bellman–Ford equation, we know that the part from node j^* to node i on the minimum cost path from node u to node i is a minimum cost path from node j^* to node i, which implies that there exists a minimum cost path from node j^* to node i that has $k - 1$ hops.

According to the induction assumption, at the end of the $(k-1)$th iteration, node j^* has obtained $\omega_{j^*i}^*$, which is sent to node u at the beginning of the kth iteration. Therefore, in the kth iteration, node u computes

$$\min\left\{c_{uj^*} + \omega_{j^*i}^*, \min_{j:j\neq j^*,(u,j)\in\mathcal{L}}\left(c_{uj} + \omega_{ji}\right)\right\} = \min\left\{\omega_{ui}^*, \min_{j:j\neq j^*,(u,j)\in\mathcal{L}} c_{uj} + \omega_{ji}\right\}$$

$$= \omega_{ui}^*,$$

where the first equality results from the definitions of ω_{ui}^* and j^*.

In a network with N nodes, any loop-free path consists of no more than $N - 1$ hops. So the theorem holds. □

Although the Bellman–Ford algorithm is a distributed and asynchronous algorithm, and does not require global knowledge, it has its own weaknesses. For example, the algorithm responds very slowly to link or node failures, and has the *counting-to-infinity* problem, i.e., it may take an infinite number of iterations for a node to find out about a link or node failure on its minimum cost path. See Exercise 7.7 at the end of this chapter. In practice, modifications to the basic algorithm are required to ensure correct operation in networks with changing topologies.

7.3   IP addressing and routing in the Internet

We conclude this chapter with a brief discussion of how nodes in the Internet are given addresses, and how packets in the Internet find their way to their destinations using these addresses.

7.3.1 IP addressing

The Internet today consists of hundreds of millions of hosts. The first challenge in routing is to assign addresses to these hosts so that they can be located. The Internet Protocol (IP) addressing scheme is a numerical method used to label all hosts that are connected to the Internet. Instead of labeling the hosts, IP addresses are assigned to network interfaces and used to identify network interfaces. If a host has multiple interfaces, e.g., both an Ethernet card and a wireless network card, then each network interface may be assigned an IP address.

In the current version of IP protocol – Internet Protocol version 4 (IPv4), an IP address consists of 32 bits. In general, an IP address is represented in *dot-decimal notation*, i.e., in the format of *A.B.C.D*, which consists of four integers that range from 0 to 255 and are separated by dots. In general, an IP address contains two parts: NetID and HostID. The NetID identifies the network and the HostID identifies a host located in the network. For example, 130.126.158.111 is an IP address where the first two integers (130.126.) form the NetID of a campus network belonging to the University of Illinois, and the last two integers (158.111) form the HostID representing a network interface that connects a computer to the campus network.

Classful IP addressing

At the early stages of IPv4, IP address space was divided into five primary classes – *A*, *B*, *C*, *D*, and *E*. Such an addressing scheme is called classful IP addressing. The classification is shown in Figure 7.6. This classful IP addressing basically defines three types of networks: Class A (large-size networks), Class B (medium-size networks), and Class C (small-size networks), and the maximum number of distinguishable IP addresses in each type of network. This addressing architecture soon proved non-scalable. A class C network can support at most $2^8 = 256$ addresses, which is too small for most organizations. So most organizations required a block of Class B addresses, which led to a quick depletion of Class B addresses.

Classless inter-domain routing protocol

The classful IP addressing architecture was later replaced by Classless Inter-Domain Routing (CIDR) protocol, which can support a NetID of arbitrary length. A *CIDR block*, which

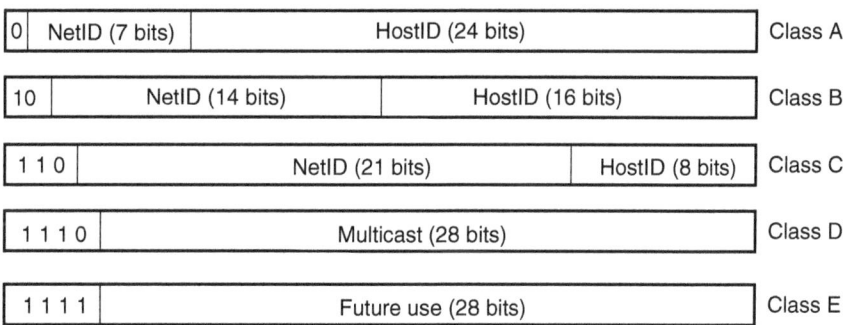

Figure 7.6 Classful IP addressing.

represents a range of IP addresses, is in the format of *A.B.C.D/X*, which consists of four integers ranging from 0 to 255, followed by a slash and a number ranging from 0 to 32. The last number, *X*, specifies the length of the NetID. For example, the previously mentioned campus network of the University of Illinois is associated with a CIDR block 130.126.0.0/16, which indicates that the first 16 bits are the NetID and the remaining 16 bits are the HostID.

The CIDR protocol does not permanently solve the problem of IPv4 address exhaustion as the size of the Internet explodes. A long-term solution is to replace IPv4 with a new version of IP protocol – IPv6 – which supports IP addresses that are 128 bits long, i.e., 2^{128} IP addresses instead of 2^{32} IP addresses.

7.3.2   Hierarchical routing

Each network interface is assigned an IP address when connected to the Internet. When a host wants to communicate with another host on the Internet, the source host needs to know the IP address associated with the destination host, and it inserts the address in the headers of the packets intended for the destination. Packets are routed by routers using routing tables. Specifically, when a router receives a packet, it checks the destination IP address of the packet contained in the packet header, and then looks up the routing table to locate the next router to which the packet should be forwarded. A routing table has at least two pieces of information fields: (i) destination and (ii) interface to the next-hop router. The destination field specifies the IP address of the destination, and the next-hop field specifies the interface to which the packet should be forwarded to reach the next-hop router, where the next-hop router is determined by the routing algorithm. Consider the network shown in Figure 7.7. A possible routing table on router A is shown in Table 7.3.

Given the size of today's Internet, it is not a scalable solution to maintain a separate entry for each host of the Internet in the routing table. Routers therefore are organized into *autonomous systems* (ASs), where each AS consists of a group of routers under the same administrative control, and is associated with one or more CIDR blocks. A packet is first routed to the destination AS using *inter-domain routing*, and then to the destination host in the AS using *intra-domain routing*. This routing mechanism is called *hierarchical routing*.

Under hierarchical routing, a router outside of an AS only needs to maintain a single entry for the AS, instead of one for each host in the AS. Only for those hosts that reside in

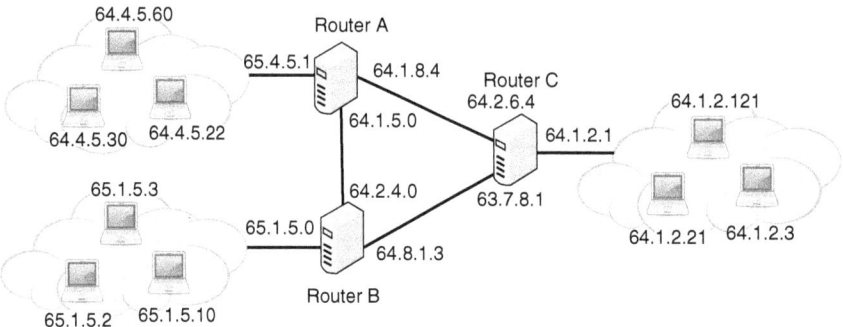

Figure 7.7 IP addressing and routing tables. Each router has three IP addresses, one for each network interface.

Table 7.3 Example of a routing table at router A

Destination	Interface
65.1.5.2	65.1.5.0
65.1.5.3	65.1.5.0
65.1.5.10	65.1.5.0
64.1.2.3	64.1.8.4
64.1.2.21	64.1.8.4
64.1.2.121	64.1.8.4

the same AS as the router, should the router have a separate entry for each of them. With the CIDR protocol, multiple ASs can also be combined into a *supernet* if they share a common prefix. For example, AS-1, with CIDR block 130.126.0.0/16, and AS-2, with CIDR block 130.127.0.0/16, can be combined into a supernet with CIDR block 130.126.0.0/15 since the first 15 bits of these two CIDR blocks are the same:

130.126.0.0/16　1 0 0 0 0 0 1 0 0 1 1 1 1 1 1 0 0 0 0 0 0 0 0 0 0 0 0 0 0 0 0 0

130.127.0.0/16　1 0 0 0 0 0 1 0 0 1 1 1 1 1 1 1 0 0 0 0 0 0 0 0 0 0 0 0 0 0 0 0

first 15 bits

In this case, a router not in AS-1 and AS-2 may maintain one entry for the supernet, which further reduces the size of the routing table. Of course, a router would maintain a single entry for AS-1 and AS-2 only if the outgoing link for both destinations is the same.

Intra-domain routing: OSPF and RIP

Routing within an AS is called intra-domain routing. The hosts within the same AS are, in general, under the same administrative control, so they use the same intra-domain routing protocol. Most intra-domain routing protocols are cost-oriented protocols, i.e., packets are routed along minimum cost paths.

Two extensively implemented intra-domain routing protocols are Open Shortest Path First (OSPF) and Routing Information Protocol (RIP). OSPF is a link-state routing protocol. Under OSPF, link-state information is broadcast to all routers in the network. Each router then constructs the complete topology of the AS and applies Dijkstra's algorithm to compute the minimum cost paths to all other routers within the same AS. RIP is a distance-vector routing protocol. Routers exchange their routing tables every 30 seconds, and then update their routing tables using the Bellman–Ford algorithm.

Inter-domain routing: BGP

Inter-domain routing decides the path from one AS to another AS. For inter-domain routing, routes are generated not only based on cost considerations, but are also decided by policies that an AS uses to route its packets to other ASs. For example, two ASs may have an agreement to route a certain fraction of their traffic through each other, and the routing tables may reflect this agreement. The Border Gateway Protocol (BGP) is the current inter-domain routing protocol for the Internet, and is much more complicated than intra-domain routing protocols such as OSPF and RIP. In the following, we present a very brief description of the basic idea behind BGP.

Under BGP, routers exchange reachability information. The messages exchanged among routers consist of two pieces of information: CIDR blocks and attributes associated with the CIDR blocks. The CIDR blocks represent the networks that are reachable through the router which sends the message, and attributes comprise detailed information on how to reach those networks. For example, one attribute is the list of all ASs on the path to the network associated with the CIDR block.

A router may receive multiple messages containing the same CIDR block, i.e., learn more than one path to the destination network. The router then needs to select the "best" path that complies with its policies. The definition of "best" is specified by a sequence of prioritized rules, such as local preference, length of paths, etc. For example, if path length has the highest priority, the path with the smallest number of hops will be selected. After selecting the "best" path, the router records the interface connecting to the next AS on the selected path, together with the destination network's CIDR block, to its routing table.

When a router receives a packet, it looks up the destination IP address of the packet in its routing table. We say an IP address matches an entry of the routing table if it matches a CIDR block, which means the first X bits of the IP address are the same as the CIDR block $A.B.C.D/X$. An IP address may match multiple CIDR blocks. For example, IP address 64.222.4.132 matches both entries 64.222.8.11/20 and 64.222.8.11/15 in the routing table shown in Table 7.4. The current IP protocol uses *longest prefix matching*, which selects a matched CIDR block with the longest NetID as it provides the most specific information about the destination's location. Again consider Table 7.4 and a packet destined to address 64.222.4.132. The router will select the entry whose destination field is 64.222.8.11/20, and send the packet to the next router via interface 65.1.3.1.

7.4   MAC layer protocols in wireless networks

In this section, we discuss the connection between the algorithms discussed in previous sections and practical MAC layer protocols.

Table 7.4 A routing table example

Destination	Interface
65.1.5.2/18	65.8.5.0
64.222.8.11/15	65.6.5.0
64.222.8.11/20	65.1.3.1
64.1.2.3	64.1.2.1
...	...

7.4.1 Proportionally fair scheduler in cellular downlink

We first present the proportionally fair scheduler, a scheduler used in the downlink of several cellular network standards, and discuss its connection to the NUM formulation. Consider a cellular downlink with N mobiles; let $r_i(t)$ denote the rate allocated to mobile i at time slot t, and $T_i(t)$ the weighted average throughput (to be defined precisely later). Assuming the base station can transmit to only one mobile at a time, the buffers for all mobiles are infinitely backlogged, and $w_i > 0$ is a constant associated with mobile i, the proportional fair scheduler is as presented in Algorithm 9.

Algorithm 9 Proportional fair scheduler

1: The base station selects mobile i^* such that
$$i^* \in \arg\max_i \frac{w_i c_i(t)}{T_i(t)},$$
where $c_i(t)$ is the maximum possible transmission rate to mobile i at time slot t. The base station transmits to the mobiles with rate $r(t)$ such that $r_i(t) = c_i(t)$ if $i = i^*$ and $r_i(t) = 0$ otherwise.

2: The base station updates the weighted average throughput such that
$$T_i(t+1) = \left(1 - \frac{1}{\alpha}\right) T_i(t) + \frac{1}{\alpha} r_i(t)$$
for some $\alpha > 1$.

We can see that this proportional fair scheduler is simply a MaxWeight scheduler, where, instead of using the queue lengths as weights, it uses the inverse of the average throughput as the weights. We will next outline an argument which shows that this scheduler maximizes the net utility

$$\sum_i w_i \log T_i(\infty),$$

where $T_i(\infty)$ is the steady-state average throughput.[1]

Recall that π_m is the probability that the network is in channel state m, $c_{m,i}$ is the maximum number of packets that can be transmitted to mobile i when the network is in state m, and $\alpha_{m,i}$ is the probability that mobile i is selected for transmission when the channel state is m. Let x_i denote the average rate allocated to mobile i, so

$$x_i = \sum_m \alpha_{m,i} c_{m,i} \pi_m.$$

Therefore, the network utility maximization problem is

$$\max \sum_i w_i \log \left(\sum_m \alpha_{m,i} c_{m,i} \pi_m \right),$$

---

1 This is the reason it is called the "proportional fair scheduler."

subject to

$$\sum_i \alpha_{m,i} \le 1, \qquad \forall m,$$

$$\alpha_{m,i} \ge 0 \qquad \forall i, m.$$

Using Lagrange multipliers, we obtain

$$\sum_i w_i \log \left(\sum_m \alpha_{m,i} c_{m,i} \pi_m \right) - \sum_m \lambda_m \left(\sum_i \alpha_{m,i} - 1 \right).$$

So the optimal solution satisfies

$$\frac{w_i}{x_i^*} \pi_m c_{m,i} - \lambda_m^* = 0 \quad \text{if} \quad \alpha_{m,i}^* > 0,$$

i.e.,

$$\frac{\lambda_m}{\pi_m} = \frac{w_i c_{m,i}}{x_i^*} \quad \text{if} \quad \alpha_{m,i}^* > 0. \tag{7.24}$$

Thus, the optimal solution has the property that $w_i c_{m,i}/x_i^*$ are equal for all i at each channel state m. The proportional fairness algorithm attempts to achieve this by scheduling the mobile with the largest value of $w_i c_{m,i}/x_i(t)$, where $x_i(t)$ is the current throughput received by mobile i. By doing this, the throughput for user i increases, thus decreasing $w_i c_{m,i}/x_i(t)$ after the current time instant, and driving the system towards the state where $w_i c_{m,i}/x_i(t)$ are equal for all i. The proof that a proportional fair scheduler achieves (7.24) will not be provided here, but we will see how proportional fairness can also be achieved using the solution presented in Chapter 6.

Under this proportional fair scheduler, the throughput allocated to mobiles is controlled by the weights $\{w_i\}$ instead of queue lengths (as in Algorithm 5). The advantage is that this provides isolation among flows: each flow is guaranteed a certain long-term rate independent of the behavior of the other users in the network.

Proportional fairness can also be achieved using an algorithm that follows from optimization decomposition, as discussed in Section 6.2, Algorithm 5. Assume that the base station maintains a counter for each mobile i, and that the value of the counter at time t is denoted by $\tilde{q}_i$. The algorithm is presented in Algorithm 10.

Note that $\tilde{q}_i$ is a counter instead of a real queue. According to the analysis in Section 6.2, Algorithm 10 maximizes $\sum w_i \log x_i$ when ϵ is small, where x_i is the long-term throughput of mobile i.

7.4.2 MAC for WiFi and ad hoc networks

WiFi access points today operate using protocols specified by the IEEE 802.11 standards. Table 7.5 summarizes the main characteristics of several popular 802.11 standards. The 802.11 standards support two modes: infrastructure mode, where mobile stations (laptops, PDAs, etc.) connect to an access point to access the network, and ad hoc mode, where mobile stations communicate directly with each other.

Algorithm 10 Proportional fair scheduler using optimization decomposition

---

1: **Virtual congestion control:** The base station generates $\tilde{x}_i(t)$ virtual arrivals to counter i such that

$$\tilde{x}_i(t) = \min\left\{\tilde{x}_{\max}, \frac{w_i}{\epsilon \tilde{q}_i(t)}\right\}.$$

2: **MaxWeight scheduling:** The base station serves mobile i^* such that

$$i^* \in \arg\max_i \tilde{q}_i(t) c_i(t).$$

Ties are broken at random. The base station transmits to the mobiles with rate $r(t)$ such that $r_i(t) = c_i(t)$ if $i = i^*$ and $r_i(t) = 0$ otherwise.

3: **The evolution of virtual queues:** The base station updates the virtual queues:

$$\tilde{q}_i(t+1) = (\tilde{q}_i(t) + \tilde{x}_i(t) - r_i(t))^+.$$

Table 7.5 Summary of 802.11 standards

Standard	Frequency (GHz)	Bandwidth (MHz)	Max. Rate (Mbps)	Ranges (in-/outdoor) (ft)
802.11a	5	20	54	115/390
802.11b	2.4	20	11	125/460
802.11g	2.4	20	54	125/460
802.11n	2.4	50	72.2	230/820
802.11n	5	50	150	230/820

WiFi operates over unlicensed spectra, so multiple mobiles may access the same spectrum at the same time. In 802.11, the interference is resolved by a random access protocol called Carrier Sensing Multiple Access/Collison Avoidance (CSMA/CA), which works as follows.

(i) When a wireless node (an access point or mobile station) has a packet to send, it first senses the channel. If the channel is sensed idle, the node transmits the packet after a short duration of time, called the Distributed Inter-Frame Space (DIFS). Otherwise, the node selects a backoff time uniformly and randomly from $[0, W_c]$ and reduces the backoff time by 1 whenever the channel is sensed idle for 1 time slot; W_c is called the contention window. When the backoff time reduces to zero and the channel is idle, the node transmits the packet. If a collision occurs, the contention window is doubled and a new backoff time is selected.

(ii) If the packet is successfully decoded at the receiver, the receiver sends back an acknowledgement after a short period of time called the Short Inter-Frame Space (SIFS).

(iii) If the acknowledgement is successfully received at the sender, the sender starts to transmit the next packet by following step (i); otherwise, the sender randomly chooses another backoff time from $[0, W_c]$ and attempts to transmit the packet again when the backoff time reaches zero.

In the mechanism described above, when the channel status changes from busy to idle, the sender waits for a random period of time and then starts to transmit, instead of transmitting immediately. This is to avoid collision. Suppose that a node transmits immediately after the channel is sensed idle. Then, if multiple nodes are waiting to access the channel, they will transmit at the same time and collide with each other. Now, if each node waits for a random backoff time and then transmits, it is likely that one node will transmit before the others, so that other nodes can detect the transmission and keep silent.

We also note that, upon receiving a packet, the receiver sends an acknowledgement after an SIFS time without even sensing the channel. In 802.11, SIFS is the smallest time interval between transmissions, e.g., DIFS is SIFS plus two slot times. So the acknowledgement is transmitted before any other transmission could occur, which gives acknowledgements the highest transmission priority in the network.

The CSMA/CA mechanism can resolve contention assuming that a wireless node can sense all interfering transmissions; this, however, is not true because of signal attenuation. Consider the scenario in Figure 7.8, where nodes a and b can both communicate with the access point, but cannot communicate with each other because of path loss. Consequently, node a cannot sense node b's transmission, while the transmissions of node a and b strongly interfere with each other at the access point. This is the so-called *hidden terminal problem*. This hidden terminal problem exists because, in CSMA/CA, it is the transmitter who senses the channel and makes transmission decisions, but the actual interference happens at the receiver side. Because of signal attenuation, the signal strength seen by the transmitter could be very different from that seen by the receiver, so the transmitter may not be able to detect all interfering transmissions.

One approach to resolve this problem is to use Request to Send (RTS) and Clear to Send (CTS). The RTS/CTS protocol works as follows. When a wireless node decides to transmit a packet after sensing that channel is idle, it first sends an RTS message to its receiver. After receiving the RTS, the receiver broadcasts a CTS message. The CTS message tells the transmitter that the receiver is ready to receive the packet, and also tells interfering

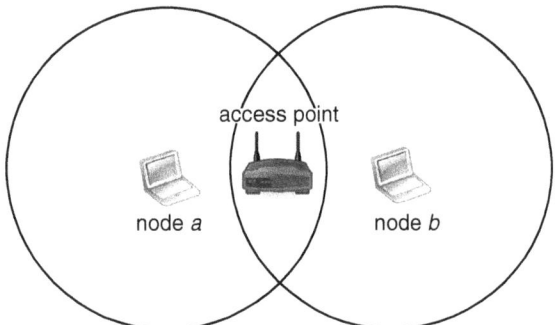

Figure 7.8 The hidden terminal problem: nodes a and b can communicate with the access point, but cannot communicate with each other.

nodes not to transmit. Consider the scenario in Figure 7.8. After the access point receives a RTS from node a, it broadcasts a CTS. Then node b, who receives the CTS, knows that someone is going to transmit data to the access point, and will keep silent.

Consider a link, say link 1. Under the RTS/CTS mechanism, any link (e.g., link 2) whose transmitter can hear the CTS from the receiver of link 1, cannot transmit at the same time with link 1. So, under the CSMA/CA, along with RTS/CTS protocol, the interference can be modeled as a conflict graph. This is the reason the interference model based on conflict graphs has been widely studied in the design of ad hoc wireless networks. The Q-CSMA protocol presented in Chapter 5 can, in principle, be combined with 802.11 protocols to achieve better performance, and this is a topic of active research.

7.5   Summary

- **TCP-Reno** TCP-Reno consists of two phases. In the slow-start phase, the window size increases by 1 each time an acknowledgement (ack) is received. When the window size hits either the slow-start threshold or ssthresh, or if a packet loss is detected, the algorithm shifts to the congestion-avoidance phase. When in the congestion-avoidance phase, the algorithm increases the window size by $1/W$ every time an ack of a successful packet transmission is received and cuts its window size by half when a packet loss is detected by the receipt of three dupacks. The TCP-Reno congestion control algorithm is approximated as a differential equation in the following manner:

$$\dot{x}_r = \frac{x_r(t - T_r)(1 - q_r(t))}{T_r^2 x_r(t)} - \beta x_r(t - T_r)q_r(t)x_r(t).$$

- **TCP-Vegas** TCP-Vegas monitors two values: the expected throughput $e(t)$ and the actual throughput $a(t)$. If $a(t)$ is less than $e(t)$, it means that the transmission rate is too high, so the algorithm cuts down the window size; otherwise, the algorithm increases the window size. The throughput of a set of sources using TCP-Vegas can be approximated by the solution to a network utility maximization problem, with

$$U_r(x_r) = \alpha T_{pr} \log x_r$$

as the utility function, where T_{pr} is the propagation delay of source r.

- **Routing** A common objective of routing algorithms is to find the route with the minimum cost from a source to its destination, where the cost of a route is simply the sum of the costs of the links in the route.

  Dijkstra's algorithm is a link-state routing algorithm, where each node knows the costs of all links in the network, and iteratively computes the minimum cost paths from itself to all other nodes in the network.

  The Bellman–Ford algorithm is a distance-vector routing algorithm. In the Bellman–Ford algorithm, nodes only communicate with immediate neighbors without knowing the complete network state, and iteratively compute the minimum cost and the next hop to each destination using the Bellman–Ford equation.

- **IP addressing and routing in the Internet** The Internet Protocol (IP) addressing scheme is a numerical method employed to label all hosts that are connected to the Internet. In the current version of the IP protocol – Internet Protocol version 4 (IPv4), an IP address consists of 32 bits. Packets are routed according to the destination IP addresses in the headers of the packets. Due to scalability consideration, routers are organized into autonomous systems (ASs). Most routing protocols within ASs, such as OSPF and RIP, are cost oriented; the current inter-AS routing protocol in the Internet is the BGP protocol.

- **Proportional fair scheduler** Under the proportional fair scheduler, the base station transmits to mobile i^* such that

$$i^* \in \arg\max_i \frac{w_i c_i(t)}{T_i(t)},$$

where $c_i(t)$ is the maximum transmission rate to mobile i at time slot t. The scheduler maximizes $\sum_i w_i \log T_i(\infty)$, so is proportionally fair.

- **Carrier sensing** Nodes sense the presence of a carrier to determine if other transmissions are taking place in their vicinity.

- **802.11 protocol** Using a random backoff timer, nodes attempt to transmit at different times to reduce the chances of a collision. In the event that a collision takes place, backoff timers are increased to avoid repeated collisions. An RTS/CTS protocol is used to avoid the hidden terminal problem.

7.6   Exercises

Exercise 7.1 (The local stability of TCP-Reno) In this exercise, we will derive the condition under which TCP-Reno is locally stable with delays. Recall that the delay differential equation of TCP-Reno for a single-source, single-link network, is given by

$$\dot{x} = x(t-T)\left(\frac{1 - f(x(t-T))}{T^2 x(t)} - \frac{f(x(t-T))x(t-T)}{2}\right).$$

(1) Linearize the delay differential equation for TCP-Reno.

(2) Obtain the sufficient condition for the local stability of TCP-Reno using Theorem 2.6.1.

Exercise 7.2 (Window sizes and throughput in TCP-Reno) Consider a single TCP-Reno connection over a link with bandwidth 15 Mbps. Assume each packet has a size of 1500 bytes and the round trip time is 80 ms. Further, we assume packet losses occur when the transmission rate exceeds the link capacity, and no time-out occurs during transmissions.

- What is the maximum possible window size (in terms of packets) for this TCP connection? Note: In reality, most versions of TCP-Reno set a maximum window size beyond which the window size cannot increase. In this problem, we assume that such an upper limit does not exist, but rather we are interested in computing the maximum window size limit that is naturally imposed by the available bandwidth and the RTT.
- What are the average window size and average throughput? Hint: The congestion window is set to one-half of the previous value when a packet loss is detected.
- How many packets were transmitted during the slow-start phase?

Exercise 7.3 (The throughput and delay in TCP-Vegas) Consider the same network as in Exercise 7.2, but assume TCP-Vegas is used. Further, assume $\alpha = \beta = 150/T_{pr}$. Compute the throughput and queueing delay at the equilibrium point.

Exercise 7.4 (An example of Dijkstra's algorithm)   Consider the network shown in Figure 7.9. Use a table similar to Table 7.1 to illustrate the computation process of Dijkstra's algorithm at node A.

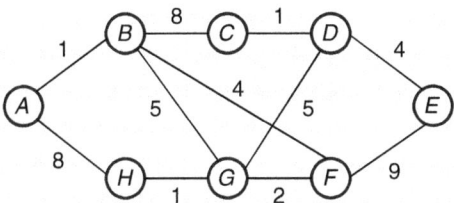

Figure 7.9 The network configuration.

Exercise 7.5 (An example of the Bellman–Ford algorithm) Consider the network in Figure 7.9. Assume node A is the only destination in the network. Use a table similar to Table 7.2 to show the computation process of the Bellman–Ford algorithm.

Exercise 7.6 (Transient loops under Dijkstra's algorithm) In Dijkstra's algorithm, nodes need to have consistent knowledge of link costs. Otherwise, packets may not be correctly routed to their destinations. For example, when the cost of a link changes, but the change is not known by all the nodes in the network, *transient loops* may occur. Consider the network in Figure 7.10, where nodes a and b know the failure of link ab, but nodes c, d, and e do not. Use Dijkstra's algorithm to construct routing tables at each node, and then show that packets from e to a will loop between nodes b and e.

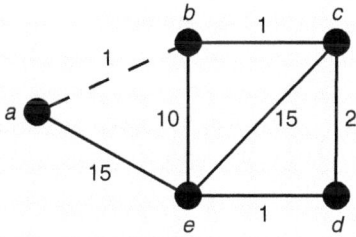

Figure 7.10 Nodes a and b know the failures of link ab, but nodes c, d, and e do not.

Exercise 7.7 (Counting to infinity under the Bellman–Ford algorithm) Consider the network shown in Figure 7.11, where each link has a cost of 1. Assume node A is the only destination, and nodes calculate the shortest paths using the Bellman–Ford algorithm. The initial routing tables at each node are illustrated in Table 7.6.

(1) Assume node A goes down. Show that, under the *synchronized* Bellman–Ford algorithm, after A goes down each node's cost of a path to A slowly increases and goes to infinity. Note: After A goes down, assume B finds out about it, i.e., B knows that the cost of link BA has to be infinity.

(2) Now assume the link between node A and node D with cost 1 fails, but there is another link between node A and node D with cost 10. Starting from the initial routing tables

(Table 7.6), how many iterations does it take for all nodes to find the alternative path to node A?

(3) **Poisoned reverse** Still consider Figure 7.11. In (1), you should have observed that it will take forever for each node to realize that it can no longer route to node A. In this part, we will explore a solution to this problem by which even the nodes that are not directly connected to a broken link can detect link failure quickly. Assume that if node x finds a path to node A which goes through its neighbor node y, node x always advertises to node y that its distance to node A is ∞ (for all other nodes, node x will advertise its true routing table entry as before). Illustrate the updates of the routing tables under the synchronized Bellman–Ford algorithm with this additional rule.

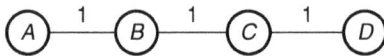

Figure 7.11 Simple line network.

Table 7.6 Routing tables at time 0

(ω_{AA}, n_{AA})	(ω_{BA}, n_{BA})	(ω_{CA}, n_{CA})	(ω_{DA}, n_{DA})
$(0, null)$	$(1, A)$	$(2, B)$	$(3, C)$

Exercise 7.8 (α fair scheduler) Using the idea behind the proportional fair scheduler (Algorithm 9), write down the α fair scheduler based on average throughput.

7.7   Notes

TCP-Reno and NewReno are improvements on TCP-Tahoe proposed in [60]. TCP's AIMD algorithm was based on [25], where it was also shown that other increase–decrease algorithms may not work well. Details of the protocols studied are available in [80, 96, 139, 158]. The differential equation for TCP without time delays was introduced in [93, 95] who also noted the connection to minimum potential delay fairness, and a delay differential equation model was introduced in [122]. The choice of the arctan utility function for TCP was proposed in [74]. The generalization of TCP-Reno was proposed in [165], and a special case was studied in [79]. TCP-Vegas was proposed in [18], and the analytical study is due to [109]. Models that capture the window dynamics of TCP in more detail than utility function based models can be found in [3, 7, 97, 136, 152].

The books [96, 139] are good references for addressing and routing protocols, and the book [10] is an excellent source for the analysis of Bellman–Ford and Dijkstra's algorithms. The limitations of the shortest-path routing algorithm are discussed in [80]. A good source for both mathematical models of networks and associated network protocols for wireline and wireless networks, including the 802.11 protocol, is [89]. A detailed mathematical model to assist in understanding the behavior of the 802.11 protocol can be found in [11]. A proportionally fair scheduler which exploits channel variations was presented in [166]. A different algorithm to exploit channel variations using an optimization formulation was presented in [106].

8 Peer-to-peer networks

As we have seen so far, the Internet is a collection of physical links and routers that communicate with each other using a common set of protocols. Many applications run on top of this physical network. Some of these applications involve a collection of computers (called peers), which form a logical overlay network on top of the physical Internet. When communication takes place between the peers, the underlying physical network is transparent to them, and so the communication appears as though it takes place between the peers without the participation of any intermediate nodes. Hence, these types of networks are called Peer-to-Peer (P2P) networks.

Typically, P2P networks are used for high bandwidth data transfer due to its scalability. In a traditional client–server system, the capacity of the system is limited by the bandwidth of the server. In a P2P network, peers upload data to other peers while receiving the data, so the network capacity increases with the number of peers. In this chapter, we will introduce three P2P applications: Distributed Hash Tables (DHTs), file-sharing applications, and streaming systems. A DHT protocol is a distributed database which stores (key, value) pairs at participating peers and facilitates the retrieval of content location. In file-sharing applications, peers who are interested in the same content (such as a movie or a music file) exchange content pieces among themselves. P2P media-streaming applications require peers to exchange data chunks originated from a streaming server, and a data chunk must be received within a certain deadline for smooth sequential playing out of the media. In this chapter, the following questions related to P2P networks will be addressed.

- *How are (key, value) pairs stored in DHTs, and what is the complexity to retrieve (key, value) pairs?*
- *How do file-sharing applications such as BitTorrent work?*
- *What is structured P2P streaming, and what is the maximum streaming rate in structured P2P streaming?*
- *What is unstructured P2P streaming, and what is the relationship between streaming rate, skip-free probability, and playout delay?*

8.1 Distributed hash tables

A Distributed Hash Table (DHT) is a P2P system that provides a lookup service for users interested in specific content (files). A DHT can be viewed as a distributed database or a search engine, which stores (key, value) pairs. The keys could be movie names, music titles, etc., and the values are the IP addresses of the peers who have the movies. The (key, value) pairs are distributed to peers instead of being stored at a central server, which makes the system scalable and resilient to a single point of failure. The functionality of a DHT is to return a (key, value) pair to a peer if the key is queried by the peer. We will discuss two DHT systems: Chord and Kademlia, where Chord uses a ring structure and Kademlia is constructed based on XOR distance.

8.1.1 Chord

Chord assigns an n-bit integer identifier to each content. This is achieved by hashing the content (such as a movie title) to produce a number between 0 and $2^n - 1$. Similarly, each node in the P2P network is also given an n-bit identifier. This is done by hashing the IP address of the node to produce a number between 0 and $2^n - 1$. Thus, both the keys and the nodes which store them have identifiers selected from the numbers 0 to $2^n - 1$. Such a scheme in which nodes and keys have the same identifier space is called consistent hashing.

Once a content has been given an identifier (called the key), the (key, value) pair associated with the content is stored in a node such that the distance from the key to the identifier of the node is the smallest. Recall that the value is simply the set of IP addresses where the content can be found. Note that original keys are the content types, not integers. So Chord hashes original keys to create integer keys. Similarly, the IP address of a node needs to be hashed to an integer identifier.

In Chord, the distance of two identifiers is defined to be the clockwise distance on the ring, which is asymmetric. For example, given $n = 3$, the distance from 4 to 5 is 1, but the distance from 5 to 4 is 7. So the node closest to a key is the node whose identifier is the same as the key or the node who is the immediate successor of the key on the ring.

Example 8.1.1

Consider a Chord network, shown in Figure 8.1, where $n = 3$. The peers in the network are associated with the following identifiers $\{0, 1, 4, 5, 6\}$. In this case, key 2 will be assigned to peer 4 and key 7 will be assigned to peer 0. □

Note that a good hash function tends to distribute the inputs "randomly" and uniformly among the 2^n identifiers. Consistent hashing, therefore, achieves load balancing, i.e., each node is responsible for roughly the same amount of keys. Consider a Chord system with K keys and N nodes. For simplicity, assume that an original key (or IP address) is hashed to an integer that is uniformly and randomly selected from $[0, 2^n - 1]$. Then the following theorem states that, with high probability, each node is responsible for no more than $2.2(\log N)K/N$ keys, deviating from the mean value (K/N) only by a factor of $2.2 \log N$.

Theorem 8.1.1 Assume that $N = o(K)$. When N is sufficiently large, with probability at least $1 - 2/N$, each peer is responsible for no more than $2.2(\log N)K/N$ keys.

Proof Consider a node with identifier m. Without loss of generality, we assume $m \geq 2^{n+1}(\log N)/N$. Note that if $m < 2^{n+1}(\log N)/N$, we can simply rotate the ring clockwise by $2^{n+1}(\log N)/N$ positions.

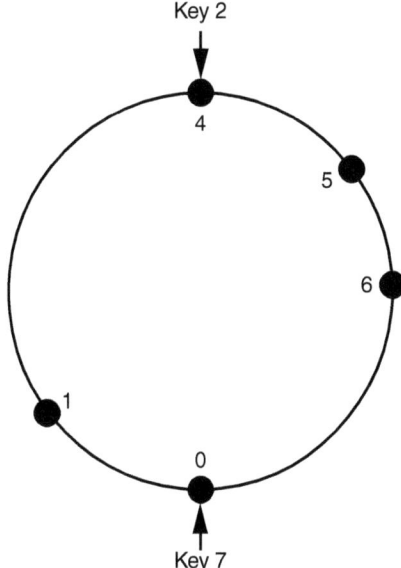

Figure 8.1 The ring structure of Chord, with $n = 3$.

We will prove this theorem based on the following two claims.

- **Claim 1:** With probability at least $1 - 1/N^2$, there is another peer with identifier k such that $0 < m - k \leq 2^{n+1}(\log N)/N$.
- **Claim 2:** With probability at least $1 - e^{-0.005K(\log N)/N}$, no more than $2.2(\log N)K/N$ keys are assigned to the range $[m - 2^{n+1}(\log N)/N + 1, m]$.

If both claims hold, using the union bound, the probability that node m is responsible for no more than $2.2(\log N)K/N$ keys is at least

$$1 - \frac{1}{N^2} - e^{-0.005K(\log N)/N}.$$

Recall that we assume $N = o(K)$, when $K \geq 400N$, we have

$$1 - \frac{1}{N^2} - e^{-0.005K(\log N)/N} \geq 1 - \frac{1}{N^2} - e^{-2\log N} = 1 - \frac{2}{N^2}.$$

Applying the union bound again, the probability that every node is responsible for no more than $2.2(\log N)K/N$ keys is at least

$$1 - N\frac{2}{N^2} = 1 - \frac{2}{N}.$$

Proof of claim 1 The probability that a node's identifier falls in the range $\left[m - 2^{n+1}(\log N)/N, m - 1\right]$ is given by

$$\frac{1}{2^n} \times \frac{2^{n+1}\log N}{N} = \frac{2\log N}{N},$$

where $1/2^n$ is the probability a node is assigned a specific identifier, and $2^{n+1}(\log N)/N$ is the number of identifiers in the range. So, the probability that none of the N identifiers is in the range $\left[m - 2^{n+1}(\log N)/N, m - 1\right]$ is given by

$$\left(1 - \frac{2\log N}{N}\right)^N \le e^{-2\log N} = \frac{1}{N^2}.$$

Hence, we prove the first claim.

Proof of claim 2 Since a key is uniformly and randomly picked in the range $[0, 2^n - 1]$, the number of keys assigned to interval $[m - \tilde{m} + 1, m]$ is a sum of K independent Bernoulli random variables, each with mean $\tilde{m}/2^n$. Using the Chernoff bound, it follows that, with a probability greater than or equal to

$$1 - e^{-\delta^2 K\tilde{m}/2^{n+2}},$$

the number of keys in range $[m - \tilde{m} + 1, m]$ is less than[1]

$$(1 + \delta)K\tilde{m}/2^n,$$

where δ is some positive constant. Claim 2 holds by choosing $\tilde{m} = 2^{n+1}(\log N)/N$ and $\delta = 0.1$. ☐

Key lookup

Recall that the main purpose of the Chord protocol is to find content. Thus, when a peer is interested in some content, it first hashes the content title and finds the content's key. Then, it initiates a query within the P2P network to find the (key, value) pair so that it can determine where the content is stored. This process of finding the (key, value) pair is called the *lookup process*. The lookup process in Chord is simple. Suppose that every node knows the IP address of only its immediate successor. By immediate successor, we mean the node with the next higher identifier. For example, node 1 knows only the IP address of node 4 for the network in Figure 8.1. When a query is initiated at a node, the query is forwarded clockwise until it reaches the node responsible for the key, i.e., the first node whose identity is larger than the key. At this point, the query is said to be resolved. This node then returns the value information to the node that initiated the query. Consider the network in Figure 8.1. If key 5 is queried at node 1, the query is forwarded along the path

$$1 \to 4 \to 5.$$

Therefore, if a node knows the IP addresses of its immediate successor, this is sufficient to answer all queries. However, if only this piece of information is available at each node, in the worst case it requires $N - 1$ message exchanges to answer the query. For example, if a query for key 7 is initiated at peer 1, the query needs to be forwarded along the path

$$1 \to 4 \to 5 \to 6 \to 0,$$

which requires four message exchanges.

To improve the lookup speed, in Chord each node keeps track of the identities of n peers instead of two. The n other peers can be thought of as n logical links outgoing from the given node. These logical links are called *fingers* in Chord. The ith finger is the peer that succeeds

---

1 This version of the Chernoff bound is obtained using the fact that $D((1 + \delta)p\|p) \ge \delta^2 p/4$.

the current node by at least 2^{i-1}; in other words, it is the node closest (in terms of clockwise distance) to identifier $(m + 2^{i-1}) \bmod (2^n)$, where m is the identity of the current node.

Example 8.1.2

Consider a Chord network, shown in Figure 8.2, where $n = 3$ and $N = 7$, and each node maintains three fingers. For example, node 0 in this figure has fingers to nodes 1, 2, and 4, respectively. Node 7 is both the first and second fingers of node 5 because node 6 is not in the network, and node 1 is the third finger. Note that we allow multiple fingers to point to the same node.

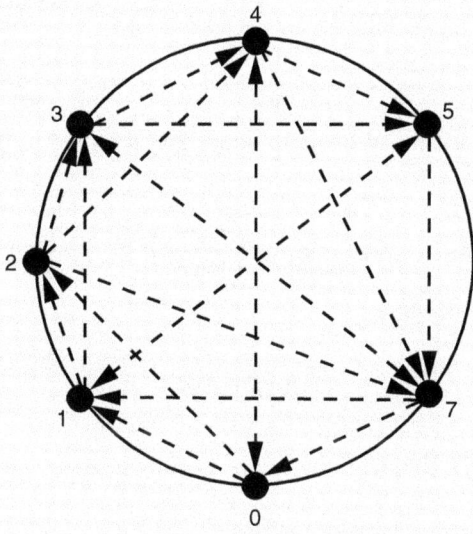

Figure 8.2 The ring structure of Chord with $n = 3$ and $N = 7$. The dashed arrows are the fingers associated with each node.

□

With n fingers at each node, after receiving a query the node not responsible for the query forwards it to the finger closest to the queried key. For example, in Figure 8.2, if key 2 is queried at node 3, the forwarding path is

$$3 \to 7 \to 1 \to 2.$$

So the query is resolved with three message exchanges instead of six message exchanges if only immediate predecessors/successors are known at each node. In general, with n fingers at each node, most queries can be resolved after $O(\log N)$ message exchanges, as shown in the following theorem. The intuition is that Chord resembles the *binary search*, and the distance between the current node and the queried key halves after every message exchange. So $\log N$ message exchanges are sufficient for resolving any query.

Theorem 8.1.2 With high probability, the number of message exchanges required to resolve a query in an N-node Chord system is $O(\log N)$.

Proof Without loss of generality, assume a query for key k is initiated at node 0. If node 0 is not responsible for key k, it forwards the query to the finger closest to key k. Assume it is node p, which is the ith finger of node 0. For simplicity, we assume each node has n distinct fingers, so $2^{i-1} < p \leq 2^i$. Further, $k \leq 2^{i+1}$, because otherwise the $(i+1)$th finger of node 0 would be closer to key k than the ith finger.

So the distance from node 0 to key k is at most 2^{i+1}, and the distance from node 0 to node p is at least 2^{i-1}. Hence, the distance from node p to key k is at most $3/4$ of the distance from node 0 to key k. This process repeats. After $\log N / \log(4/3)$ forwardings, the distance from the current node to key k reduces to at most $1/N$ of the distance from node 0 to k, which is at most[2] $2^n/N$.

Now consider the interval from $k - 2^n/N + 1$ to k. The probability a node's identifier falls in this interval is given by

$$\frac{2^n}{N} \times \frac{1}{2^n} = \frac{1}{N}.$$

Given N nodes, the probability that $\log N$ or more nodes are assigned to this interval is upper bounded by

$$\binom{N}{\log N}\left(\frac{1}{N}\right)^{\log N} \leq \frac{1}{(\log N)!} \leq \left(\frac{e}{\log N}\right)^{\log N} \leq \frac{1}{N},$$

where the second inequality holds because

$$\frac{m^m}{m!} < \sum_{i=1}^{\infty} \frac{m^i}{i!} = e^m.$$

Therefore, after another $O(\log N)$ contacts, the query is forwarded to the node that is responsible for the key. □

Peer churn

In P2P networks, nodes may join and leave dynamically; this is called *peer churn*. To handle peer churn, a node needs to know the IP address of its immediate predecessor, in addition to the IP addresses of its immediate successor and the nodes in its finger table. When peer churn occurs, nodes need to:

(1) update the identities of their predecessors and successors;
(2) update their finger tables in response to peer churn; and
(3) update their (key, value) databases.

When a node joins/leaves, it needs to notify its immediate successor and predecessor so that they can update their immediate successors/predecessors. This step ensures the connectivity of Chord systems. Specifically, when a node, say node m, leaves the system, it notifies its immediate successor, say node m_s, and predecessor, say node m_p. Then node m_s changes its immediate predecessor to m_p and node m_p changes its immediate successor to m_s. The update process is shown in Figure 8.3.

---

2 A careful, but more complicated, analysis can show that the distance actually halves after each message exchange, but that is omitted here.

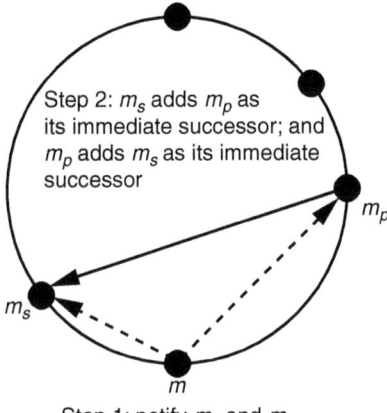

Step 2: m_s adds m_p as its immediate successor; and m_p adds m_s as its immediate successor

m_p

m_s

m

Step 1: notify m_s and m_p

Figure 8.3 The process of updating immediate predecessors/successors when a node leaves the network. Dashed lines represent message exchanges and solid lines represent connections.

When node m wants to join the network, it needs to know the IP address of a node already in the system. Node m then queries for key m. The node responsible for key m, denoted by p, is the immediate successor of peer m. Node p notifies node m of its own (identifier, IP address) and the (identifier, IP address) of its immediate predecessor, denoted by p_p. Node m records node p as its immediate successor and node p_p as its immediate predecessor. Node p changes its immediate predecessor to node m. Finally, node m notifies node p_p of its presence, and node p_p updates its immediate successor to node m. The update process is shown in Figure 8.4.

To update finger tables, each node, say node m, periodically initiates queries for keys

$$\left\{ \left(m + 2^{i-1} \right) \bmod \left(2^n \right) \right\}_{i=1,\ldots,n}$$

to find out the correct fingers.

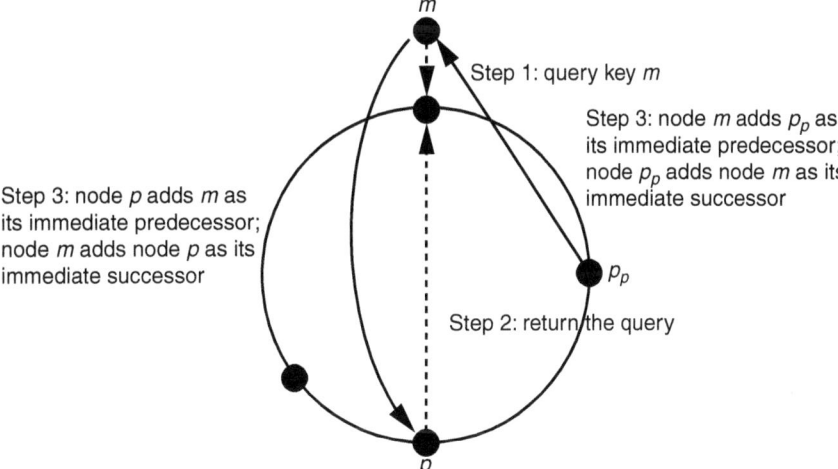

m

Step 1: query key m

Step 3: node m adds p_p as its immediate predecessor; node p_p adds node m as its immediate successor

Step 3: node p adds m as its immediate predecessor; node m adds node p as its immediate successor

p_p

Step 2: return the query

p

Figure 8.4 The process of updating immediate predecessors/successors in Chord when a node joins the network.

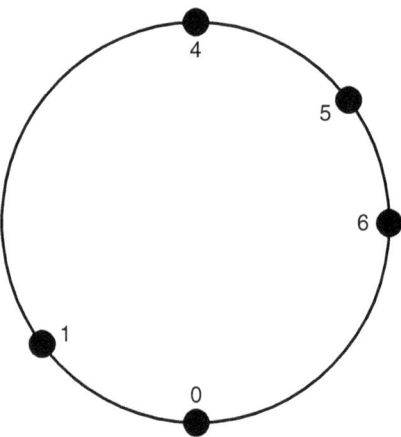

Figure 8.5 Chord network with five nodes.

When a node leaves the network, it needs to transfer its (key, value) pairs to its immediate successor because its immediate successor is now responsible for those pairs. When a peer joins the network, its immediate successor needs to transfer those (key, value) pairs closer to the new node than itself to the new node. Consider the network shown in Figure 8.5. When node 5 leaves the system, it needs to transfer (key, value) pairs associated with key 5 to node 6. When node 3 joins the network, node 4 needs to transfer (key, value) pairs associated with keys 2 and 3 to node 3. Since each node is responsible for $O((\log N)K/N)$ keys with high probability, a node arrival or departure only requires $O((\log N)K/N)$ (key, value) transfers.

8.1.2   Kademlia

Kademlia is another DHT system. Similar to Chord, Kademlia uses consistent hashing to achieve load balancing. The fundamental difference between the two DHTs is that Kademlia uses XOR distance instead of circular distance. The distance between two nodes in Kademlia is the integer value of the XOR of the two identifiers. For example, consider two nodes with identifiers 001 and 010. XORing the two identifiers yields 011, and hence the distance between these two nodes is 3. Note that the XOR distance is symmetric whereas the circular distance is asymmetric. A (key, value) pair is stored at a node which is closest in XOR distance to the key. In fact, as will be explained later, to improve reliability in the case of peer churn (i.e., node arrivals and departures), Kademlia stores a (key, value) pair at multiple nodes that are close to the key under the XOR distance metric.

To facilitate the lookup processes, each node, say node m, in Kademlia maintains n buckets. Each bucket consists of k entries, and each entry in the ith bucket is the IP address of a node whose most significant $i - 1$ bits (i.e., the left-most $i - 1$ bits) are the same as those of node m, but the ith most significant bit is not.

Example 8.1.3

Consider a Kademlia network with three-bit node identifiers, i.e., $n = 3$. Therefore, there can be a maximum of eight nodes in this network. Suppose that there are only six nodes in this network, and let the node identifiers of these six nodes be 000, 001, 010, 100, 110, and 111.

Recall that each node maintains n buckets, which means, in this example, each node will maintain three buckets. Consider the buckets maintained by node 100. The first bucket will consist of all other nodes whose left-most bit is 0. Thus, nodes 000, 001, and 010 are eligible to go in bucket 1. If the maximum number of entries in a bucket, which we denoted earlier by k, is three, then all three nodes can be in bucket 1. Otherwise, any k of the eligible entries will be put into bucket 1. Bucket 2 consists of all nodes whose left-most bit is 1 and the next bit is 1. Thus, bucket 2 can contain the nodes 110 and 111. Again, the actual contents of bucket 2 are determined by the value of k (and other considerations to be mentioned later). The third bucket should contain nodes whose left-most two bits are 10 and the next bit is 1. But 101 is not a node in this network, so the third bucket is empty. □

Key lookup
The lookup process of Kademlia is different from that of Chord. We first introduce the lookup process of Kademlia for $k = 1$, i.e., each bucket contains at most one entry.

(1) Key y is queried at node m.
(2) Node m selects the node closest to key y from the set of nodes including itself and all nodes in its buckets. Call the selected node p.
(3) Node m contacts node p. Node p selects the node closest to key y from the set of nodes including itself and all nodes in its buckets, and returns the node to node m. Call the returned node node q.
(4) If node q is closer to the key than node p, set $p \leftarrow q$ and repeat step (3). Otherwise, the lookup process terminates, and node p is claimed to be the node that stores the key.

In the following analysis, we will show that it takes at most n contacts to resolve a query in Kademlia. We first introduce the following notation: for a given node m, we represent its identifier as $b_{n-1}^m b_{n-2}^m \cdots b_0^m$. Thus, b_0^m is the right-most bit in the identifier and b_{n-1}^m is the left-most bit in the identifier. Define

$$\tau(m, n) = \max \left\{ j : b_j^m \neq b_j^n \right\},$$

i.e, the most significant bit (the left-most bit) in which the identifiers of m and n differ. We first present a useful lemma about XOR distances. The proof is an immediate consequence of the definition of XOR distance, so we do not provide a proof. Instead, after the lemma, we present an example to illustrate the idea.

Lemma 8.1.3 Consider three n-bit identifiers x, y, and z; $d(x, z) > d(y, z)$ if and only if $b_{\tau(x,y)}^y = b_{\tau(x,y)}^z$. In other words, y is closer to z than x if and only if y and z agree at the left-most position where x and y differ. □

According to Lemma 8.13, to compare $\text{XOR}(x, z)$ and $\text{XOR}(y, z)$, we can simply look at the $\tau(x, y)$th bits of x, y, and z.

Example 8.1.4

Consider three binary numbers $x = 11000110$, $y = 11110001$, and $z = 10100110$; $XOR(x, z) = 01100000$ and thus $d(x, z) = 96$. Similarly, $XOR(y, z) = 01010111$ and thus $d(y, z) = 87$. Note that x and y differ in the third most significant bit where y and z agree.   □

Theorem 8.1.4 It takes at most n contacts to resolve a query in Kademlia, i.e., after at most n contacts, Kademlia correctly identifies the node closest to the queried key.   □

We will prove this theorem by establishing the following lemmas.

Lemma 8.1.5 When the lookup process terminates and returns a node p in response to a query y, it implies there is no other node w such that

$$d(w, y) < d(p, y). \tag{8.1}$$

Proof By contradiction. Suppose that there is a node w satisfying (8.1). This implies, by the property of XOR distances given in Lemma 8.1.3, that the left-most bit where w and p are different is a bit where w and y agree. Assume it is the ith most significant bit. The ith bucket of node p must then contain a node which is closer to y than p. For example, consider node p with identifier 100100, node y with identifier 100010, and node w with identifier 100001. Then, $i = 4$ and the fourth bucket of node $p = 100100$ must contain a node with identifier $1000**$, which is closer to $y = 100010$ than $p = 100100$. Note that the fourth bucket of node p cannot be empty since w is a node that can be placed in the bucket. But, as we assumed that the lookup process terminated and returned p, which means that p cannot find any node in its buckets which is closer to y than itself, this leads to a contradiction.   □

Lemma 8.1.6 Suppose that node m is queried for identifier y. Further, suppose that node w in m's ith bucket is closer to y than node m and any of the other nodes in m's buckets. Then, if there is any node u in the network that is closer to y than w, u's i most significant bits are identical to those of w.

Proof By contradiction. Suppose there exists a node u which is closer to y than node w, but that one of u's first i most significant bits is different from the corresponding bit of w. Let u and w differ in the jth most significant bit and agree on the $j - 1$ most significant bits. Then, by Lemma 8.1.3, the jth most significant bit of u agrees with the jth most significant bit of y because

$$d(u, y) < d(w, y).$$

Consider an example in which $u = 1101**$, $w = 1111**$ and $m = 1110**$, as shown in Figure 8.6. The length of an arc represents the XOR distance of the two nodes. In this example, node w is in node m's fourth bucket, so $i = 4$; u and w agree on the two most significant bits and differ in the third most significant bit, so $j = 3$. Since node u is closer to y than node w, the third most significant bit of y must agree with the corresponding bit of u, so it must be 0.

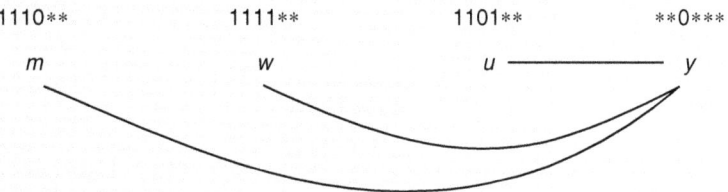

Figure 8.6 Example to illustrate the proof.

Now assume $j < i$ and consider a node z in m's jth bucket. Recall that node w is in node m's ith bucket, so node m and node w agree on the $i - 1$ most significant bits. Therefore, the jth most significant bit of z agrees with the corresponding bit of y and differs from that of w, and node z and w agree on the $j - 1$ most significant bits. From Lemma 8.1.3, node z is closer to y than w, which contradicts the fact that w is closer to y than all the other nodes in m's buckets. Note that u is a potential candidate for node m's jth bucket, so the bucket must be non-empty. In the example, node z has an identity $110***$, so is closer to $y = **0***$ than $w = 1111**$.

If $j = i$, it can be shown that node m is closer to y than node w, which contradicts the fact that w is closer to y than m. The proof is left as an exercise. ☐

From Lemma 8.1.6, each time a node returns another node from its ith bucket in response to a query, in further steps of the lookup the i most significant bits of any node that is returned cannot be changed. The process stops when no more nodes closer to the query can be found. Thus, the query process must terminate after n contacts, and Theorem 8.1.4 holds.

We considered the case $k = 1$ because of its simplicity. In reality, in Kademlia, each bucket of a Kademlia node contains more than one entry, i.e., $k > 1$. Further, a (key, value) pair is stored at the k nodes closest to the key. We will now describe the lookup process for this more general case. Assume that node m receives a query for key y. The lookup process is described below.

(1) Node m selects the k closest nodes to key y among all nodes in its buckets, including itself, and then chooses α nodes at random from the k nodes.

(2) Node m contacts the α nodes. Each of these α nodes returns a list of k nodes closest to the key according to its buckets and itself. Thus, node m receives αk nodes.

(3) Node m selects k nodes closest to key y from the returned αk nodes and those in the original k-node list to form a new k-node list. This finishes one iteration.

(4) Node m selects α nodes that have not been contacted from the k-node list, and repeats steps (2) and (3). The process terminates after all the k nodes in the list have been contacted and the list remains unchanged. These k nodes are the ones that store the target (key, value) pair.

Example 8.1.5

Consider the Kademlia system described in Example 8.1.3. Assume $k = \alpha = 2$. Consider that node 100 receives a query for key 010. The buckets of node 100 are as in Table 8.1.

Table 8.1 The buckets of node 100

First bucket	Second bucket	Third bucket
000, 001	110, 111	

Node 100, therefore, first gets a two-node list {000, 001}, and contacts both of them. Assume node 000's buckets are as in Table 8.2. Node 000 returns nodes 010 and 000 to node 100.

Table 8.2 The buckets of node 000

First bucket	Second bucket	Third bucket
100, 111	010	001

Assume node 001's buckets are as in Table 8.3. Node 011 returns nodes 010 and 000.

Table 8.3 The buckets of node 001

First bucket	Second bucket	Third bucket
100, 110	010	000

After receiving the responses, node 100 updates the two-node list to {010, 000}. Since 000 has been contacted, node 100 contacts 010 only. Assume node 010's buckets are as in Table 8.4. Node 010 returns nodes 010 and 000.

Table 8.4 The buckets of node 010

First bucket	Second bucket	Third bucket
110, 111	000, 001	

The two-node list of node 100 remains unchanged and the lookup process terminates. It is easy to verify that 010 and 000 are the two closest nodes to key 010, so both nodes are responsible for key 010. □

Bucket maintenance

Kademlia exploits query processes to initialize and update its buckets. Suppose that node m receives a query from node p. By identifying the left-most bit where the identifier of node p differs from it, node m can find the bucket to which p may belong. Call this bucket i. The bucket, which is a list of nodes, is updated as follows.

- If node p is already in the bucket, move node p to the tail of the list.
- If the ith bucket is not full, and node p is not in the bucket, node p is added to the tail of the list.
- If node p is not in the bucket and the bucket is full, node m contacts the least recently seen node, say node h. If node h responds, node p is discarded and node h is moved to the tail of the list; otherwise, node h is removed from the bucket and node p is added to the tail of the list.

Note that this update process is a *least recently seen eviction policy,* but live nodes (nodes still in the network) are never evicted from a bucket. Also note that the buckets are updated when nodes receive actual queries. No extra queries are needed for maintaining the bucket information.

Peer churn

To handle peer churn, Kademlia uses redundant storage. Each node periodically finds the k closest nodes for each key the node is responsible for using the key lookup process, and replicates the (key, value) pair to the k nodes. So, when a node leaves or fails, (key, value) pairs that the node is responsible for are available at other nodes. When a node joins the network, the node first queries its own identifier in the network. From all returned node lists during the lookup process, the node can start to construct its buckets. This process also notifies other nodes of the presence of the new node. After querying its own identifier, the new node queries a random key in the ith bucket range for each i to construct its buckets further. For example, node 000 may query key 100. The lookup process will return the k closest nodes to 100. Some of these k nodes can be placed in the first bucket of node 000 if their identities are in the form of $1 * *$.

Kademlia has been used in several file-sharing networks, including Kad Network, Bit-Torrent, and Gnutella. Next we will discuss one of these networks, namely BitTorrent, which is currently the most popular P2P file-sharing system.

8.2   P2P file sharing

In traditional file-sharing systems, users interested in a file (such as a video or webpage) contact a server (or a fixed number of servers) and download the file. Thus, as the number of users increases, the server's upload capacity has to be shared by many users, and, thus, the download rate of each user will go down. On the other hand, once a user (a peer in P2P terminology) in a P2P network receives a file (either fully or partially), it starts uploading the file to others who may not have it. Thus, as the number of users increases, not only does demand increase, but also the uploading capacity of the network itself increases.

Consider a network with a single server, with upload bandwidth c_u, and n peers, each with download bandwidth c and upload bandwidth μ. Then the average download speed per peer under the client/server model is c_u/n, which decreases rapidly as the number of clients increases. The average download speed per peer under the P2P model, however, is

$$\min\left\{\frac{c_u}{n} + \mu, c\right\},$$

assuming that the P2P network has been well designed such that the entire upload capacity of a peer can be fully utilized. At least in principle, as $n \to \infty$, each peer's download rate is lower bounded by $\min\{c, \mu\}$. So the P2P model is a scalable solution for file-sharing. There are many P2P file sharing programs, such as Kazza, Gnuttella, eDonkey/overnet, and BitTorrent, to name a few. In this section, we will study the behavior of BitTorrent, which is the most popular P2P file-sharing protocol.

8.2.1   The BitTorrent protocol

Pictorially, the BitTorrent system can be described as in Figure 8.7. In the BitTorrent system, the set of peers distributing the same file is called a *swarm*. The peers in a swarm are further classified into two categories: *seeds*, the peers who have the complete copy of the file; and *leechers*, the peers who have not yet downloaded the complete file. Each file is divided into chunks, and one chunk is transferred at a time between two peers. In Figure 8.7, the file is divided into four chunks. Peers 1 and 2 have the complete copy, so they are seeds; peers 3, 4, and 5 only have a subset of the chunks, so they are leechers. The tracker is a central server that maintains the list of peers in the swarm.

To share a file in BitTorrent, the publisher first needs to create a .torrent file, which contains the information of the file and the tracker that is responsible for the file. A peer, who wants to download the file, needs to download the .torrent file, and then registers at

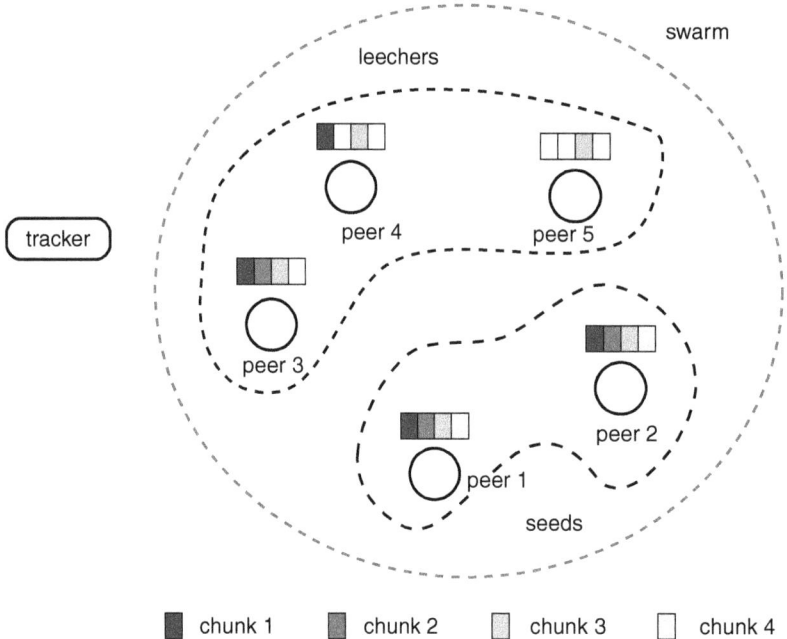

Figure 8.7 BitTorrent system.

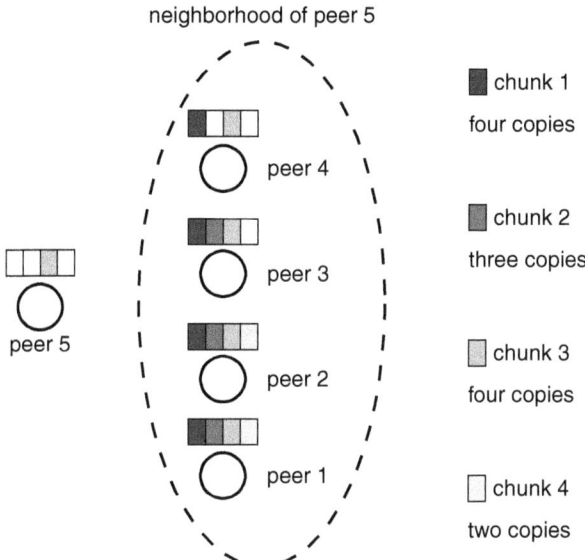

Figure 8.8 The peer first requests chunk 4, which has only two copies in its neighborhood.

the corresponding tracker to find out the set of peers in the swarm. The peer then connects to a subset of peers in the swarm to download the file.

When downloading the file, a peer asks each of its neighbors (connected peers) for a list of available chunks, and requests missing chunks. When requesting missing chunks, BitTorrent uses the *rarest-first policy*: requesting those chunks that are the rarest in a peer's neighborhood. Considering Figure 8.8 and assuming all peers are connected to each other, peer 5 finds four copies of chunks 1 and 3 in its neighborhood, three copies of chunk 2, and two copies of chunk 4. Under the rarest-first policy, the peer requests chunk 4 first. By reducing the rarity of chunks, this policy attempts to ensure that a peer that needs only one more chunk does not have to wait very long due to the fact that the chunk is rare.

In uploading the files, BitTorrent adopts a tit-for-tat strategy to incentivize uploading. Under the tit-for-tat policy, a peer uploads to m neighbors who are uploading to it at the highest rates, called the top m list. Hence, a peer who is uploading with high speed has a better chance of being selected by other peers and gets a high download speed. The uploading rates of the neighbors are periodically re-evaluated.

A strict tit-for-tat strategy, however, has two drawbacks: (i) newly joined peers will not receive any data because they do not have any chunk to upload; and (ii) two peers that can exchange chunks at a high rate may not discover each other because neither of them initiates the exchange. To overcome these two issues, BitTorrent implements *optimistic unchoking*. Every 30 seconds, each peer will contact a randomly selected new peer and starts to upload chunks. If this randomly selected new peer provides a good upload rate and can be moved to the top m list under the tit-for-tat strategy, it will be moved to the top m list in the next round, and the two peers keep exchanging chunks. Otherwise, in the next round, the new peer will be choked and another random peer is selected. This optimistic unchoking allows newly joined peers to receive data without uploading any data, and helps peers discover each other.

Since the tracker in BitTorrent is a central server, the system is vulnerable to single-point failure. Some BitTorrent systems now use Kademlia as the database to resolve this issue. In BitTorrent, peers upload chunks to each other, so the total capacity available to upload files increases proportionally to the number of peers in the network, assuming that a peer always contains chunks that are useful to others and a peer with missing chunks can always find another peer from which it can download one of the missing chunks. If this is the case, each peer can download files at a constant rate, and the time required by a peer to download a file does not increase with the number of peers in the network. This is in contrast to the client/server model, where the upload capacity is fixed, so the download time increases with the number of clients in the system. To understand the performance of BitTorrent analytically, we need to model the system as a continuous-time Markov chain. The analysis is deferred to Chapter 9, after continuous-time Markov chains are introduced.

8.3   Structured P2P streaming

Traditionally, much of the P2P traffic consisted of file transfers, including video and music file sharing. As P2P file-sharing networks have matured, many ideas from file sharing have been transplanted to media-streaming applications, some with considerable success. Streaming media using P2P imposes a more exacting set of constraints than file sharing because each chunk in the streamed file must be received within a certain deadline so that the content can be played out smoothly. A natural approach for designing streaming networks is to create a content-delivery tree amongst the source and peers to *push* chunks on the tree, which we call *structured P2P streaming*. In this section, we will look at the capacity of structured P2P streaming, and describe a simple yet optimal method to construct trees for streaming purposes.

We consider a P2P network with a single source, denoted by s, and N nodes. Let $T_1, T_2, T_3, \ldots, T_k$ denote the possible spanning trees with s as the root, where a spanning tree is a tree that includes all the nodes in the network. For example, consider a network with a single source and two nodes, as shown in Figure 8.9. The three spanning trees are shown in the same figure. Each tree represents a possible path from the source to all the other nodes in the network. Given the trees, such as the ones in Figure 8.9, different chunks from the streaming content may be distributed over different trees. Our goal is to find the maximum possible streaming rate from the source to all the nodes. To do this, we will compute the maximum rate at which content can be delivered over each tree, taking into account the fact that each node appears in all the trees. Note that this model for delivery of streaming content is very general. Suppose we do not use trees to transfer content, but simply track the path of a single chunk through the network. If we ignore transmissions that lead to loops, as such transmissions are redundant, the path taken by a chunk to reach all nodes is a tree.

We introduce the following notation:

- r_t the rate (packets/second) at which streaming data is transmitted over tree T_t;
- λ the rate at which streaming data is sent out from the source;
- u_s the upload bandwidth of the source;

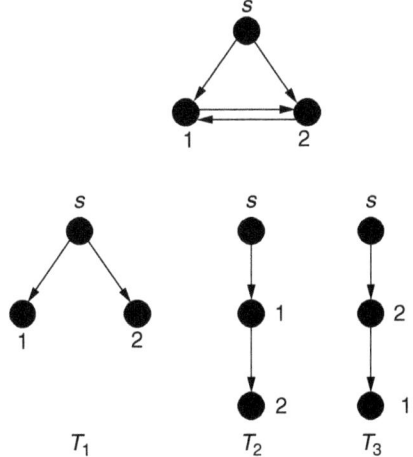

Figure 8.9 The three spanning trees associated with the three-node network.

- u_i the upload bandwidth of peer i;
- d_i the download bandwidth of peer i.

To compute the maximum streaming rate possible, we note that if content is streamed at rate r_t over tree t, each of the intermediate nodes in the tree has to upload content at this rate, but the leaf nodes, being receivers, do not have to upload any content along this tree. Keeping this in mind, the constraints required for streaming rate λ to be supportable in the network are the following:

$$\lambda \leq \sum_t r_t, \tag{8.2}$$

$$\sum_t r_t \leq u_s, \tag{8.3}$$

$$\sum_t r_t \leq d_i, \qquad \forall i, \tag{8.4}$$

$$\sum_t c_t(i)r_t \leq u_i, \qquad \forall i, \tag{8.5}$$

where $c_t(i)$ is the number of children of node i in tree t. Note that constraint (8.2) states that the streaming rate from the source must be less than or equal to the total streaming rate over all trees, constraint (8.3) is the upload capacity constraint of the server; constraints (8.4) and (8.5) are the download and upload capacity constraints, respectively, at each peer. Consider the three-node network in Figure 8.9, and assume each link can support 1 packet/time slot. To support streaming rate $\lambda = 2$, a possible solution is $r_1 = 0$, $r_2 = r_3 = 1$. In other words, the source pushes data over trees T_2 and T_3, each with rate one packet/time slot.

Given a P2P network, computing the maximum streaming rate λ is to find

$$\max \lambda, \quad \text{subject to } (8.2)–(8.5). \tag{8.6}$$

Because a general graph may have a large number of spanning trees, e.g., a complete graph with N nodes has N^{N-2} spanning trees, (8.6) is a very complicated high-dimensional linear programming problem. But it has a surprisingly simple solution. In what follows, we will show that

$$\max \lambda = \min \left\{ u_s, d_{\min}, \frac{\sum_{i=1}^n u_i + u_s}{N} \right\}, \tag{8.7}$$

where $d_{\min} = \min_i d_i$ is the minimum download bandwidth among all peers. We will also show that this maximum streaming rate can be achieved by utilizing $N + 1$ simple trees.

First, the upper bound

$$\lambda \leq \min \left\{ u_s, d_{\min}, \frac{\sum_{i=1}^n u_i + u_s}{N} \right\}$$

is intuitive.

(i) The streaming rate λ cannot be greater than the server's upload capacity u_s.
(ii) It cannot be larger than the download bandwidth of any peer.
(iii) The overall upload capacity of the network is $u_s + \sum_{i=1}^N u_i$. So, even if this entire capacity is used to serve all the peers in the network, the maximum value for λ can only be

$$\frac{\sum_{i=1}^N u_i + u_s}{N}.$$

Theorem 8.3.1 The streaming rate

$$\lambda = \min \left\{ u_s, d_{\min}, \frac{\sum_{i=1}^n u_i + u_s}{N} \right\}$$

is achievable.

Proof We will prove that this upper bound is achievable by considering the following three cases.

Case 1: Assume that

$$u_s \leq \min \left\{ d_{\min}, \frac{\sum_{i=1}^N u_i + u_s}{N} \right\},$$

i.e., the upload bandwidth of the source is the bottleneck. In this case, consider the N streaming trees, as shown in Figure 8.10. Note that the ith tree consists of sending a packet to node i from the server; node i, in turn, delivers the packet to the rest of the nodes.

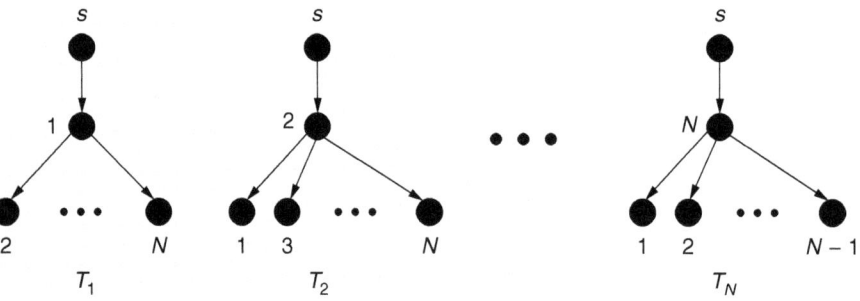

Figure 8.10 The N spanning trees used for case 1.

We let the streaming rate over the ith tree be

$$r_i = \frac{u_i}{\sum_{j=1}^{N} u_j} u_s.$$

Clearly, the total streaming rate over the N trees is

$$\sum_{i=1}^{N} r_i = u_s = \min\left\{u_s, d_{\min}, \frac{\sum_{i=1}^{N} u_i + u_s}{N}\right\}.$$

We will now verify that all of the constraints are satisfied. Constraints (8.2), (8.3), and (8.4) are obviously satisfied. To verify (8.5), note that each peer only uploads in one of the trees. Specifically, peer i uploads in the ith tree, whose rate is r_i. So peer i has to upload at rate r_i to each of the rest of the $N - 1$ peers. The required upload bandwidth, therefore, is given by

$$(N - 1)r_i = \frac{(N - 1)u_i}{\sum_{j=1}^{N} u_j} u_s.$$

According to the assumption,

$$u_s \leq \frac{u_s + \sum_{j=1}^{N} u_j}{N}$$

or

$$u_s \leq \frac{1}{N - 1} \sum_{j=1}^{N} u_j,$$

so

$$(N - 1)r_i \leq \frac{(N - 1)u_i}{\sum_{j=1}^{N} u_j} \times \frac{1}{N - 1} \sum_{j=1}^{N} u_j = u_i,$$

and constraint (8.5) is satisfied.

Case 2: Assume that

$$\frac{u_s + \sum_{j=1}^{N} u_j}{N} \leq \min\{d_{\min}, u_s\},$$

i.e., the total upload capacity becomes the bottleneck. In this case, consider $(N + 1)$ trees: the N trees in Case 1 and an $(N + 1)$th tree, as in Figure 8.11.

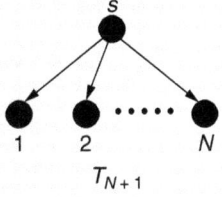

Figure 8.11 The $(N + 1)$th spanning tree.

Let the streaming rate over the ith tree be $r_i = u_i/(N-1)$ and

$$r_{N+1} = \frac{1}{N}\left(u_s - \frac{\sum_{j=1}^{N} u_j}{N-1}\right),$$

so the streaming rate is

$$\lambda = \sum_{i=1}^{N+1} r_i = \frac{\sum_{i=1}^{N} u_i}{N-1} + \frac{1}{N}\left(u_s - \frac{\sum_{j=1}^{N} u_j}{N-1}\right) = \frac{u_s}{N} + \frac{\sum_{j=1}^{N} u_j}{N},$$

which matches the upper bound.

We will now verify that all the constraints are satisfied. Constraint (8.2) simply states that any rate less than or equal to $\sum_{i=1}^{N} r_i + r_{N+1}$ is supportable. To check constraints (8.3) and (8.4), note that

$$\lambda = \frac{u_s}{N} + \frac{1}{N}\sum_{j=1}^{N} u_j \le \min\{u_s, d_{\min}\}$$

by our assumption for Case 2.

Constraint (8.5) is satisfied because peer i only uploads in the ith tree, with upload rate

$$(N-1)\frac{u_i}{N-1} = u_i.$$

Case 3: Assume

$$d_{\min} \le \min\left\{u_s, \frac{u_s + \sum_{j=1}^{N} u_j}{N}\right\},$$

i.e., the download capacity of one peer becomes the bottleneck. We further divide this case into two subcases.

Case 3a: For the first subcase, we assume $d_{\min} \le \sum_{j=1}^{N} u_j/(N-1)$. In this subcase, we use the trees in Case 1 with

$$r_i = d_{\min}\frac{u_i}{\sum_{j=1}^{N} u_j}.$$

Therefore,

$$\sum_{i=1}^{N} r_i = d_{\min} \le u_s.$$

So the server's upload capacity constraint and the peers' download capacity constraints are satisfied. Further, the upload rate of peer i is

$$(N-1)r_i = \frac{(N-1)u_i}{\sum_{j=1}^{N} u_j}d_{\min} \le \frac{(N-1)u_i}{\sum_{j=1}^{N} u_j} \times \frac{\sum_{j=1}^{N} u_j}{N-1} = u_i,$$

so constraint (8.5) is satisfied.

Case 3b: For this subcase, we assume $d_{\min} > \sum_{j=1}^{N} u_j/(N-1)$. We use the trees in Case 2 with the following rates:

$$r_i = \frac{u_i}{N-1}, \qquad \text{for } i \le N,$$

$$r_{N+1} = d_{\min} - \frac{\sum_{i=1}^{N} u_i}{N-1}.$$

The streaming rate is

$$\lambda = \sum_{i=1}^{N+1} r_i = d_{\min} \leq u_s,$$

so conditions (8.2)–(8.4) hold. The upload rate of peer i is

$$(N-1)\frac{u_i}{N-1} = u_i,$$

so condition (8.5) is satisfied. $\square$

Structured P2P networks have the advantage that, since the trees are pre-specified and the rates on each tree are also pre-specified, tight control can be exercised over the delays in delivering packets to all the peers. However, the main drawback is that there is a significant overhead to maintaining the trees, especially when there are arrivals and departures of peers.

8.4   Unstructured P2P streaming

An alternative approach to structured P2P streaming is to obtain chunks from randomly selected peers, as in BitTorrent. We call such an approach *unstructured P2P streaming*. Several systems exist that use this approach to content streaming, and notable examples are CoolStreaming, which was one of the first systems to employ this approach, and some highly popular in East Asia, such as PPLive, QQLive, and TVAnts.

Note that in P2P streaming networks, chunks need to be played out sequentially, and the chunk to be played next should be available when required. This sequential-playout constraint makes unstructured P2P streaming different from P2P file sharing. In this section, we consider a simplified unstructured P2P streaming system, where peers are connected to each other. The system is time slotted such that, during each time slot, a peer can randomly select another peer and download one (and at most one) chunk. We do not limit the number of chunks a peer can upload. In other words, the P2P streaming system is assumed to be download bandwidth limited. Since the download bandwidth of each peer is 1 chunk/time slot, the streaming rate is at most 1 chunk/time slot.

When chunks are randomly distributed in the network, the time a chunk arrives at a specific peer is a random variable. So chunks may arrive out of order, and each peer needs to maintain a buffer to store those chunks that have arrived but are not required to be played out. We assume all peers have a buffer of size m, with the buffer positions indexed by $i \in \{1,...,m\}$.

We further assume chunks are indexed from 0 to ∞, where chunk t is the chunk sent out from the source at time t. At the beginning of time slot t, the chunk in buffer position i is the one that was sent out from the source at time $t - i$. An unstructured P2P streaming network works as follows.

(i) At the beginning of time slot t, the source randomly selects a peer and uploads chunk t to the selected peer.

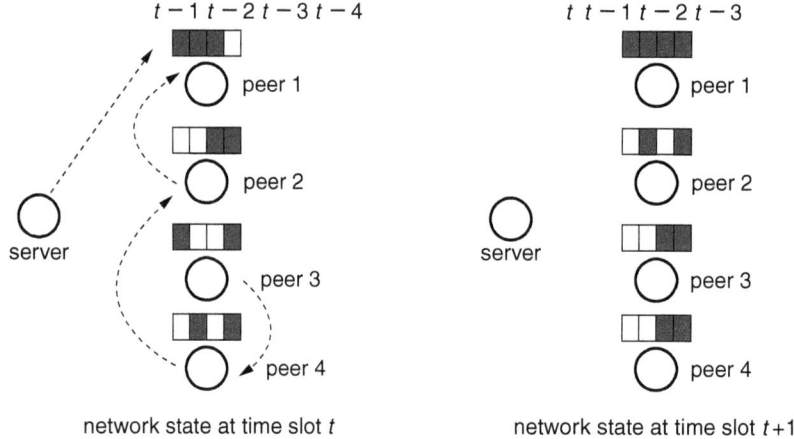

Figure 8.12 Unstructured P2P streaming. Dashed lines represent peer selections.

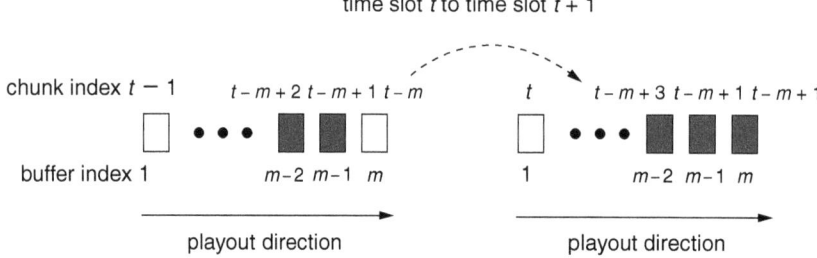

Figure 8.13 Playout process under the unstructured P2P streaming.

(ii) Each peer, if not selected by the source, chooses another peer at random, and may request a chunk from it according to some *chunk-selection policy,* as shown in Figure 8.12.

(iii) At the end of time slot t, a peer attempts to play out the chunk at buffer position m, causing a rightward shift of the buffer contents by one position, as shown in Figure 8.13. To simplify analysis, here we will assume that all peers are time synchronized and that they all attempt to play out the *same chunk*. The chunk is then removed from the system by all peers. So, if the buffer position m is empty, there is a gap in playout, and the missing chunk is never recovered.

The probability that a chunk is not available when required is called *skip-free probability,* which measures the quality of the streaming video. We assume a target skip-free probability of q over all peers. Further, we are interested in the *maximum streaming rate* and *playback delay,* where the maximum streaming rate is the maximum rate the system can support, and

$$\text{playback delay} = \text{buffer size } m \times \text{playout time of one chunk,}$$

which measures the delay from when a chunk is sent out from the source to when the chunk is played out at the peers.

It is clear that the maximum streaming rate is 1, as the download capacity of a peer is 1. We are interested in the playback delay the network has to tolerate to support streaming rate 1 and skip-free probability q. Since the playback delay is a linear function of buffer size m, we next calculate the minimum buffer size required to support skip-free probability q and stream rate 1, which determines the minimum playback delay.

We consider a specific chunk-selection policy, called *rarest-first policy*, which selects a missing chunk that is most recently sent out from the source. In other words, the policy requests chunks in buffer positions with small indices first. To study the performance of the rarest-first policy, we assume the system is in steady state. Specifically, we assume that the buffer-occupancy probabilities have a steady-state distribution, where the steady-state probability that buffer space i is occupied is denoted by $p(i)$. We assume that this distribution is identical and independent across peers. As we will see shortly, this i.i.d. assumption makes the system analytically tractable.

Under the above assumption, we can write down a simple relation between the steady-state buffer-occupancy probabilities as follows:

$$p(i+1) = p(i) + s(i)p(i)(1 - p(i)), \qquad \forall\, i \geq 1, \tag{8.8}$$

$$p(1) = \frac{1}{N}. \tag{8.9}$$

In the above, because buffer position $i+1$ is filled by a rightward shift from buffer position i, its steady-state probability is the probability that buffer position i was already filled at the beginning of the last time slot, plus the probability that i was filled during the last time slot. The latter term is derived by considering that $p(i)(1 - p(i))$ is the probability that a peer, say peer α, does not possess the chunk in buffer position i, but the selected peer, say peer β, does; and $s(i)$ is the probability that peer α chooses to fill buffer position i given that peer α does not possess the chunk in buffer i while peer β does.

Note that the rarest-first policy considers buffer positions with small indices first. Therefore,

$$s(i) = \left(1 - \frac{1}{N}\right)\prod_{j=1}^{i-1}(1 - p(j)(1 - p(j))), \tag{8.10}$$

where $(1 - 1/N)$ is the probability that the peer is not selected by the source so that it can download a chunk from other peers; and $1 - p(j)(1 - p(j))$ is the probability that either buffer j at peer α is filled, or that both peers do not possess the chunk in buffer position j. Note that we have made the assumption that the probabilities $\{p(j)\}$ (where $p(j)$ is the probability of having a chunk in position j) are independent across j. We note that this assumption, as well as the earlier assumption that the chunk availability probabilities are identical and independent across peers, are difficult to justify. However, the models resulting from these assumptions work reasonably well in predicting performance in simulations reported in the literature, and so we assume them here.

By induction, it can be shown that, under the rarest-first policy,

$$s(i) = 1 - p(i). \tag{8.11}$$

First, it is obvious that $s(1) = 1 - 1/N$. Now assume that (8.11) holds for any $i \leq k$. Then we can obtain

$$s(k+1) = \left(1 - \frac{1}{N}\right) \prod_{j=1}^{k} (1 - p(j)(1 - p(j)))$$

$$= {}_{(a)} (1 - p(k)(1 - p(k)))s(k)$$

$$= s(k) - s(k)p(k)(1 - p(k))$$

$$= {}_{(b)} 1 - p(k) - s(k)p(k)(1 - p(k))$$

$$= {}_{(c)} 1 - p(k+1),$$

so claim (8.11) holds. Note that equality (a) holds due to (8.10), (b) holds due to the induction assumption (8.11), and (c) holds due to (8.8).

Now, the buffer-occupancy probabilities satisfy

$$p(i+1) = p(i) + p(i)(1 - p(i))s(i)$$

$$= p(i) \left(1 + (1 - p(i))^2\right).$$

Note that this expression is the same for all m's as long as $i + 1 \leq m$. Therefore we define M_q to be the last buffer position such that $p(M_q) < q$, i.e.,

$$M_q = \max\{i : p(i) < q\}.$$

Next we will establish upper bounds on M_q and show $M_q = O(\log N)$. So, under the rarest-first policy, the minimum playback delay peers have to tolerate to achieve skip-free probability q is $O(\log N)$.

Since $p(\cdot)$ is increasing in i, for $i \leq M_q$, we have

$$p(i+1) = p(i) \left(1 + (1 - p(i))^2\right)$$

$$\geq p(i) \left(1 + (1 - q)^2\right),$$

which implies that

$$p(i) \geq \frac{(1 + (1 - q)^2)^{i-1}}{N}.$$

From this inequality, we obtain

$$M_q \leq \frac{\log q + \log N}{\log(1 + (1 - q)^2)} + 1,$$

because, otherwise,

$$p(M_q) \geq q,$$

contradicting the definition of M_q. Note that $p(M_q + 1) \geq q$ according to the definition of M_q. So, under the rarest-first policy, the buffer size required to achieve skip-free probability q for any constant q is $O(\log N)$.

Theorem 8.4.1 Given any target skip-free probability q, the buffer size required to achieve this skip-free probability with the rarest-first policy is $O(\log N)$. □

We remark that, in this model, we assumed that the system is download bandwidth limited. In fact, similar results hold if the system is upload bandwidth limited, but not download bandwidth limited. In that case, the number of peers having a specific chunk can at most double in every time slot, so the time required to deliver a chunk to the N peers is at least $\log N$, and a playback delay of $O(\log N)$ is order-wise optimal.

8.5   The gossip process

In the preceding section, we made a number of assumptions to develop a simple model to understand the performance of unstructured P2P streaming systems. However, more precise models are possible in some instances. The central idea behind these models is to relate P2P file transfers to a well-studied stochastic model called the gossip process. In a gossip process, a single packet has to be transferred from a source to all peers. The connection between this model and a P2P network is difficult to develop and is beyond the scope of this book. In this section, we present only an analysis of the gossip model to provide an interested reader with the necessary background to analyze P2P streaming networks further. Some useful references for this purpose are provided at the end of the chapter.

The gossip process is defined as follows. Consider a town with N people. Initially, there is only one person in the town who is in possession of a rumor. At each time slot, each person who knows the rumor randomly contacts someone from the town, including himself/herself, to tell the rumor. Thus, the person contacted could be someone who already knows the rumor. This completes the description of the gossip process. The question of interest in such a model is the following: how many time slots does it take for all N people to know the rumor?

Before we provide a precise answer to the above question, we present some intuition behind the analysis that leads to the answer. The analysis is divided into two phases.

- **Phase I**: When the number of people who know the rumor is relatively small, with high probability anyone they contact will not know the rumor. Thus, the number of people who know the rumor will roughly double in each time slot. In the analysis, we will find it easier to show that, as long as the number of people who know the rumor is less than $N/6$, the number of people who know the rumor increases by a factor of 1.5, with high probability.
- **Phase II**: Next, since the number of people who know the rumor is a non-negligible fraction of N, it can be shown that each person that does who know a rumor will be contacted with a non-negligible probability by a person who knows the rumor in each time slot. This fact will be used, along with the union bound, to show that all remaining people who do not know the rumor will hear the rumor within $O(\log N)$ steps, with high probability.

We now analyze the two phases separately, and combine the results of these analyses to obtain our main result on the gossip process.

Phase I: It is trivial to note that the number of people who know the rumor is 2 after the first time slot, with probability $1 - 1/N$. Now, suppose that the number of people who know the rumor at the beginning of a time slot is i. We will now derive a lower bound on the probability that the number of people who know the rumor by the end of the time slot is at least $1.5i$. Note that, if there are i people that know the rumor at the beginning of a time slot, exactly i contacts will be made within the time slot to try to spread the rumor. A key insight is that the number of people who know the rumor at the end of the time slot is the same, independent of whether the contacts are made simultaneously or sequentially in a particular order. The latter viewpoint is more convenient for analysis. Thus, the number of people who know the rumor can potentially change within the time slot from i to $i+1$ to $i+2$ and so on. However, people who newly acquire the rumor in the time slot will not be making contacts till the next time slot.

Denote by W_i the number of contacts required to increase the number of people who know the rumor from i to $i+1$ within the time slot. Since each contact spreads the rumor to a new person with probability $1-i/N$, W_i is a geometrically distributed random variable with parameter $1 - i/N$, and $\{W_i\}$ are independent across i. Therefore, the following equalities hold.

$$\Pr(W_i = k) = \left(\frac{i}{N}\right)^{k-1}\left(1 - \frac{i}{N}\right), \qquad k \geq 1,$$

and

$$E\left[e^{\theta W_i}\right] = \frac{N-i}{Ne^{-\theta} - i}, \qquad \text{for } \theta < \log\frac{N}{i}. \tag{8.12}$$

We now calculate the probability that the number of people who know the rumor does not increase from i to at least $1.5i$ in this time slot, where exactly i contacts occur. The event that the number of people who know the rumor does not increase from i to at least $1.5i$ is the same as the event $W_i + \cdots + W_{\lfloor 1.5i \rfloor} > i$. According to inequality (8.12), for any θ such that $e^\theta < N/1.5i$, we have

$$\Pr\left(W_i + \cdots + W_{\lfloor 1.5i \rfloor} > i\right) \leq e^{-i\theta} \prod_{j=i}^{\lfloor 1.5i \rfloor} \frac{N-j}{Ne^{-\theta} - j}$$

$$= e^{-(i-1)\theta} \prod_{j=i}^{\lfloor 1.5i \rfloor} \frac{N-j}{N - je^\theta}$$

$$\leq e^{-(i-1)\theta} \left(\frac{N - \lfloor 1.5i \rfloor}{N - \lfloor 1.5i \rfloor e^\theta}\right)^{\lfloor 1.5i \rfloor - i + 1}$$

$$\leq e^{-(i-1)\theta} \left(\frac{N - \lfloor 1.5i \rfloor}{N - \lfloor 1.5i \rfloor e^\theta}\right)^{0.5i+1}.$$

Choosing θ such that $e^\theta = N/2\lfloor 1.5i \rfloor$, we have

$$\Pr\left(W_i + \cdots + W_{\lfloor 1.5i \rfloor} > i\right) \leq e^{-(i-1)\theta}(\sqrt{2})^i 2 \leq \left(\frac{3\sqrt{2}i}{N}\right)^{i-1} 2\sqrt{2}.$$

We now argue that the above probability is of the order of $1/\sqrt{N}$. To see this, first consider $2 \leq i \leq \sqrt{N}$. Then

$$\Pr\left(W_i + \cdots + W_{\lfloor 1.5i \rfloor} > i\right) \leq \left(\frac{3\sqrt{2}i}{N}\right)^{i-1} 2\sqrt{2} \leq \left(\frac{3\sqrt{2}\sqrt{N}}{N}\right)^{i-1} 2\sqrt{2} \leq \frac{12}{\sqrt{N}}.$$

On the other hand, if $\sqrt{N} < i \leq N/6$, we have

$$\Pr\left(W_i + \cdots + W_{\lfloor 1.5i \rfloor} > i\right) \leq \left(\frac{1}{\sqrt{2}}\right)^{\sqrt{N}-4} \leq \frac{12}{\sqrt{N}}.$$

Thus, we have shown that the number of people who know the rumor increases by a factor of at least 1.5 in Phase I with high probability. Because the number of people who know the rumor increases exponentially, we can expect Phase I to last at most an order of $\log N$ time slots with high probability. To make this precise, let T_1 be the duration of Phase I. Thus, from the union bound,

$$\Pr\left(T_1 \geq \frac{1}{\log 1.5}\log\frac{N}{6} + 1\right) \leq \left(\frac{1}{\log 1.5}\log\frac{N}{6} + 1\right)\frac{12}{\sqrt{N}}.$$

In other words, the probability that Phase I lasts longer than $O(\log N)$ time slots goes to zero as $N \to \infty$.

Phase II: Note that at least $N/6$ contacts will be made in each time slot in this phase. A person who does not know the rumor is selected with probability $1/N$ during each contact, so the probability that the person is not contacted in a time slot is given by

$$\left(1 - \frac{1}{N}\right)^{N/6} \leq e^{-1/6}.$$

Therefore, the probability that a particular person is not contacted for $7\log N$ time slots is upper bounded by

$$\left(e^{-1/6}\right)^{7\log N} = N^{-7/6}.$$

Thus, using the union bound, the probability that at least one of the remaining $5N/6$ persons is not contacted within $7\log N$ time slots in Phase II is upper bounded by

$$\frac{5N}{6}N^{-7/6} = \frac{5}{6N^{1/6}}.$$

Putting together the results from the analysis of the two phases gives us the following theorem.

Theorem 8.5.1 Let T_f be the amount of time taken for the gossip process to spread the rumor to all N persons. Then,

$$P\left(T_f < \left(\frac{1}{\log 1.5} + 7\right)\log N - \frac{\log 6}{\log 1.5} + 1\right) \geq 1 - \left(\frac{1}{\log 1.5}\log\frac{N}{6} + 1\right)\frac{12}{\sqrt{N}} - \frac{5}{6N^{1/6}}.$$

$\square$

Thus, we have proved that the gossip process ends within $O(\log N)$ time slots with high probability.

8.6   Summary

- **Chord** Chord is an architecture for DHTs in which users are organized in a ring structure. The name of each content is hashed to an identifier (called the key), and the (key, value) pair associated with the content is stored in a node whose identifier is closest to the key. The value refers to the IP address of the node that has the content. Consider a Chord system with K keys and N users. Exploiting the randomness of hash functions, it can be shown that each user is responsible for no more than $2.2(\log N)K/N$ keys, with high probability. To improve the key lookup speed, each user maintains connections to $\log N$

peers, called fingers. The ith finger is the first peer who succeeds the current node by at least 2^{i-1}. By navigating through this network of fingers, it can be shown that a query for the location of a content can be resolved with $O(\log N)$ message exchanges with high probability.

- **Kademlia** Kademlia is another DHT system, which uses XOR distance instead of circular distance. The distance between two nodes in Kademlia is the integer value of the XOR of the two identifiers. A (key, value) pair is stored at multiple nodes that are close in XOR distance to the key. To facilitate the lookup processes, each node, say node m, maintains n buckets. Each bucket consists of k routing entries, and each entry in the ith bucket is the routing information to a node whose most significant $i - 1$ bits (i.e., the left-most i bits) are the same as those of node m, but the ith significant bit is not. It takes at most n contacts to resolve a query in Kademlia.

- **BitTorrent** BitTorrent is a P2P file-sharing system. In BitTorrent, each file is divided into chunks. The chunks are first distributed among the peers in the network. Then, the peers download the chunks they are missing from each other. Thus, instead of overloading a single server to get the file, the peers can download from each other. While uploading chunks, BitTorrent adopts a tit-for-tat strategy to incentivize uploading, and uses a process called optimistic unchoking to allow newly joined peers to receive data without uploading any data and help peers discover each other.

- **Structured P2P streaming** In contrast to file-sharing systems, streaming systems deliver data with strict delay guarantees. When a specific topology is used to connect the peers, the resulting P2P network is called a structured network; otherwise, it is called an unstructured network. The maximum streaming rate under structured P2P streaming is

$$\min \left\{ u_s, d_{\min}, \frac{\sum_{i=1}^{n} u_i + u_s}{N} \right\}.$$

This rate can be achieved using at most $N + 1$ spanning trees, where N is the number of peers in the network.

- **Unstructured P2P streaming** The rarest-first policy in P2P streaming selects a missing chunk that was most recently sent from the source. Given any target skip-free probability q, the buffer size required to achieve this skip-free probability with the rarest-first policy is $O(\log N)$, where N is the number of peers in the network.

8.7   Exercises

Exercise 8.1 (An example of Chord) Consider a Chord system with identifiers in the range $[0, 31]$. Suppose there are ten nodes with identifiers 0, 6, 8, 14, 16, 17, 22, 23, 26, and 30.

(1) Assume there are 12 keys: 1, 2, 7, 9, 11, 12, 13, 20, 21, 23, 27, and 28. For each node in the network, identify the keys that each node is responsible for.

(2) A query for key 27 is initiated at node 0. Write down the sequence of nodes the query will be forwarded to, and the finger tables of these nodes.

Exercise 8.2 (Peer churn in Chord) Consider the Chord system in Exercise 8.1, and assume that a node with identifier 10 joined the system. After node 10 has been added to the network, for each node in the network identify the keys that it is responsible for, and the finger table of node 10.

Exercise 8.3 (Fingers in Chord) Prove that, with high probability, every node is the ith finger of $O(\log N)$ other nodes in Chord, where N is the number of nodes in the network.

Exercise 8.4 (The XOR distance in Kademlia) Consider three nodes with identifiers $11\star\star\star\star$, $100\star\star\star$, and $101\star\star\star$, where $\star$ denotes that the corresponding bit is unknown. In a Kademlia system, which node is the closest one to node 011010, and which node is the second closest?

Exercise 8.5 (An example of Kademlia) Consider a Kademlia system with $k = 2$ and identifiers in the range $[0, 15]$. Suppose there are six nodes with identifiers 0, 1, 2, 6, 8, 14.

(1) Assume there are five keys: 1, 2, 7, 9, and 11. For each node in the network, identify the keys that each node is responsible for.

(2) Assume $\alpha = 1$; the 2-buckets at the six nodes are shown in Table 8.5. A query for key 0 is initiated at node 14. Write down the sequence of nodes that are contacted and the information returned by these nodes.

Exercise 8.6 (Peer churn in Kademlia) Consider the Kademlia system in Exercise 8.5, and assume that a node with identifier 4 joined the system. Identify the keys that each node is responsible for in the network and the 2-buckets of node 4.

Exercise 8.7 (Key queries in Kademlia) Consider a Kademlia system. Suppose that node m is queried for identifier y. Further, suppose that node w in m's ith bucket is closer to y than node m. Prove by contradiction that there does not exist a node u such that u is closer to y than w, and that u and w differ in the ith significant bit and agree on the $i - 1$ most significant bits.

Exercise 8.8 (BitTorrent) In a BitTorrent system, assume each leecher can connect to K other leechers. Further assume that the number of chunks a leecher has is uniformly distributed in $\{0, \ldots, Z - 1\}$, where Z is the total number of chunks, and the chunks a leecher has are uniformly and randomly chosen from the Z chunks. The set of chunks a leecher has is independent from those of other leechers. Note that the efficiency η can be viewed as the probability that a leecher has at least one chunk that is missing in its neighbors. Compute η.

Table 8.5 Buckets at all six nodes

	Node 0	Node 1	Node 2	Node 6	Node 8	Node 14
First bucket	8, 14	8, 14	8, 14	8, 14	1, 2	0, 6
Second bucket	6	6	6	0, 2	14	8
Third bucket	2	2	0, 1			
Fourth bucket	1	0				

Exercise 8.9 (An example of structured P2P streaming) Consider the P2P network shown in Figure 8.14. The upload/download capacities of the four nodes are summarized in Table 8.6.

(1) If node 1 is the streaming source and the other nodes are peers, what is the maximum stream rate? Write down the set of trees that can be used to achieve the streaming rate.

(2) Repeat (1) when node 2 becomes the streaming source and node 1 becomes a peer.

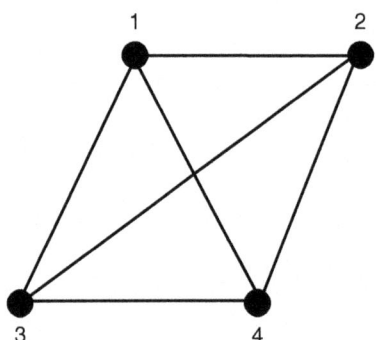

Figure 8.14 Network with four nodes.

Table 8.6 Upload/download capacities of the four nodes

	Node 1	**Node 2**	**Node 3**	**Node 4**
Upload	3	4	3	5
Download	2	6	4	4

Exercise 8.10 (A tighter upper bound on the buffer size under the rarest-first policy) In Section 8.4, we have shown that the buffer size required to achieve skip-free probability q with the rarest-first policy in the unstructured P2P streaming is at most

$$\frac{\log q + \log N}{\log(1 + (1 - q)^2)} + 2.$$

When q is close to 1, i.e., $\epsilon = 1 - q$ is small, $\log(1 + (1-q)^2) \approx \epsilon^2$, so the upper bound on the buffer size becomes

$$\frac{\log q + \log N}{\epsilon^2} + 2 = O\left(\frac{1}{\epsilon^2} \log N\right).$$

In this exercise, we will derive a better bound, and prove that, for any $\delta \in (0, q)$, the buffer size required is at most

$$\frac{\log \delta + \log N}{\log(1 + (1 - \delta)^2)} + \frac{\log q - \log \delta}{\log(1 + (1 - q)^2)} + 3 = O\left(\frac{1}{\epsilon^2} + \log N\right).$$

(1) Define M_δ to be the last buffer position such that $p(M_\delta) < \delta$, i.e.,

$$M_\delta = \max\{i : p(i) < \delta\}.$$

Prove that

$$M_\delta \leq \frac{\log \delta + \log N}{\log(1 + (1 - \delta)^2)} + 1.$$

(2) Prove that

$$M_q - M_\delta \leq \frac{\log q - \log \delta}{\log(1 + (1 - q)^2)} + 1.$$

Exercise 8.11 (A greedy policy for unstructured P2P streaming) In this exercise, we consider a different chunk-selection policy for unstructured P2P streaming, called the greedy policy, which selects chunks with priority $m - 1 > m - 2 > \cdots > 2$. As in Section 8.4, we assume that the buffer-occupancy probabilities have a steady-state distribution and the steady-state distribution is identical and independent across peers.

(1) Prove that $s(i) = p(i + 1) + 1 - p(m) - 1/N$.

(2) Prove that, under the greedy policy, to guarantee that $p(m) \geq q \geq 4/N$, it requires

$$m \geq \log N + \log q - 1 + \frac{1}{\log (2 - q + 3/N)}.$$

8.8   Notes

The Chord and Kademlia protocols were presented in [99, 159] and [116], respectively. The BitTorrent file-sharing protocol is described in [26]. The throughput bounds for structured P2P streaming networks were obtained in [91, 92, 126, 127]. The analysis of unstructured P2P streaming networks is based on the results in [151], which in turn is based on the model developed by [178]. The analysis of the gossip process is based on the results in [42]. However, for the sake of simplicity, we have presented bounds that are much looser than the results in [42]. While we have not presented the connection to P2P streaming systems, the connection between the classical gossip process and the throughput and delay of P2P systems has been studied in [13, 147].

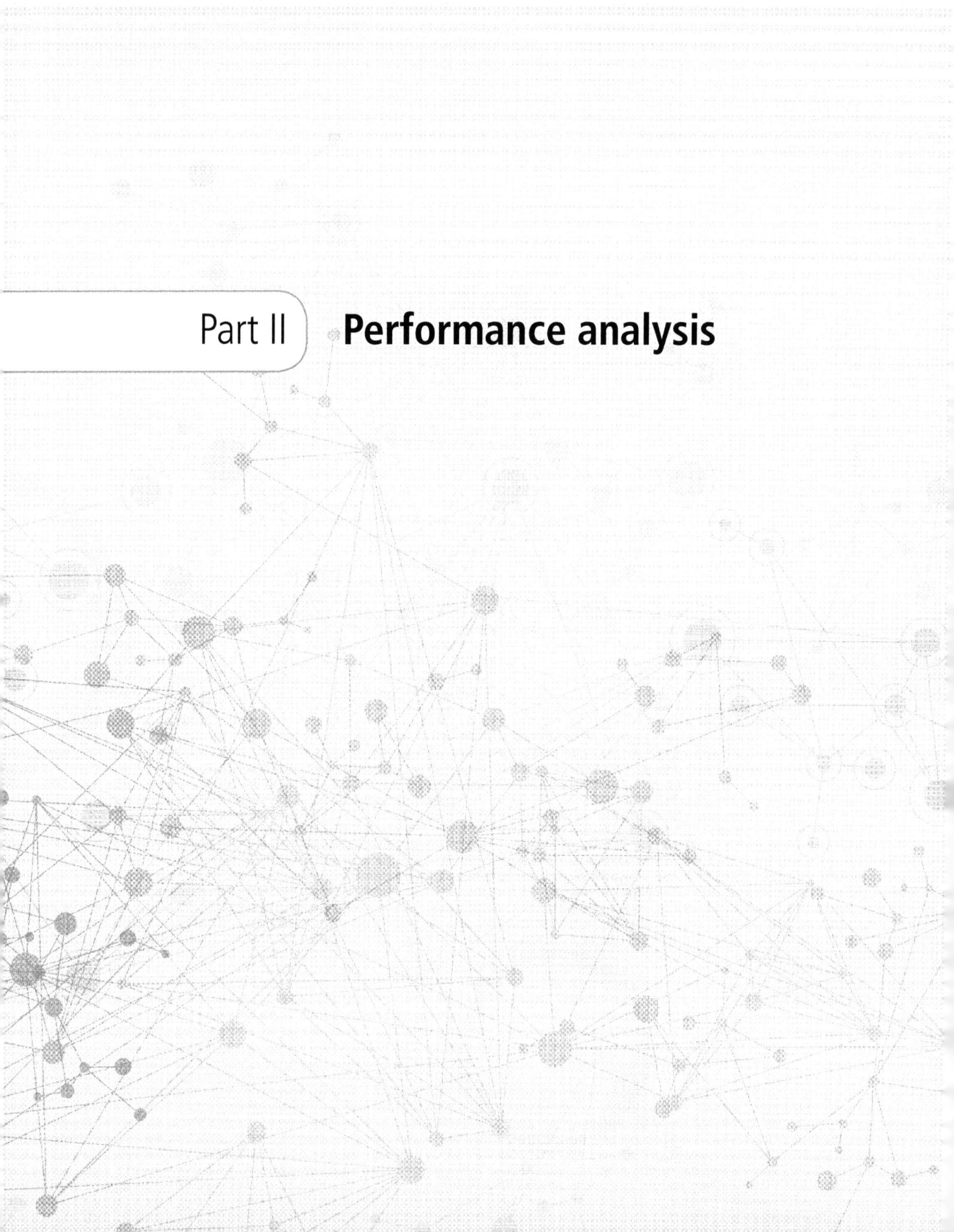

Part II **Performance analysis**

Queueing theory in continuous time

We have learned about Discrete-Time Markov Chains (DTMCs) and discrete-time queueing systems in Chapter 3. In this chapter, we will look at continuous-time queueing systems, which were first developed to analyze telephone networks but has broader applications in communication networks, manufacturing, and transportation. The focus of this chapter is on the basics of continuous-time queueing theory and its application to communication networks. As in the case of discrete-time queueing systems, continuous-time queueing systems will be analyzed by relating them to Markov chains. For this purpose, we will first introduce Continuous-Time Markov Chains (CTMCs) and related concepts, such as the global balance equation, the local balance equation, and the Foster–Lyapunov theorem for CTMCs. Then, we will introduce and study simple queueing models including the M/M/1 queue, the M/GI/1 queue, the GI/GI/1 queue, and the Jackson network, as well as important concepts such as reversibility and insensitivity to service time distributions. Finally, we will relate CTMCs and queueing theory to the study of connection-level stability in the Internet, distributed admission control, telephone networks, and P2P file-sharing protocols. The following questions will be answered in this chapter.

- *Under what conditions does a CTMC have a stationary distribution, and how should it be computed if it exists?*
- *What are the mean delays and queue lengths of simple queueing models?*
- *What is the reverse chain of a CTMC, and how does the concept of reversibility help the calculation of the stationary distributions of queueing networks?*
- *How should CTMCs and queueing theory be used to model and analyze communication networks, including the Internet, telephone networks, and P2P networks?*

9.1 Mathematical background: continuous-time Markov chains

A stochastic process $\{X(t)\}$ is a *Continuous-Time Markov Chain (CTMC)* if the following conditions hold:

(i) t belongs to some interval of $\mathcal{R}$;
(ii) $X(t) \in \mathcal{S}$, where $\mathcal{S}$ is a countable set; and
(iii) $\Pr(X(t+s)|X(u), u \leq s) = \Pr(X(t+s)|X(s))$, i.e., the conditional probability distribution at time $t+s$, given the complete history up to time s, depends only on the state of the process at time s.

A CTMC is said to be *time homogeneous* if $\Pr(X(t+s)|X(s))$ is independent of s, i.e., the conditional probability distribution is time independent. Define $\gamma_i(t)$ to be the amount of time the process stays in state i given the process is in state i at time t. In other words,

$$\gamma_i(t) = \inf\{s > 0 : X(t + s) \neq X(t) \text{ and } X(t) = i\}.$$

The following theorem says that $\gamma_i(t)$ is an exponentially distributed random variable if the CTMC is time homogeneous.

Theorem 9.1.1 The time spent in any state $i \in S$ of a time-homogeneous CTMC is exponentially distributed, i.e., $\gamma_i(t)$ is exponentially distributed with a parameter depending only on state i.

Proof Assume the Markov chain is in state i at time 0, i.e., $X(0) = i$. We consider the probability that the CTMC stays in state i for more than $u + v$ units of time, given that it has stayed in state i for v units of time. Since $\{X(t)\}$ is a CTMC, we have

$$\Pr(\gamma_i(0) > u + v | \gamma_i(0) > v) = \Pr(X(t) = i, v < t \leq u + v | X(s) = i, 0 \leq s \leq v)$$
$$= \Pr(X(t) = i, v < t \leq u + v | X(v) = i).$$

Further, because $X(t)$ is time homogeneous, we have

$$\Pr(\gamma_i(0) > u + v | \gamma_i(0) > v) = \Pr(X(t) = i, 0 < t \leq u | X(0) = i)$$
$$= \Pr(\gamma_i(0) > u),$$

which implies that

$$\Pr(\gamma_i(0) > u + v) = \Pr(\gamma_i(0) > u)\Pr(\gamma_i(0) > v).$$

Therefore, the probability distribution of $\gamma_i(0)$ is memoryless. Since the exponential distribution is the only distribution on the positive real line with this property, the theorem holds. $\square$

In this book, we will assume that the number of transitions in a finite time interval is finite with probability 1. One can construct CTMCs with strange behavior if we do not assume this condition. This condition is called *non-explosiveness*.

A time-homogeneous CTMC can be described by a jump process, where a jump is a state transition. Let $Y(k)$ denote the state of the CTMC after k jumps, and let T_k denote the time duration between the $(k-1)$th and the kth jumps, as shown in Figure 9.1. According to Theorem 9.1.1, T_k is an exponentially distributed random variable with its parameter depending on $Y(k-1)$ only, so a time-homogeneous CTMC can also be described as follows.

(i) During each visit to state i, the CTMC spends an exponentially distributed amount of time in this state. We let $1/q_i$ denote the mean time spent in state i.
(ii) After spending an exponentially distributed amount of time in state i, the CTMC moves to state $j \neq i$ with probability P_{ij}.

Therefore, the CTMC is completely specified by the parameters $\{q_i\}$ and $\{P_{ij}\}$, and the probability distribution of the state at time $t = 0$.

An important quantity of interest is

$$p_l(t) = \Pr(X(t) = l).$$

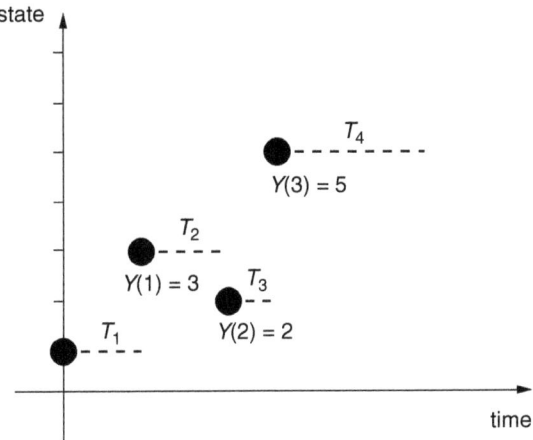

Figure 9.1 A time-homogeneous CTMC as a jump process.

Now we will derive a differential equation to describe the evolution of $p_i(t)$. First, note that

$$p_i(t + \delta) = \Pr(X(t + \delta) = i)$$

$$= \sum_j \Pr(X(t + \delta) = i | X(t) = j) \Pr(X(t) = j)$$

$$= \Pr(X(t + \delta) = i | X(t) = i) \Pr(X(t) = i)$$

$$+ \sum_{j \neq i} \Pr(X(t + \delta) = i | X(t) = j) \Pr(X(t) = j)$$

$$= e^{-q_i \delta} p_i(t) + \sum_{j \neq i} \left(1 - e^{-q_j \delta}\right) P_{ji} p_j(t) + O(\delta^2).$$

In the final step above, the terms that are not of the order δ^2 correspond to no or only one state transition occurring in $[t, t + \delta]$. The probability that more than one transition occurs in $[t, t + \delta]$ is $O(\delta^2)$. When δ is sufficiently small, the above equality can be approximated as follows:

$$p_i(t + \delta) \approx (1 - q_i \delta) p_i(t) + \sum_{j \neq i} \delta q_j P_{ji} p_j(t).$$

Moving $p_i(t)$ on the right-hand side to the left-hand side and dividing both sides by δ, we obtain

$$\frac{p_i(t + \delta) - p_i(t)}{\delta} = -q_i p_i(t) + \sum_{j \neq i} q_j P_{ji} p_j(t),$$

which implies that

$$\dot{p}_i(t) = -q_i p_i(t) + \sum_{j \neq i} P_{ji} q_j p_j(t).$$

Defining $Q_{ji} = P_{ji} q_j$ and $Q_{ii} = -q_i$, we have

$$\dot{p}_i(t) = Q_{ii} p_i(t) + \sum_{j \neq i} Q_{ji} p_j(t)$$

or

$$\dot{p}(t) = p(t)\mathbf{Q},$$

where $\mathbf{Q}$ is an $|\mathcal{S}| \times |\mathcal{S}|$ matrix, called the *transition rate matrix* or *generator matrix* of the CTMC. We remark that Q_{ij} can be interpreted as the rate of transition to state j, given that the CTMC is in state i. In other words, for very small δ, $Q_{ij}\delta$ is the probability of transitioning to state $j, j \neq i$, from state i in a small interval of time of duration δ.

Note that $\mathbf{Q}$ completely describes the CTMC, along with $p(0)$. So, given $\mathbf{Q}$, we need not specify $\{q_i\}$ and $\{P_{ij}\}$. It follows from the definition of $\mathbf{Q}$ that it has the following two properties:

(i) $Q_{ij} \geq 0$ for $i \neq j$, and
(ii) $\sum_j Q_{ij} = 0$.

The first property is easy to verify because $Q_{ij} = q_i P_{ij}$ when $i \neq j$. The second property holds because

$$\sum_j Q_{ij} = \sum_{j \neq i} q_i P_{ij} - q_i$$

$$= q_i \left(\sum_{j \neq i} P_{ij} - 1 \right)$$

$$= 0.$$

Similar to DTMCs, we are interested in the following questions.

- Does there exist a $\pi \geq 0$ such that $0 = \pi\mathbf{Q}$ and $\sum_i \pi_i = 1$? Note that if such a π exists, $p(t) = \pi$ for all t when $p(0) = \pi$. Therefore, such a π is called a stationary distribution.
- If there exists a unique stationary distribution, does $\lim_{k \to \infty} p(t) = \pi$ for all $p(0)$? In other words, does the distribution of the Markov chain converge to the stationary distribution starting from any initial state?

Similar to DTMCs, it is possible to construct examples to show that a CTMC may not possess a unique stationary distribution if it is not irreducible. The definition of irreducibility for CTMCs is presented below.

Definition 9.1.1 (Irreducibility) A CTMC is said to be irreducible if, given any two states i and j,

$$\Pr\left(X(t) = j | X(0) = i\right) > 0$$

for some finite t.   □

Note that there is no concept of aperiodicity for CTMCs because state transitions can happen at any time. The following theorem states that a CTMC has a unique stationary distribution if it is irreducible and has a finite state space.

Theorem 9.1.2 A finite-state-space, irreducible CTMC has a unique stationary distribution π and $\lim_{t \to \infty} p(t) = \pi$, $\forall p(0)$.   □

For a finite-state-space, irreducible CTMC, the stationary distribution can be computed by finding a $\pi \geq 0$ such that $\pi Q = 0$ and $\sum_i \pi_i = 1$. If the state space is infinite, irreducibility is not sufficient to guarantee that the CTMC has a unique stationary distribution.

Example 9.1.1

Consider a CTMC $X(t)$ with the state space to be integers. The transition rate matrix $\mathbf{Q}$ is as follows:

$$Q_{ij} = \begin{cases} 1, & \text{if } j = i+1, \\ 1, & \text{if } j = i-1, \\ -2, & \text{if } j = i, \\ 0, & \text{otherwise.} \end{cases}$$

It is easy to verify that $\Pr(X(t) = j | X(0) = i) > 0$ for some finite t, so $X(t)$ is irreducible. If a stationary distribution π exists, it has to satisfy $0 = \pi Q$, which can be rewritten as

$$0 = \pi_{k-1} - 2\pi_k + \pi_{k+1}, \quad \forall k.$$

Thus, we have to solve for π that satisfies

$$2\pi_k = \pi_{k-1} + \pi_{k+1}, \quad \forall k,$$

$$\sum_{k=-\infty}^{\infty} \pi_k = 1,$$

$$\pi_k \geq 0, \quad \forall k.$$

We have shown in Example 3.3.4 in Chapter 3 that no π can satisfy all three equalities, so a stationary distribution does not exist. $\square$

Similar to DTMCs, we now introduce the notion of recurrence and conditions beyond irreducibility to ensure the existence of stationary distributions.

Definition 9.1.2 Assuming $X(0) = i$, the *recurrence time* is the first time the CTMC returns to state i after it leaves the state. Recall that the amount of time spent in state i is defined as

$$\gamma_i = \inf\{t > 0 : X(t) \neq i \text{ and } X(0) = i\}.$$

The recurrence time τ_i is defined as

$$\tau_i = \inf\{t > \gamma_i : X(t) = i \text{ and } X(0) = i\}.$$

State i is called *recurrent* if

$$\Pr(\tau_i < \infty) = 1,$$

and transient otherwise.

A recurrent state is *positive recurrent* if $E[\tau_i] < \infty$, and is *null recurrent* if $E[\tau_i] = \infty$. $\square$

The following lemma and theorems are stated without proofs.

Lemma 9.1.3 For an irreducible CTMC, if one state is positive recurrent (null recurrent), all states are positive recurrent (null recurrent). Further,

$$\lim_{t \to \infty} p_i(t) = \frac{1}{E[\tau_i](-Q_{ii})},$$

which holds even when $E[\tau_i] = \infty$. $\qquad\square$

Theorem 9.1.4 Consider an irreducible and non-explosive CTMC. A unique stationary distribution $\pi \geq 0$ (i.e., $\sum_i \pi_i = 1$ and $\pi Q = 0$) exists if and only if the CTMC is positive recurrent. $\qquad\square$

This theorem states that if we can find a vector $\pi \geq 0$ such that $\pi Q = 0$ and $\sum_i \pi_i < \infty$, and the CTMC is non-explosive, the CTMC is positive recurrent. Note that if $\sum_i \pi_i \neq 1$, we can define $\tilde{\pi}_i = \pi_i / \sum_j \pi_j$, and $\tilde{\pi}$ is a stationary distribution.

Theorem 9.1.5 If there exists a π such that $\pi Q = 0$ and $\sum_i \pi_i = \infty$ for an irreducible CTMC, the CTMC is not positive recurrent and

$$\lim_{t \to \infty} p_i(t) = 0$$

for all i. $\qquad\square$

Now we have characterized the conditions under which the CTMC is positive recurrent and not positive recurrent. The following lemma presents an alternative characterization of the equation that has to be satisfied by the stationary distribution of a CTMC.

Lemma 9.1.6 $\pi Q = 0$ is equivalent to

$$\sum_{i \neq j} \pi_i Q_{ij} = \pi_j \sum_{i \neq j} Q_{ji}, \quad \forall j. \tag{9.1}$$

Equation (9.1) is often called the *global balance equation*. Note that the equation simply states that, given a state j, the rate at which transitions take place into state j is equal to the rate at which transitions take place out of state j.

Proof Assume that equality (9.1) holds. For fixed j, we have

$$\sum_i \pi_i Q_{ij} = \pi_j \sum_{i \neq j} Q_{ji} + \pi_j Q_{jj}$$

$$= \pi_j \sum_i Q_{ji}$$

$$= 0,$$

where the last equality holds due to the definition of Q.

Now assume $\pi Q = 0$. Then we have

$$\sum_i \pi_i Q_{ij} = 0$$

for any j, which implies that

$$\sum_{i \neq j} \pi_i Q_{ij} = -\pi_j Q_{jj}.$$

Since $\sum_i Q_{ji} = 0$, we further conclude that

$$\sum_{i \neq j} \pi_i Q_{ij} = -\pi_j Q_{jj} = \pi_j \sum_{i \neq j} Q_{ji}.$$

$\square$

Example 9.1.2

Consider a CTMC $X(t)$ with the state space to be non-negative integers. The transition rate matrix $\mathbf{Q}$ is assumed to be as follows:

$$\mathbf{Q} = \begin{pmatrix} -1 & 1 & 0 & 0 & 0 & 0 & \dots \\ 2 & -3 & 1 & 0 & 0 & 0 & \dots \\ 0 & 2 & -3 & 1 & 0 & 0 & \dots \\ & & & \dots & & & \end{pmatrix}.$$

Consider $\pi_i = 2^{-(i+1)}$. First, we have

$$\sum_{i=0}^{\infty} \pi_i = \sum_{i=0}^{\infty} 2^{-(i+1)} = \frac{1/2}{1 - 1/2} = 1.$$

Further, for $j > 0$,

$$\sum_{i \neq j} \pi_i Q_{ij} = \pi_{j-1} Q_{j-1,j} + \pi_{j+1} Q_{j+1,j} = 2^{-j} + 2^{-j-2} \times 2 = 3 \times 2^{-j-1}$$

and

$$\pi_j \sum_{i \neq j} Q_{ji} = \pi_j \left(Q_{j,j-1} + Q_{j,j+1} \right) = 2^{-j-1} \times (1 + 2) = 3 \times 2^{-j-1},$$

so the global balance equation holds. It is easy to verify that the global balance equation also holds when $j = 0$. So $X(t)$ has a unique stationary distribution $\pi_i = 2^{-(i+1)}$. $\square$

Unlike the above example, there are many instances where one cannot easily verify the global balance equation. Next we introduce a sufficient condition under which the global balance equation holds.

Lemma 9.1.7 The global balance equation holds if

$$\pi_i Q_{ij} = \pi_j Q_{ji}, \qquad \forall i \neq j. \tag{9.2}$$

Equation (9.2) is called the *local balance equation*. If the local balance equation is satisfied, then, given a pair of states i and j, the rate at which transitions take place from state i to state j is equal to the rate at which transitions take place from state j to state i.

Proof For fixed j, we

$$\sum_{i \neq j} \pi_i Q_{ij} = \sum_{i \neq j} \pi_j Q_{ji} = \pi_j \sum_{i \neq j} Q_{ji}. \qquad \square$$

Example 9.1.3

The local balance equation is often much easier to verify than the global balance equation. Consider the CTMC in Example 9.1.2. Since $Q_{ij} = 0$ when $|i - j| > 1$, the local balance equations are given by

$$\pi_i Q_{i,i+1} = \pi_{i+1} Q_{i+1,i}, \qquad \forall i.$$

It is straightforward to verify that $\pi_i = 2^{-(i+1)}$ satisfies the above equality. $\qquad \square$

Note that the local balance equation is only a sufficient condition for a CTMC to be positive recurrent. In particular, it is possible to have a stationary distribution π that satisfies the global balance equation but not the local balance equation.

Often it is difficult to find the π to satisfy either the global or the local balance equation. Nevertheless, in the applications it is important to know whether π exists, even if we cannot find it explicitly. Therefore, similar to the Foster–Lyapunov theorem for DTMCs, we present the following Foster–Lyapunov theorem for CTMCs, which provides another sufficient condition for a CTMC to be positive recurrent.

Theorem 9.1.8 (Foster–Lyapunov theorem for CTMCs) Suppose $X(t)$ is irreducible and non-explosive. If there exists a function $V : \mathcal{S} \to \mathcal{R}^+$ such that

(1) $\sum_{j \neq i} Q_{ij}(V(j) - V(i)) \leq -\epsilon$ if $i \in \mathcal{B}^c$, and
(2) $\sum_{j \neq i} Q_{ij}(V(j) - V(i)) \leq M$ if $i \in \mathcal{B}$,

for some $\epsilon > 0, M < \infty$, and a bounded set $\mathcal{B}$, then $X(t)$ is positive recurrent.

Proof We omit the details, and just present the intuition based on DTMCs. For sufficiently small δ, the probability that $X(t)$ makes the transition from state i to state j in time interval $[t, t + \delta]$ is approximately $Q_{ij}\delta$, so the expected drift of the $V(t)$ given $X(t) = i$ is

$$E\left[\frac{V(X(t + \delta)) - V(X(t))}{\delta} \,\middle|\, X(t) = i \right]$$

$$= \frac{\sum_{j \neq i} Q_{ij}\delta V(j) + (1 + Q_{ii}\delta)V(i) - V(i)}{\delta}$$

$$= \sum_{j \neq i} Q_{ij}V(j) + Q_{ii}V(i).$$

Substituting $Q_{ii} = -\sum_{j \neq i} Q_{ij}$ into the above equation, we obtain

$$E\left[\left.\frac{V(X(t+\delta)) - V(X(t))}{\delta}\right| X(t) = i\right] = \sum_{j \neq i} Q_{ij}(V(j) - V(i)),$$

which is less than $-\epsilon$ when $i \in \mathcal{B}^c$ and less than M otherwise. $\qquad \square$

9.2 Queueing systems: introduction and definitions

We consider queueing systems in which customers (or packets) arrive at a queue according to some arrival process, the service time of each customer is some random variable, and the number of servers in the system could be one or more. A pictorial illustration of a queueing system with one server and one queue is shown in Figure 9.2.

Customers who find all servers busy form a queue. In general, the order in which the customers in the queue has to be specified to complete the description of the queueing system. Unless otherwise stated, we will assume that the customers are served in FIFO (first-in, first-out) order. We will encounter two types of queues: ones in which there is infinite waiting space and ones where the number of waiting spaces is limited. When the number of waiting spaces is limited, we will assume that a customer who finds no empty waiting space is *blocked* and *lost* forever. In other words, we assume that blocked customers do not attempt to join the queue at a later time.

We will use λ to denote the arrival rate of customers (the average number of arriving customers per second) into the system and $1/\mu$ to denote the average service time of each customer. Thus, μ is the rate at which a server can serve the customers, i.e., the average number of customers served by a single server per second is μ.

Little's law, $L = \lambda W$, relates the average number of customers in the system, L, to the average time spent in the system, W, through arrival rate λ. This relation applies to the system as a whole (queue + servers), the queue alone, or just the servers.

The intuition behind Little's law can be explained as follows: suppose that the customers pay the system \$1 for each second spent in the system. Since a customer spends an average of W seconds in the system, the average rate of revenue for the system is λW \$/s. But it is also true that the average number of customers in the system is L. Thus, the rate of revenue is also L \$/s. Therefore, L has to be equal to λW.

We would like to comment that Little's law is true in great generality with very few assumptions on the arrival process, service times, the number of servers, the number of waiting spaces, etc. The proof is similar to the proof for discrete-time queues, and is omitted here.

Figure 9.2 Pictorial illustration of a simple queueing system.

A queueing system can be specified by specifying the arrival process, the service process, the number of servers, and the buffer space. Therefore, we use the following standard notation.

Definition 9.2.1 (M/M/*s*/*k* queue) The first M denotes the fact that inter-arrival times are exponential (memoryless, hence the M); the second M denotes that service times are exponential (memoryless); s is the number of servers; and k is the total number of customers allowed in the system at any time. Thus, $k - s$ is the number of waiting spaces in the queue, also known as buffer space. When the queue is full, any arriving customer is blocked from entering the system and lost. Unless specified, we assume that the service order is FIFO. □

If, instead of M, we use G, it denotes either general inter-arrival times or service times, depending on whether the G is in the first or second position in the above notation. The notation GI is used to indicate that the inter-arrival times (or service times) are independent. We also use D to denote constant (or deterministic) inter-arrival or service times.

An arrival process with exponentially distributed i.i.d. inter-arrival times is called the Poisson process. We next present the definition and some properties of Poisson processes.

Definition 9.2.2 (Poisson process) $N(t)$ $(t \geq 0)$ is a Poisson process with parameter λ if the following conditions hold:

(i) $N(0) = 0$;
(ii) N is a counting process, i.e., $N(t)$ can increase at most by 1 at any time instant;
(iii) $N(t)$ is an independent increment process; and
(iv) $N(t) - N(s) \sim \text{Poi}(\lambda(t - \delta))$. □

Since $N(t) - N(s)$ follows a Poisson distribution with parameter $\lambda(t - s)$,

$$E[N(t) - N(s)] = \lambda(t - s) \quad \text{and} \quad Var(N(t) - N(s)) = \lambda(t - s).$$

Next we present three equivalent definitions of Poisson processes. For all three, we assume $N(0) = 0$.

(i) $N(t)$ is a counting process and the inter-arrival times are i.i.d. exponentially distributed with parameter λ.
(ii) $N(t) - N(s) \sim \text{Poi}(\lambda(t - s))$; given $N(t) - N(s) = n$, the event times are uniformly distributed in $[s, t]$.
(iii) $N(t)$ has stationary and independent increments, $\Pr(N(\delta) = 1) = \lambda\delta + o(\delta)$, and $\Pr(N(\delta)) \geq 2) = o(\delta)$. Here, $o(\delta)$ denotes a function $g(\delta)$ with the property $\lim_{\delta \to 0} g(\delta)/\delta = 0$.

We do not prove the fact that the above three definitions are equivalent to Definition 9.2.2. The proof can be found in many textbooks on random processes.

Below are two well-known properties of Poisson processes that will be used in this chapter. The proofs of these results are left as exercises at the end of the chapter.

Result 9.2.1 If $N_1(t), \ldots, N_K(t)$ are independent Poisson processes with parameters $\lambda_1, \ldots, \lambda_K$, respectively, $\sum_{i=1}^{K} N_i(t)$ is a Poisson process with parameter $\sum_{i=1}^{K} \lambda_i$. ☐

Result 9.2.2 Assume that $N(t)$ is a Poisson process. We can generate K random processes $N_1(t), \ldots, N_K(t)$ as follows: when there is an arrival according to $N(t)$, make it an arrival for process $N_k(t)$, with probability p_k, where $\sum_{k=1}^{K} p_k = 1$. Then, $N_1(t), \ldots, N_K(t)$ are independent Poisson processes with parameters $\lambda p_1, \ldots, \lambda p_K$, respectively. ☐

9.3   The M/M/1 queue

The M/M/1 queue is a short-hand notation for an M/M/1/∞ queue. Let $q(t)$ denote the number of customers in the system, which forms a time-homogeneous CTMC. Let $\pi(n)$ be the (steady-state) probability that there are n customers in the system. Note that $P_{ij} = 0$ if $j \neq i - 1$ or $i + 1$, and the inter-arrival times and service times are exponential, so the transition rate matrix $\mathbf{Q}$ is given by

$$
Q_{ij} = \begin{cases} \lambda, & \text{if } j = i + 1, \\ \mu, & \text{if } j = i - 1, \\ -\lambda - \mu, & \text{if } j = i, \\ 0, & \text{otherwise.} \end{cases}
$$

To see this, consider a small interval of time $[t, t + \delta)$. Then,

$$\Pr(q(t \mid \delta) = i \mid 1 | q(t) = i)$$

$$= \Pr(\{\text{one arrival and no departures in } \delta \text{ time units}\} \text{ or}$$

$$\{\text{two arrivals and one departure in } \delta \text{ time units}\} \text{ or} \ldots)$$

$$= \lambda \delta + o(\delta),$$

where the last equality follows from one of the definitions of a Poisson process. Thus,

$$Q_{i,i+1} = \lim_{\delta \to 0} \Pr(q(t + \delta) = i + 1 | q(t) = i)/\delta = \lambda.$$

The other entries of the $\mathbf{Q}$ matrix can be derived similarly. The Markov chain is illustrated in Figure 9.3. In general, a CTMC can be represented by such a figure, where the nodes represent the state of the Markov chain and the arrows represent possible transitions. The numbers on the arrows represent the corresponding entries of the transition rate matrix $\mathbf{Q}$.

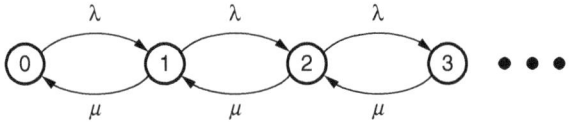

Figure 9.3 Markov chain for an M/M/1/∞ queue.

Recall that, if we can find a stationary distribution π which satisfies the local balance equation (9.2), then this π will also satisfy $\pi Q = 0$. The local balance equation for the M/M/1 queue is given by $\lambda \pi(n) = \mu \pi(n+1)$, or

$$\pi(n+1) = \rho \pi(n), \tag{9.3}$$

where $\rho = \lambda/\mu$. We note that ρ is dimensionless, but it is often expressed in units of erlangs. Repeatedly using (9.3), we obtain

$$\pi(1) = \rho \pi(0),$$

$$\pi(2) = \rho \pi(1) = \rho^2 \pi(0),$$

$$\vdots$$

$$\pi(n) = \rho \pi(n-1) = \rho^n \pi(0).$$

Since $\sum_{n=0}^{\infty} \pi(n) = 1$, we have $\pi(0) \sum_{n=0}^{\infty} \rho^n = 1$.

Suppose $\rho < 1$, then

$$\sum_{n=0}^{\infty} \rho^n = \frac{1}{1-\rho},$$

which yields

$$\pi(n) = \rho^n (1-\rho).$$

We note that $\rho < 1$ implies $\mu > \lambda$, i.e., the service rate of the server is larger than the arrival rate. Intuitively, this is required; otherwise, the number of customers in the system will grow forever. In other words, $\rho < 1$ is required to ensure the stability of the queueing system.

Using $\pi(n) = \rho^n (1-\rho)$ and Little's law, we can derive the following results.

- The mean number of customers in the system: $L = \sum_{n=0}^{\infty} n\pi(n) = \rho/(1-\rho)$.
- From Little's law, the mean delay (i.e., waiting time in the queue + service time) of a customer: $W = L/\lambda = 1/(\mu - \lambda)$.
- The mean waiting time of a customer in the queue: $W_q = W - 1/\mu = \rho/(\mu - \lambda)$.
- The mean number of customers in the queue (again, by Little's law): $L_q = \lambda W_q = \rho^2/(1-\rho)$.

Next, we study the fraction of time the server is busy, i.e., the mean server utilization. Note that

fraction of time the server is busy = Pr(there is at least one customer in the system),

so

$$U \triangleq \text{mean server utilization} = 1 - \pi_0 = \rho.$$

As the server utilization becomes close to 1, $L \to \infty$. Thus, the mean delay, the waiting time, and the number of customers in the queue, all go to infinity.

One can also see $U = \rho$ as a consequence of Little's law. Let us apply Little's law, $L = \lambda W$, to the *server*. Note that at most one customer can be in the server, so

$$L_s = \text{average number of customers in the server}$$

$$= \text{fraction of time that the server is busy}$$

$$= U.$$

Considering the waiting time, we have

$$W_s = \text{average amount of time spent by a customer in service}$$
$$= 1/\mu.$$

Thus, applying Little's law to the server alone, we obtain

$$U = L_s = \lambda W_s = \lambda/\mu = \rho.$$

Note that, in applying Little's law to compute U, we did not use the fact that the queue is an M/M/1 queue. Indeed, the result applies to any general i.i.d. inter-arrival and service time distributions, i.e., $L_s = \lambda W_s$ $U = \rho$ applies to any GI/GI/1 queue.

9.4 The M/M/*s*/*s* queue

Consider a M/M/*s*/*s* loss model, where the queueing system has s servers and *no* buffer. We are interested in the blocking probability, i.e., the probability that an arriving customer is blocked and lost due to the fact that all servers are busy. As in the case of the M/M/1 queue, the number of customers in the M/M/*s*/*s* queue evolves as a Markov chain, and is shown in Figure 9.4. The transition rates can be derived as in the case of the M/M/1 queue.

Next, we derive the stationary distribution π of the number of customers in the queue for the M/M/*s*/*s* loss model.

According to the local balance equation, we have that

$$\lambda \pi(0) = \mu \pi(1),$$
$$\lambda \pi(1) = 2\mu \pi(2),$$
$$\vdots$$
$$\lambda \pi(s-1) = s\mu \pi(s).$$

Thus, for $0 \leq n \leq s$,

$$\pi(n) = \frac{\rho^n}{n!} \pi(0),$$

where, as before, $\rho = \lambda/\mu$. Using the fact that $\sum_{k=0}^{s} \pi(k) = 1$, we obtain

$$\pi(0) = \frac{1}{\sum_{k=0}^{s} \rho^k/k!},$$

and

$$\pi(n) = \frac{\rho^n/n!}{\sum_{k=0}^{s} \rho^k/k!}.$$

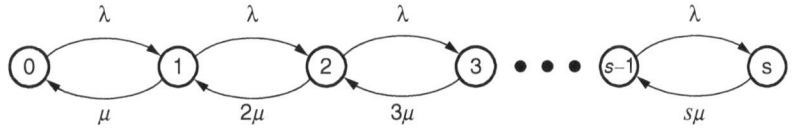

Figure 9.4 Markov chain for an M/M/*s*/*s* queue.

Thus, the probability that all servers are busy is given by

$$\pi(s) = \frac{\rho^s/s!}{\sum_{k=0}^{s} \rho^k/k!}. \tag{9.4}$$

9.4.1 The PASTA property and blocking probability

It turns out that $\pi(s)$, the probability that all servers are busy, is also equal to the blocking probability. To see that the two quantities are equal, we have to show that, when the arrival process is Poisson, computing the average of any quantity by sampling the system at customer arrival times is equal to the average of the same quantity computed over all time. In the case of the M/M/s/s model, the customer average is the average fraction of time that an arriving customer sees that all servers are busy, i.e., the blocking probability. The time average is the fraction of time that all servers are busy, i.e., the probability that all servers are busy. The fact that these two quantities are equal is a consequence of a more general property called Poisson Arrivals See Time Averages (PASTA). We now present the intuition behind the PASTA property, which is nearly identical to the BASTA property discussed in Section 3.4.3 for discrete-time queues. Let $q(t)$ be the number of busy servers at time t, and let $a(t, t+\delta)$ be the event that there is an arrival in the time interval $[t, t+\delta]$. Then, the probability that an arriving customer is blocked is given by $\Pr(q(t) = s|a(t, t+\delta) = 1)$. This probability can be calculated as follows:

$$\Pr(q(t) = s|a(t, t+\delta) = 1) = \frac{\Pr(a(t, t+\delta) = 1|q(t) = s)\Pr(q(t) = s)}{\Pr(a(t, t+\delta) = 1)} = \Pr(q(t) = s),$$

where the last step follows from the fact that, because the arrival process is Poisson, the number of arrivals in $[t, t+\delta]$ is independent of the past and, hence, independent of $q(t)$. Thus, the blocking probability is equal to $\Pr(q(t) = s)$, which is π_s in the steady state.

Formula (9.4) for the blocking probability is called the *Erlang-B formula* in honor of A. K. Erlang, who was the father of queueing theory and derived the blocking probability formula. The Erlang-B formula also holds when the service times are non-exponential. The proof of this insensitivity to service-time distribution is similar to the insensitivity proof for a different model provided in Section 9.10.

To see that non-Poisson arrivals may not see time averages, consider the following example. Consider a D/D/1/1 model, i.e., deterministic (or constant) inter-arrival times, deterministic service times, one server and no buffer. Let the inter-arrival times and service times be 10 s. Then, the server is always busy, but no customer is blocked because each arrival occurs immediately after the previous customer has been served. Thus, the fraction of time for which the system is full is 1, whereas the fraction of blocked customers is 0.

9.5 The M/M/s queue

Consider the M/M/s model, where the queueing system has s servers and infinite buffer. The time-homogeneous Markov chain describing the number of customers in the system is shown in Figure 9.5. From the local balance equation, we have

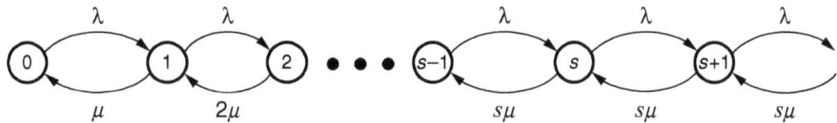

Figure 9.5 Markov chain for an M/M/s queue.

$$\lambda\pi(0) = \mu\pi(1),$$

$$\lambda\pi(1) = 2\mu\pi(2),$$

$$\vdots$$

$$\lambda\pi(s-1) = s\mu\pi(s),$$

$$\lambda\pi(s) = s\mu\pi(s+1),$$

$$\vdots$$

Thus,

$$\pi(n) = \frac{(s\rho)^n}{n!}\pi(0), \qquad n \le s,$$

$$\pi(n) = \frac{s^s\rho^n}{s!}\pi(0), \qquad n > s,$$

where $\rho = \lambda/s\mu < 1$. Using the fact that $\sum_{n=0}^{\infty}\pi(n) = 1$, we obtain

$$\pi(0) = \left[\sum_{n=0}^{s-1}\frac{(s\rho)^n}{n!} + \frac{(s\rho)^s}{s!(1-\rho)}\right]^{-1}.$$

The probability that an arrival finds all servers busy is given by

$$P_Q = \sum_{n=s}^{\infty}\pi(n) = \frac{\pi(0)(s\rho)^s}{s!(1-\rho)}.$$

This is called the *Erlang-C formula*.

9.6   The M/GI/1 Queue

Consider a queueing system with exponential inter-arrival times (mean $1/\lambda$) and *general but identical and independent service times* (mean $1/\mu$), as shown in Figure 9.6. Note that, for general service times, the amount of time a customer has been served determines the remaining service time of the customer. So, unless the service times are exponentially

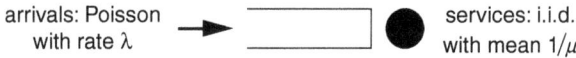

Figure 9.6 M/GI/1 queue.

distributed, the number of customers in the system is not a Markov chain. To compute the stationary distribution of M/GI/1 queues, we sample the system at departure times. Denote by q_k the number of customers in the system immediately after the kth departure, and let A_{k+1} be the number of arrivals during the service time of the $(k + 1)$th customer.

When $q_k > 0$, it is easy to see that

$$q_{k+1} = q_k + A_{k+1} - 1.$$

When $q_k = 0$, the number of arrivals between the kth departure and the $(k + 1)$th departure is $1 + A_{k+1}$, where the 1 accounts for the arrival of the $(k + 1)$th customer, and A_{k+1} is the number of arrivals when the $(k + 1)$th customer is served. Therefore, when $q_k = 0$,

$$q_{k+1} = q_k + 1 + A_{k+1} - 1.$$

Therefore, the sampled queue can be described as the following dynamical system:

$$q_{k+1} = q_k + A_{k+1} - U_{k+1},$$

where $U_{k+1} = 1$ when $q_k > 0$ and $U_{k+1} = 0$ when $q_k = 0$. Since the arrival process is Poisson, q_k is a DTMC.

Denote by S_{k+1} the service time of the $(k + 1)$th customer. Given $S_{k+1} = s_{k+1}$, A_{k+1} follows a Poisson distribution with parameter λs_{k+1}. We can therefore compute the first moment and second moment of A_{k+1} given S_{k+1} as follows:

$$E[A_{k+1}|S_{k+1}] = \lambda S_{k+1},$$

$$E[A_{k+1}^2|S_{k+1}] = \lambda S_{k+1} + \lambda^2 S_{k+1}^2.$$

Recall that $E[S_{k+1}] = 1/\mu$. Therefore,

$$E[A_{k+1}] = \lambda \frac{1}{\mu} = \rho,$$

$$E[A_{k+1}^2] = \frac{\lambda}{\mu} + \lambda^2 E[S^2] = \rho + \lambda^2 E[S^2].$$

We ignore the subscript $k + 1$ of S_{k+1} since the service times are i.i.d.

Note that $\{A_k\}$ are i.i.d. because the inter-arrival times are exponential and the service times are i.i.d. So $\{q_k\}$ is similar to DTMCs that we have analyzed before. If $\rho < 1$, $\{q_k\}$ is positive recurrent and has a well-defined stationary distribution. As in previous chapters, we also say that the queueing system is *stable* if $q(t)$ has a well-defined stationary distribution. Now the question is: does $\{q(t)\}$ have a well-defined stationary distribution at $t \to \infty$? This question is answered by the following theorem.

Theorem 9.6.1 Consider the M/GI/1 queue introduced at the beginning of this section. When $\rho < 1$, the queueing system has a well-defined stationary distribution, the same distribution as that of $\{q_k\}$.

Proof We present the key ideas behind the proof here. First, we sample the system at arrival times and denote by Q_k the number of customers in the queue before the kth arrival. We claim that Q_k and q_k have the same stationary distribution.

To prove this fact, let $F_n(t)$ denote the number of up crossings from n to $(n+1)$ in $[0, t]$, let $G_n(t)$ denote the number of down crossings from $(n+1)$ to n in $[0, t]$, let $A(t)$ denote the number of arrivals in $[0, t]$, and let $D(t)$ denote the number of departures in $[0, t]$. Note that $F_n(t)$ is the number of samples that the system is in state n when the system is sampled at arrival times, so

$$\pi_Q(n) = \lim_{t \to \infty} \frac{F_n(t)}{A(t)},$$

which is the steady-state probability that an arrival sees a queue with length n. Similarly, $G_n(t)$ is the number of samples that the system is in state n when the system is sampled at departure times, so

$$\pi_q(n) = \lim_{t \to \infty} \frac{G_n(t)}{D(t)},$$

which is the steady-state probability that a departure sees a queue with length n.

Note that $|F_n(t) - G_n(t)| \leq 1$ because the number of customers in the system has to come down before it can go up again, and vice versa; so,

$$\lim_{t \to \infty} \frac{|F_n(t) - G_n(t)|}{A(t)} = 0.$$

Therefore, we obtain

$$\lim_{t \to \infty} \frac{F_n(t)}{A(t)} = \lim_{t \to \infty} \frac{G_n(t)}{A(t)} = \lim_{t \to \infty} \frac{G_n(t)}{D(t)} \frac{D(t)/t}{A(t)/t}.$$

As $\{q_k\}$ is stable, we have

$$\lim_{t \to \infty} \frac{D(t)}{t} = \lim_{t \to \infty} \frac{A(t)}{t} = \lambda,$$

which implies that

$$\lim_{t \to \infty} \frac{F_n(t)}{A(t)} = \lim_{t \to \infty} \frac{G_n(t)}{D(t)},$$

and

$$\pi_Q(n) = \pi_q(n).$$

Next, by the PASTA property, $\pi_Q(n)$ must also be the steady-state probability distribution of having n customers in the system:

$$\pi_Q(n) = \lim_{t \to \infty} \Pr(q(t) = n | A(t, t+\delta) = 1)$$

$$= \lim_{t \to \infty} \Pr(q(t) = n) \frac{\Pr(A(t, t+\delta) = 1 | q(t) = n)}{\Pr(A(t, t+\delta) = 1)}$$

$$= \lim_{t \to \infty} \Pr(q(t) = n)$$

$$= \pi(n).$$

The theorem, therefore, holds.   □

9.6.1   Mean queue length and waiting time

Next, we compute the mean queue length of the M/GI/1 queue. Since the stationary distribution of $q(t)$ is the same as that of q_k, the system sampled at departure times, we compute $E[q_k]$. We first compute the second moment of q_k as follows:

$$E\left[q_{k+1}^2\right] = E[(q_k + A_{k+1} - U_{k+1})^2]$$
$$= E[q_k^2] + E[A_{k+1}^2] + E[U_{k+1}^2] + 2E[q_k(A_{k+1} - U_{k+1})] - 2E[A_{k+1}U_{k+1}].$$
$$(9.5)$$

Since $U_k \in \{1, 0\}$, we have

$$E[U_{k+1}^2] = E[U_{k+1}] = \Pr(q_k > 0) = \rho,$$

where the final step follows by applying PASTA and Little's law to the server.

If $\rho < 1$, q_k has a stationary distribution, and

$$\lim_{k \to \infty} E[q_{k+1}^2] = \lim_{k \to \infty} E[q_k^2].$$

Recall that

$$E[A_{k+1}] = \rho \qquad \text{and} \qquad E[A_{k+1}^2] = \rho + \lambda^2 E[S^2].$$

Moving $E[q_k^2]$ in (9.5) to the left-hand side and letting k go to infinity, we obtain

$$0 = \rho + \lambda^2 E[S^2] + \rho + 2L\rho - 2L - 2\rho^2,$$

where $L = \lim_{k \to \infty} E[q_k]$ and we used the facts that $q_k U_{k+1} = q_k$ and that A_{k+1} and q_k are independent. Rearranging the terms in the above equation, we obtain

$$0 = \lambda^2 E[S^2] + 2(L - \rho)(\rho - 1),$$

which leads to

$$L = \rho + \frac{\lambda^2 E[S^2]}{2(1 - \rho)}.$$

Next, we can compute the number of customers in the queue and the waiting time in the queue using Little's law:

$$L_q = L - \rho = \frac{\lambda^2 E[S^2]}{2(1 - \rho)},$$

$$W_q = \frac{L_q}{\lambda} = \frac{\lambda E[S^2]}{2(1 - \rho)},$$

where the second equality is the *Pollaczek–Khinchine formula,* also called the P-K formula.

Note that, if S is exponential,

$$E[S^2] = \frac{1}{\mu^2} + \frac{1}{\mu^2} = \frac{2}{\mu^2},$$

which implies that

$$W_q = \lambda \frac{2}{\mu^2} \frac{1}{2(1 - \rho)} = \frac{\rho}{\mu(1 - \rho)}.$$

Thus, we have recovered the average waiting time for M/M/1 queueing systems.

9.6.2   Different approaches taken to derive the P-K formula

We now introduce two alternative approaches taken to derive the P-K formula. Let $R(t)$ denote the remaining service time of the customer in the server at time t. Figure 9.7 illustrates a sample path of $R(t)$, where s_i is the service time of customer i.

Let R denote the average remaining service time of the customer in the server when a new customer joins the queue. Due to the PASTA property,

$$W_q = \frac{L_q}{\mu} + R,$$

where R is the time required to serve the customer in the server and L_q/μ is the amount of time required to serve all customers in the queue. According to Little's law, $L_q = \lambda W_q$, so

$$W_q(1 - \rho) = R,$$

or

$$W_q = \frac{R}{1 - \rho}. \tag{9.6}$$

We index the customers according to the sequence in which they depart the system, and let $D(t)$ denote the number of customers who have departed up to and including time t. Further, we define

$$\overline{R}_t = \frac{1}{t} \int_0^t R(s)ds.$$

Assume customer i started to receive service at time t_i and its service time is s_i, then

$$\int_{t_i}^{t_i+s_i} R(s)ds = \frac{s_i^2}{2},$$

as shown in Figure 9.7, which yields

$$\frac{1}{t} \sum_{i=1}^{D(t)} \frac{s_i^2}{2} \le \overline{R}_t \le \frac{1}{t} \sum_{i=1}^{D(t)} \frac{s_i^2}{2} + \frac{1}{t} \frac{s_{D(t)+1}^2}{2}.$$

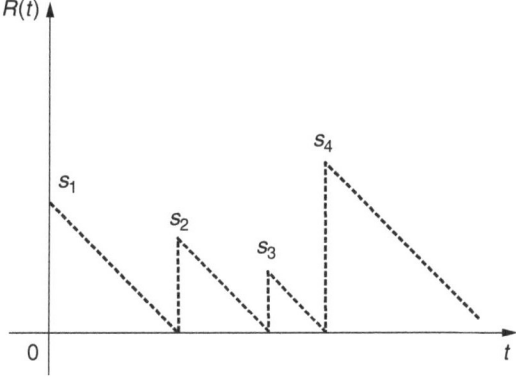

Figure 9.7 Sample path of $R(t)$.

Note that, as $t \to \infty$,

$$\frac{1}{t}\frac{s_{D(t)+1}^2}{2} \to 0,$$

so

$$R = \lim_{t \to \infty} \overline{R}_t = \lim_{t \to \infty} \frac{D(t)}{t}\frac{1}{D(t)}\sum_{i=1}^{D(t)}\frac{1}{2}s_i^2 = \frac{1}{2}\lambda E[S^2].$$

Combining the above equality with equality (9.6), we obtain the P-K formula:

$$W_q = \frac{\lambda E[S^2]}{2(1-\rho)}.$$

Another way to derive the P-K formula is to compute the entire distribution of $q(t)$. Consider the probability generating function for q_k:

$$Q(z) \triangleq E[z^{q_k}] = \sum_{i=0}^{\infty} z^i \pi(i).$$

Assuming that q_k is in steady state, we have

$$Q(z) = E\left[z^{q_{k-1}-U_k+A_k}\right]$$
$$= E\left[z^{q_{k-1}-U_k}\right]A(z),$$

where $A(z) = E\left[z^{A_k}\right]$. Recall that $U_k = 1$ if $q_{k-1} > 0$ and $U_k = 0$ otherwise, so

$$Q(z) = \left(\pi(0) + \sum_{i=1}^{\infty} \pi(i)z^{i-1}\right)A(z)$$
$$= \left(\pi(0) + \frac{1}{z}\sum_{i=1}^{\infty} \pi(i)z^i\right)A(z)$$
$$= \left(\pi(0) + \frac{1}{z}(Q(z) - \pi(0))\right)A(z)$$
$$= \left(\pi(0)\left(1 - \frac{1}{z}\right) + \frac{Q(z)}{z}\right)A(z).$$

Moving the term containing $Q(z)$ to the left-hand side, we obtain

$$Q(z)\left(1 - \frac{A(z)}{z}\right) = \pi(0)A(z)\left(1 - \frac{1}{z}\right),$$

which implies that

$$Q(z) = \frac{\pi(0)A(z)(z-1)}{(z-A(z))}.$$

Recall that $\pi(0) = 1 - \rho$, so $Q(z)$ can be obtained after $A(z)$ is known. According to the definition of A_k, we have

$$A(z) = E\left[z^{A_k}\right]$$

$$= E\left[E\left[z^{A_k}\Big|\text{ service time} = t\right]\right]$$

$$= E\left[\sum_{i=0}^{\infty} z^i \frac{(\lambda t)^i e^{-\lambda t}}{i!}\right]$$

$$= E\left[e^{\lambda z t}e^{-\lambda t}\right]$$

$$= E\left[e^{\lambda(z-1)t}\right]$$

$$= M(\lambda(z-1)),$$

where $M(\theta) = E[e^{\theta t}]$ is the moment generating function of the random variable t.

Assuming the closed-form expression of $Q(z)$ is known, we can obtain L as follows:

$$L = E[q_k] = \frac{d}{dz}E\left[z^{q_k}\right]\bigg|_{z=1} = Q'(1).$$

Then, the queue length and the waiting time in the queue can be computed using

$$L_q = L - \rho \qquad \text{and} \qquad W_q = \lambda L_q.$$

You will be asked to compute $Q(z)$ and recover the P-K formula in Exercise 9.4.

9.7 The GI/GI/1 queue

We now consider a general queueing system where inter-arrival times are general i.i.d. random variables and service times are general i.i.d. random variables. This queueing system is either written as the G/G/1 or the GI/GI/1 if one wants to emphasize that the inter-arrival times are independent and the service times are independent. Here we use the GI/GI/1 terminology. We introduce the following notation:

- W_k: waiting time in the queue of the kth customer;
- S_k: service time of the kth customer;
- I_{k+1}: inter-arrival time between the kth customer and the $(k+1)$th customer.

Then the evolution of W_k can be written as follows:

$$W_{k+1} = (W_k + S_k - I_{k+1})^+;$$

see Figure 9.8.

We define $X_k = S_k - I_{k+1}$. Then the evolution of W_k can be re-written as follows:

$$W_{k+1} = (W_k + S_k - I_{k+1})^+$$

$$= (W_k + X_k)^+$$

$$= \max\{0, W_k + X_k\}$$

$$= \max\{0, \max\{0, W_{k-1} + X_{k-1}\} + X_k\}$$

$$= \max\{0, W_{k-1} + X_{k-1} + X_k, X_k\}.$$

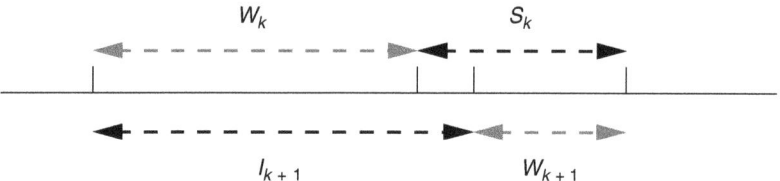

Figure 9.8 The evolution of W_k. The waiting time of the $(k+1)$th customer equals the overall time the kth customer spends in the system minus the inter-arrival time between the kth customer and the $(k+1)$th customer.

Iterating backwards and noting $W_1 = 0$, we obtain

$$W_k = \max\{0, X_1 + X_2 + \cdots + X_k,$$
$$X_2 + X_3 + \cdots + X_k,$$
$$\vdots$$
$$X_k\}.$$

Recall that $\{X_k\}$ are i.i.d., so W_k is distributionally equivalent to

$$\widetilde{W}_k = \max\{0, X_1, X_1 + X_2, \ldots, X_1 + X_2 + X_3 + \cdots + X_k\}.$$

As $\widetilde{W}_k$ is a non-decreasing sequence, it has a limit in $[0, \infty]$. Suppose $E[X_1] = E[S_k - I_{k+1}] < 0$, or $1/\mu < 1/\lambda$, or $\lambda < \mu$; then, by the Strong Law of Large Numbers (SLLN),

$$\lim_{N \to \infty} \frac{1}{N} \sum_{k=0}^{N} X_k = E[X_1],$$

with probability 1. In other words,

$$\lim_{N \to \infty} \sum_{k=0}^{N} X_k = -\infty,$$

with probability 1, which implies

$$\widetilde{W}_\infty < \infty,$$

with probability 1, where

$$\widetilde{W}_\infty = \lim_{k \to \infty} \widetilde{W}_k.$$

Therefore, we can conclude that

$$\Pr(W_\infty < \infty) = 1,$$

i.e., W_∞ is a "proper" random variable; the random variable is called "defective" otherwise. Thus,

$$W_k \xrightarrow{d} W_\infty$$

if $\lambda < \mu$, where d means convergence in distribution.

9.8   Reversibility

In this section, we introduce an important concept that is quite useful in the study of CTMCs – reversibility. Reversibility is a powerful tool used to obtain stationary distributions of complicated CTMC models. Consider a stationary CTMC X, i.e., $p(t) = \pi$ for all $t \in (-\infty, \infty)$. We define another CTMC Y by

$$Y(t) = X(-t),$$

which is called the reversed chain of X. Note that Y is stationary because X is stationary. We can view Y as a movie of X running backwards. Below we show that Y is also a CTMC.

Lemma 9.8.1 Y is a CTMC, and the transition rate matrix of Y, denoted by $\mathbf{Q}^*$, satisfies $Q_{ji}^* = Q_{ij}(\pi_i/\pi_j)$ for $i \neq j$.

Proof First, we prove that Y is a CTMC. Consider a sequence of times, such that $t \geq t_1 \geq t_2 \geq \cdots \geq t_n$, and the following conditional probability:

$$\Pr(Y(t)|Y(t_1), Y(t_2), \ldots, Y(t_n)) = \Pr(X(-t)|X(-t_1), X(-t_2), \ldots, X(-t_n)).$$

Note that $-t \leq -t_1 \leq -t_2 \leq \cdots \leq -t_n$. If we can prove

$$\Pr(X(-t)|X(-t_1), X(-t_2), \ldots, X(-t_n)) = \Pr(X(-t)|X(-t_1)), \tag{9.7}$$

then

$$\Pr(Y(t)|Y(t_1), Y(t_2), \ldots, Y(t_n)) = \Pr(X(-t)|X(-t_1)) = \Pr(Y(t)|Y(t_1)),$$

and Y is a CTMC.

According to the definition of conditional probability, we have

$$\Pr(X(-t)|X(-t_1), X(-t_2), \ldots, X(-t_n)) = \frac{\Pr(X(-t), X(-t_1)|X(-t_2), \ldots, X(-t_n))}{\Pr(X(-t_1)|X(-t_2), \ldots, X(-t_n))}$$

$$= \frac{\Pr(X(-t), X(-t_1)) \Pr(X(-t_2), \ldots, X(-t_n)|X(-t), X(-t_1))}{\Pr(X(-t_2), \ldots, X(-t_n))}$$

$$\times \frac{1}{\Pr(X(-t_1)|X(-t_2), \ldots, X(-t_n))}$$

$$= \frac{\Pr(X(-t), X(-t_1)) \Pr(X(-t_2), \ldots, X(-t_n)|X(-t_1))}{\Pr(X(-t_1)) \Pr(X(-t_2), \ldots, X(-t_n)|X(-t_1))}$$

$$= \frac{\Pr(X(-t), X(-t_1))}{\Pr(X(-t_1))},$$

so equality (9.7) holds.

Next, we show the relationship between $\mathbf{Q}$ and $\mathbf{Q}^*$. Recall that $\mathbf{Q}$ is the transition rate matrix of $X(t)$ and $\mathbf{Q}^*$ is the transition rate matrix of $Y(t)$. For $i \neq j$ and sufficiently small δ, we have

$$\Pr(Y(t+\delta) = i|Y(t) = j) = \Pr(Y(t) = j|Y(t+\delta) = i)\frac{\Pr(Y(t+\delta) = i)}{\Pr(Y(t) = j)}$$

$$= \Pr(X(-t) = j|X(-t-\delta) = i)\frac{\pi_i}{\pi_j}$$

$$= Q_{ij}\delta\frac{\pi_i}{\pi_j}.$$

Therefore, we have

$$Q_{ji}^* = Q_{ij}\frac{\pi_i}{\pi_j} \qquad \text{for } i \neq j. \qquad \square$$

The following theorem is a very useful result for computing the stationary distribution of a CTMC.

Theorem 9.8.2 Let X be a CTMC with transition rate matrix $\mathbf{Q}$. If there exists a transition rate matrix $\mathbf{Q}^*$ and probability vector π such that

$$Q_{ij}^*\pi_i = Q_{ji}\pi_j, \qquad \forall i,j, \qquad (9.8)$$

then $\mathbf{Q}^*$ must be the transition rate matrix of the reversed chain and π must be the stationary distribution of both the forward and the reversed chains.

Proof First, π must be the stationary distribution because

$$\sum_j \pi_j Q_{ji} = \sum_j \pi_j \frac{Q_{ij}^*}{\pi_j}\pi_i$$

$$= \sum_j Q_{ij}^*\pi_i$$

$$= 0,$$

where the last equality holds because the row sum of a transition rate matrix equals zero.

Further, $\mathbf{Q}^*$ must be the transition rate matrix of the reversed chain, as we have already shown that, if $\mathbf{Q}^*$ is the transition rate matrix of the reversed chain, then

$$Q_{ij}^*\pi_i = Q_{ji}\pi_j, \qquad \forall i \neq j. \qquad \square$$

The above theorem is often used in the following manner. If we have a guess for π, we can guess the elements of the transition rate matrix $\mathbf{Q}^*$ of the reversed chain from (9.8). Next, if we can verify that $\mathbf{Q}^*$ is indeed a transition rate matrix by checking that the off-diagonal terms are positive, the diagonal terms are negative, and each row sum is equal to zero, then our guess for π must be correct.

Next, we introduce the concept of reversible CTMCs and a lemma stating that the local balance equation implies that a CTMC is reversible.

Definition 9.8.1 (Reversibility) A stationary CTMC X is time reversible (or simply reversible) if

$$Q_{ij}^* = Q_{ij}.$$

Note that, since $\mathbf{Q}^*$ is the transition rate matrix of the reversed chain of X, the above relationship states that the statistics of the forward and the reverse chains are identical. □

Lemma 9.8.3 A CTMC is reversible if and only if the local balance equation is satisfied, i.e.,

$$\pi_i Q_{ij} = \pi_j Q_{ji} \qquad \text{for } i \neq j.$$

Proof Recall that

$$Q_{ij}^* = \frac{Q_{ji}\pi_j}{\pi_i}.$$

According to the definition, a CTMC is reversible if

$$Q_{ij}^* = Q_{ij},$$

which implies that

$$\pi_i Q_{ij} = \pi_j Q_{ji}.$$

Now, if $\pi_i Q_{ij} = \pi_j Q_{ji}$ holds, we have

$$Q_{ij}^* = \frac{Q_{ji}\pi_j}{\pi_i} = Q_{ij}.$$

So the CTMC is reversible. □

Next, we discuss several applications of the concept of reversibility, including applications to the M/M/1 queue and the tandem M/M/1 queue.

9.8.1   The M/M/1 queue

Consider an M/M/1 queue with arrival rate λ and service rate μ. Assume that the queue is in steady state, which is important because it is part of the definition of reversibility.

For the M/M/1 queue, the local balance equation holds, so an M/M/1 queue in steady state is reversible, which implies that the reversed Markov chain is indistinguishable from the forward Markov chain. Note that, in the forward chain, the queue size goes up by 1 at each arrival; in the reversed chain, the queue size goes up by 1 at each departure. As the Markov chain is reversible, the following claims hold in steady state.

(i) Since the time instants at which arrivals occur form a Poisson process, so the time instants at which departures occur must also form a Poisson process with the same rate, i.e., the departure process is also Poisson of rate λ.

(ii) Since future arrivals are independent of current queue size, the queue size must be independent of past departures.

Results (i) and (ii) together are called *Burke's theorem,* which also holds for M/M/*s* and M/M/∞ queues.

9.8.2 The tandem M/M/1 queue

Using our observations about M/M/1 queues, we can study a tandem M/M/1 queue, which is shown in Figure 9.9.

Figure 9.9 Tandem M/M/1 queue.

We know that the departure process from queue 1 is Poisson in steady state. So the second queue is also an M/M/1 queue. Further, the queue length at the second queue depends only on the past arrivals to the second queue and departures from the second queue. But the past arrivals to the second queue are the past departures from the first queue, which are independent of the current length of the first queue. Therefore, the first queue and the second queue are independent in steady state. If $\rho = \lambda/\mu_1 < 1$ and $\rho_2 = \lambda/\mu_2 < 1$,

$$\pi(i,j) = \Pr(q_1 = i, q_2 = j \text{ in the steady state})$$

$$= \rho_1^i(1 - \rho_1)\rho_2^j(1 - \rho_2).$$

Such a solution is called a product-form solution,

$$\pi(i,j) = \pi_1(i)\pi_2(j),$$

where π_1 and π_2 are the steady-state distributions of the first and second queues, respectively.

The above example motivates us to ask the following question: are there more complicated networks for which the product-form solution holds, i.e., where the steady-state distribution at each queue can be multiplied to get the overall steady-state distribution for the network? We now present several such examples.

9.9 Queueing systems with product-form steady-state distributions

Motivated by the tandem M/M/1 queueing system, in this section we present two examples of queueing systems whose steady-state distributions have product forms. The first model we consider is called a *Jackson network*, which is a direct generalization of the tandem M/M/1 queueing system. The Jackson network consists of exponential server queues (also called nodes), which are interconnected to form a queueing network. Customers move from node to node in a probabilistic manner to be described later. The main result is that the number of customers in each queue has the same steady-state distribution as that of an independent M/M/1 queue, and the overall steady state of the number of customers in the network is the product of the individual steady-state distributions at each queue. The second model we consider is an M/M/1 queue in which there are many classes of customers. This model is not a network model in the sense that there is only node in the system, so it is not related to the tandem queue in the preceding section. However, it also has a product-form distribution in a sense to be described later in this section. We do not directly use the concepts of a reversed chain or reversibility in establishing the product-form results. We simply guess that the steady-state distribution is product form and verify that this is indeed the case by checking that the global balance equations are satisfied. An alternative verification technique would have been to guess the form of the reversed chain and use Theorem 9.8.2. We briefly outline this technique when we consider the Jackson network.

9.9.1   The Jackson network

We consider a network where external arrivals into node i comprise a Poisson process with rate r_i. A customer moves from node i to node j with probability P_{ij}, after the service is completed, and leaves the network with probability $1 - \sum_j P_{ij} = P_{i0}$. Therefore, the aggregated arrival rate to node i is given by

$$\lambda_i = r_i + \sum_j \lambda_j P_{ji}.$$

It is shown in Exercise 9.12 that a unique solution λ exists to the above equations under a fairly general assumption that, for every node j, there exists a node i such that $P_{i0} > 0$ and a route $(m_1, m_2, \ldots, m_k)$ such that $P_{jm_1} P_{m_1 m_2} \cdots P_{m_{k-1} m_k} P_{m_k i} > 0$. This assumption implies that every customer will eventually leave the network with probability 1. This network is called the *Jackson network*. Two nodes of a Jackson network are shown in Figure 9.10.

Theorem 9.9.1 Assume that $\lambda_i < \mu_i$ for all i. The Jackson network with K nodes has a product-form stationary distribution given by

$$\pi(n) = \prod_{i=1}^{K} \rho_i^{n_i}(1 - \rho_i), \tag{9.9}$$

where $n = (n_1, \ldots, n_K)$, n_i is the number of customers at node i, and $\rho_i = \lambda_i/\mu_i$.

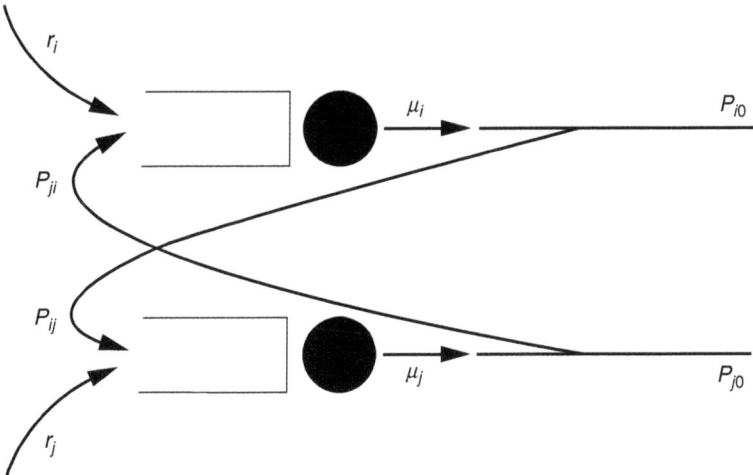

Figure 9.10 Two nodes of a Jackson network.

Proof To prove this theorem, we first introduce the following notation:

$$T_i(n) = (n_1, \ldots, n_i - 1, n_{i+1}, \ldots, n_K),$$

$$T^i(n) = (n_1, \ldots, n_i + 1, n_{i+1}, \ldots, n_K),$$

$$T_i^j(n) = (n_1, \ldots, n_i - 1, \ldots, n_j + 1, \ldots, n_K).$$

Note that state $T_i(n)$ is the state after one customer of node i left the network when the network was in state n, and state $T^i(n)$ is the state after one external customer arrived at node i when the network was in state n, and state $T_i^j(n)$ is the state after one customer moved from node i to node j when the network was in state n.

We now check the global balance equation:

$$\pi(n) \left(\sum_{i=1}^{K} \mu_i P_{i0} \mathbb{I}_{n_i > 0} + \sum_{i=1}^{K} r_i + \sum_{i=1}^{K} \sum_{j=1, j \neq i}^{K} \mu_i P_{ij} \mathbb{I}_{n_i > 0} \right)$$

$$= \sum_{i=1}^{K} \pi(T_i(n)) \mathbb{I}_{n_i > 0} r_i + \sum_{i=1}^{K} \pi(T^i(n)) \mu_i P_{i0} + \sum_{i=1}^{K} \sum_{j=1, j \neq i}^{K} \pi(T_i^j(n)) \mu_j P_{ji} \mathbb{I}_{n_i > 0}.$$

Substituting $\pi(n)$ from (9.9) into the above equation, we can verify the equality holds. Further, $\sum_n \pi(n) = 1$, so the theorem holds. □

We now present an alternative proof.

Proof Let $\tilde{Q}$ be the transition rate matrix of a Jackson network with external arrival rates

$$\tilde{r}_i = \lambda_i \left(1 - \sum_{j=1}^{K} P_{ij} \right),$$

and assume that the probability of moving from node i to node j is given by

$$\tilde{P}_{ij} = \frac{\lambda_j P_{ji}}{\lambda_i}.$$

We can verify that this is indeed the reversed chain and that (9.9) is in fact the stationary distribution using Theorem 9.8.2. □

9.9.2   The multi-class M/M/1 queue

Consider an M/M/1 queue where arrivals consist of K classes. The arrival rate of class-k customers is λ_k, and the mean service time of a class-k customer is $1/\mu_k$. Assume that the network is in state x such that $x = (c_1, \ldots, c_n)$, where c_i is the class of the ith customer in the queue, and the first customer is the one at the head of the queue. We next show that if $\mu_k = \mu$ for all k, the steady-state distribution of this multi-class M/M/1 queue has a product form and

$$\pi(x) = (1 - \rho) \prod_{i=1}^{n} \rho_{c_i}, \tag{9.10}$$

where $\rho = \sum_{k=1}^{K} \rho_k$ and $\rho_k = \lambda_k / \mu_k = \lambda_k / \mu$.

Given the queue is at state $x = (c_1, \ldots, c_n)$, the system moves to state $(c_1, \ldots, c_n, c_{n+1})$ at rate $\lambda_{c_{n+1}}$ (with an arrival belonging to class c_{n+1}) and moves to state $(c_2, \ldots, c_n)$ at rate μ_{c_1} (with the departure of the customer at the head of the queue). For $n > 0$, we first prove the following equality:

$$\pi(x)\left(\sum_{k=1}^{K} \lambda_k + \mu_{c_1}\right) = \pi(c_1, \ldots, c_{n-1})\lambda_{c_n} + \sum_{k=1}^{K} \pi(k, c_1, \ldots, c_n)\mu_k, \tag{9.11}$$

as illustrated in Figure 9.11.

Substituting $\pi(x)$ from (9.10) into (9.11), we obtain

$$(1-\rho)\left(\prod_{i=1}^{n} \rho_{c_i}\right)\left(\sum_{k=1}^{K} \lambda_k + \mu_{c_1}\right) = (1-\rho)\left(\prod_{i=1}^{n-1} \rho_{c_i}\right)\lambda_{c_n} + \sum_{k=1}^{K}(1-\rho)\rho_k\left(\prod_{i=1}^{n} \rho_{c_i}\right)\mu_k.$$

Canceling out the common terms on both sides, we have

$$\rho_{c_n}\left(\sum_{k=1}^{K} \lambda_k + \mu_{c_1}\right) = \lambda_{c_n} + \sum_{k=1}^{K} \rho_k \rho_{c_n} \mu_k,$$

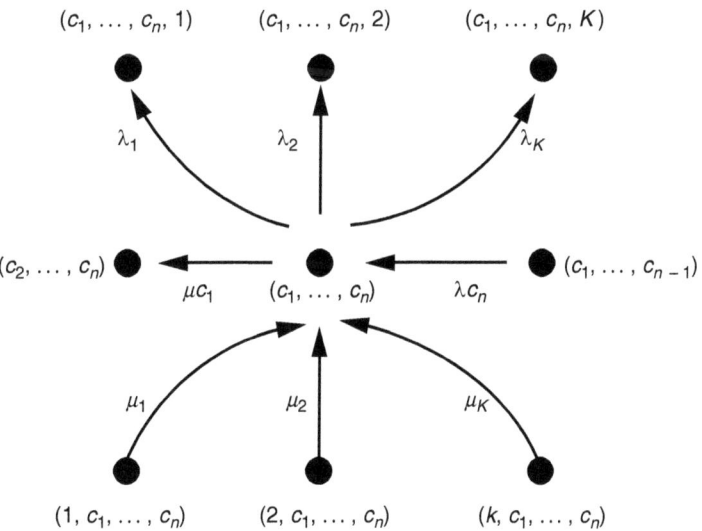

Figure 9.11 Illustration of local transitions when the queue is in state $(c_1, \ldots, c_n)$.

which is equivalent to

$$\rho_{c_n} \left(\sum_{k=1}^{K} \lambda_k + \mu_{c_1} \right) = \lambda_{c_n} + \rho_{c_n} \sum_{k=1}^{K} \rho_k \mu_k$$

$$= \lambda_{c_n} + \rho_{c_n} \sum_{k=1}^{K} \lambda_k,$$

which implies that

$$\rho_{c_n} \left(\sum_{k=1}^{K} \lambda_k + \mu_{c_1} \right) = \rho_{c_n} \left(\mu_{c_n} + \sum_{k=1}^{K} \lambda_k \right).$$

If $\mu_{c_n} = \mu$ for all $n > 0$, the above inequality is satisfied, and (9.11) holds when $n > 0$.

Now consider the case when $n = 0$. That equality (9.11) holds requires

$$\pi(\emptyset) \left(\sum_{k=1}^{K} \lambda_k \right) = \sum_{k=1}^{K} \pi(k) \mu_k.$$

Substituting (9.10) into the above equation, we obtain

$$(1 - \rho) \left(\sum_{k=1}^{K} \lambda_k \right) = (1 - \rho) \sum_{k=1}^{K} \rho_k \mu_k,$$

which is equivalent to

$$\sum_{k=1}^{K} \lambda_k = \sum_{k=1}^{K} \rho_k \mu_k = \sum_{k=1}^{K} \lambda_k.$$

So (9.11) holds for $n = 0$.

According to Lemma 9.1.6, equality (9.11) means the global balance equation holds, so $\pi \mathbf{Q} = 0$. It is easy to verify that $\sum_x \pi(x) = 1$, so we can conclude that $\pi(x)$ is the stationary distribution of the CTMC, which is in product form.

9.10   Insensitivity to service-time distributions

In previous sections, we studied the stationary distributions of FIFO queues with different arrival and service models, and we have seen that the stationary distributions depend on both the arrival process and the service-time distribution. However, certain queueing models have stationary distributions that are *insensitive* to service-time distributions, i.e., the stationary distribution depends only on the mean of the service time and not on the entire distribution. In this section, we will study one such queueing model.

The queueing system that we consider in this section is the M/M/1 processor-sharing model. Under processor sharing, if there are n customers in the queue, the capacity of the queue is equally divided between the n customers. This is different from the M/M/1 queue studied earlier, where the queue is operated in FIFO (first-in, first-out) fashion. In a FIFO queue, the server devotes its entire capacity to the first customer in the queue. Processor-sharing models have been widely studied because they are used to model two well-known applications.

- Consider files arriving at the Internet and suppose that the files have a common bottle-neck node. Further, suppose that some congestion control mechanism is used to divide the capacity of the bottleneck node equally among the files. If we view the files as customers, the resulting model is the processor-sharing model described above. However, this model may not be appropriate if there are multiple bottlenecks in the network or if the bottleneck node does not allocate its capacity equally to all files using it. A better model for modeling such situations will be presented in Section 9.11. Nevertheless, a simple processor-sharing model as described above is a useful first step.
- Web servers process requests from many jobs simultaneously. One resource allocation algorithm used by such a server is to allocate its upload capacity equally to all jobs in the system. This is another example of a processor-sharing service discipline.

9.10.1   The M/M/1-PS queue

The Markov chain for the M/M/1 processor-sharing (M/M/1-PS) queue is the same as the Markov chain for the M/M/1-FIFO queue because each customer departs at rate μ/n in the PS queue, but, because there are n customers, the total departure rate is μ.

Therefore, the stationary distribution of the number of customers in the M/M/1-PS queue is

$$\pi(n) = \rho^n(1 - \rho).$$

Next, we consider the M/GI/1-PS queue. The goal is to show that the stationary distribution of the number of customers is not affected by the service-time distribution.

9.10.2   The M/GI/1-PS queue

The Markov chains for M/GI/1-FIFO and M/GI/1-PS queues are different. We first consider a special case, where the service times are Erlang distributed, and then show that the stationary distribution of the number of customers in an M/GI/1-PS queue is insensitive to the parameters of the Erlang distribution as long as the mean service time is the same.

Definition 9.10.1 (Erlang distribution) X is an Erlang-distributed random variable with K phases if

$$X = X_1 + X_2 + \cdots + X_K,$$

where $X_i \sim \exp(K\mu)$ and the X_i's are i.i.d.   □

Note that, for an Erlang-distributed random variable,

$$E[X] = \frac{1}{\mu} \quad \text{and} \quad Var(X) = \frac{1}{K\mu^2}.$$

Further, as $K \to \infty$, X approximates a constant service time $1/\mu$ because the variance goes to zero. Later, we will use this fact to argue that any distribution on the positive real line can be approximated by a mixture of Erlang distributions. For now, let us consider a PS-queue with service times that are Erlang with K phases. A customer with Erlang-K service time distribution can be thought as having K phases of service, each requiring $\exp(K\mu)$ time: the customer completes the ith phase and then moves to the $(i + 1)$th phase.

Let $(c_1, c_2, \ldots, c_n)$ be the phases of the n customers in the system. When a new customer arrives, it randomly picks a number $\{0, 1, 2, 3, \ldots, n+1\}$ as its index. Then, the arriving customer inserts itself into the queue at the position indicated by its index, and customers already in the system at positions greater than or equal to the index of the arriving customer increase their index by 1. A customer departs when it has completed K phases. If the ith customer departs, we relabel customers $(i+1, i+2, \ldots, n)$ as $(i, i+1, \ldots, n-1)$, respectively.

Theorem 9.10.1 The stationary distribution of the $M/GI/1$-PS queue satisfies

$$\pi(c_1, c_2, \ldots, c_n) = \frac{\rho^n(1-\rho)}{K^n}, \qquad (9.12)$$

where $\rho = \lambda/\mu$. Thus, the steady-state probability that there are n customers in the queue is given by

$$\sum_{c_1=1}^{K} \sum_{c_2=1}^{K} \cdots \sum_{c_n=1}^{K} \pi(c_1, \ldots, c_n) = \rho^n(1-\rho),$$

which is the same as the $M/M/1$-PS queue.

Proof To prove (9.12), we use the concept of a reversed chain. In the forward chain,

$$(c_1, \ldots, c_i, \ldots, c_n) \to T^{i+}(c) \triangleq (c_1, \ldots, c_i+1, \ldots, c_n)$$

at rate $K\mu/n$ if $c_i < K$, which occurs when the ith customer moves from the c_ith phase to the (c_i+1)th phase;

$$(c_1, \ldots, c_{i-1}, K, c_{i+1}, \ldots, c_n) \to T_i(c) \triangleq (c_1, \ldots, c_{i-1}, c_{i+1}, \ldots, c_n)$$

at rate $K\mu/n$, which occurs when the ith customer leaves the queue;

$$(c_1, \ldots, c_i, c_{i+1}, \ldots, c_n) \to T^{i,1}(c) \triangleq (c_1, \ldots, c_i, 1, c_{i+1}, \ldots, c_n)$$

at rate $\lambda/(n+1)$, which occurs when a customer joins the system and is placed between the ith customer and the $(i+1)$th customer. Note that there are $n+1$ positions a joining customer can select, so the transmission rate is $\lambda/(n+1)$.

In the reversed chain, we have

$$(c_1, \ldots, c_i, \ldots, c_n) \to T^{i-}(c) \triangleq (c_1, \ldots, c_i-1, \ldots, c_n),$$

$$(c_1, \ldots, c_{i-1}, 1, c_{i+1}, \ldots, c_n) \to T_i(c) \triangleq (c_1, \ldots, c_{i-1}, c_{i+1}, \ldots, c_n),$$

$$(c_1, \ldots, c_i, c_{i+1}, \ldots, c_n) \to T^{i,K}(c) \triangleq (c_1, \ldots, c_i, K, c_{i+1}, \ldots, c_n).$$

Note that c in the reversed chain can be interpreted as the vector of the remaining phases of each packet. To verify that (9.12) is the stationary distribution, we utilize Theorem 9.8.2. We will construct $\mathbf{Q}^*$ such that

$$Q^*_{c,\hat{c}} = Q_{\hat{c},c} \frac{\pi(c)}{\pi(\hat{c})},$$

and show that $\mathbf{Q}^*$ is the transition rate matrix of the reversed chain and (9.12) is the stationary distribution of the CTMC. We consider the following three cases.

(1) We know that if $c_i < K$

$$Q_{c,T^{i+}(c)} = \frac{K\mu}{n}$$

and

$$\pi(c) = \pi(T^{i+}(c)) = \frac{\rho^n(1-\rho)}{K^n},$$

so

$$Q^*_{T^{i+}(c),c} = Q_{c,T^{i+}(c)} \frac{\pi(c)}{\pi(T^{i+}(c))} = \frac{K\mu}{n}, \text{ or equivalently, } Q^*_{c,T^{i-}(c)} = \frac{K\mu}{n}.$$

(2) We know that if $c_i = K$

$$Q_{c,T_i(c)} = \frac{K\mu}{n}$$

and

$$\pi(c) = \frac{\rho^n(1-\rho)}{K^n} \quad \text{and} \quad \pi(T_i(c)) = \frac{\rho^{n-1}(1-\rho)}{K^{n-1}},$$

so we have

$$Q^*_{T_i(c),c} = Q_{c,T_i(c)} \frac{\pi(c)}{\pi(T_i(c))} = \frac{\lambda}{n}, \text{ which is equivalent to } Q^*_{c,T^{i,K}(c)} = \frac{\lambda}{n+1}.$$

(3) We have that

$$Q_{c,T^{i,1}(c)} = \frac{\lambda}{n+1}$$

and

$$\pi(c) = \frac{\rho^n(1-\rho)}{K^n} \quad \text{and} \quad \pi(T^{i,1}(c)) = \frac{\rho^{n+1}(1-\rho)}{K^{n+1}},$$

so we obtain

$$Q^*_{T^{i,1}(c),c} = Q_{c,T^{i,1}(c)} \frac{\pi(c)}{\pi(T^{i,1}(c))} = \frac{K\mu}{n+1}.$$

Next we show that

$$\sum_{\hat{c} \neq c} Q_{c,\hat{c}} = \sum_{\hat{c} \neq c} Q^*_{c,\hat{c}}.$$

Consider the case where $1 < c_i < K$ for all i. In this case, no departure will occur in the forward chain because $c_i < K$ for all i; no departure will occur in the reversed chain because $c_i > 1$ for all i. The above equality holds because

$$\sum_{\hat{c} \neq c} Q_{c,\hat{c}} = \sum_{i=1}^{n} Q_{c,T_i^+(c)} + \sum_{i=0}^{n} Q_{c,T^{i,1}(c)}$$

$$= n\frac{K\mu}{n} + \lambda$$

$$= \lambda + K\mu$$

and

$$\sum_{\hat{c} \neq c} Q^*_{c,\hat{c}} = \sum_{i=0}^{n} Q^*_{c,T^{(i,K)}(c)} + \sum_{i=1}^{n} Q^*_{c,T^{i-}(c)}$$

$$= \frac{\lambda}{n+1}(n+1) + \frac{K\mu}{n}n$$

$$= \lambda + K\mu.$$

As $\sum_{\hat{c} \neq c} Q_{c\hat{c}} + Q_{cc} = 0$ and $Q_{cc} = Q_{cc}^*$, we have that $\sum_{\hat{c}} Q_{c\hat{c}}^* = 0$. The other cases can be checked similarly.

We therefore have that $\mathbf{Q}^*$ is also a transition rate matrix. According to Theorem 9.8.2, we conclude that $\mathbf{Q}^*$ is the transition rate matrix of the reversed chain and (9.12) is the stationary distribution. $\qquad\square$

Next, let us consider an M/GI/1-PS queue with the service-time distribution being a mixture of Erlang distributions. A mixture of Erlang-K distributions is defined as follows.

Definition 9.10.2 (Mixture of Erlang-K distributions) With probability p_j, the service time follows Erlang$\{K_j, \mu_j\}$, such that

$$\sum_{j \geq 1}^{\infty} p_j = 1 \qquad \text{and} \qquad \frac{1}{\mu} \triangleq \sum_{j=1}^{\infty} \frac{p_j}{\mu_j}. \qquad\square$$

The usefulness of Erlang mixtures is due to the fact that any distribution over the positive real line can be approximated by a mixture of Erlang-K distributions. Specifically, the cumulative distribution function of any random variable X taking values in $[0, \infty)$ can be approximated by a mixture of Erlang-K distributions. We present the basic intuition behind this result. Consider a distribution with Cumulative Distribution Function (CDF) $F(x)$ as shown in Figure 9.12, and assume that $\Pr(y \leq x \leq y + \Delta) = p_y$. We then define a mixture of Erlang-K distributions such that, with probability p_y, it is Erlang$(K, 1/y)$, i.e., K phases each with the mean service time y/K. It is easy to see that, as $K \to \infty$, we are approximating a random variable that takes the value y (a constant). Thus, as K becomes large, the mixture random variable takes the value y with probability p_y. Thus, for small Δ, we are approximating the desired CDF.

Consider an M/GI/1-PS system with the service-time distribution given in Definition 9.10.2. Suppose there are n customers in the system. The state of the system is

$$x \triangleq \{(t_1, c_1), (t_2, c_2), \ldots, (t_n, c_n)\},$$

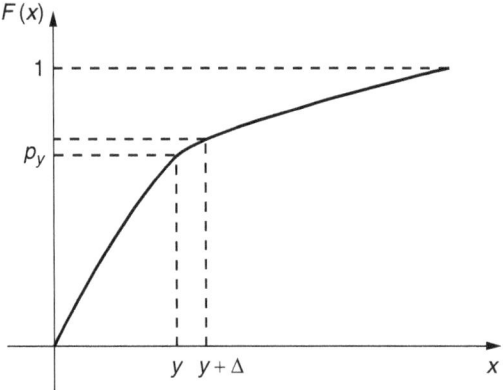

Figure 9.12 General CDF, $F(x)$.

where t_i is the type of Erlang distribution chosen at arrival using the probabilities $\{p_1, p_2, \ldots\}$, and c_i is the phase that the packet is currently in.

Our guess for the steady-state distribution is

$$\pi(x) = \frac{\hat{\pi}(t_1, c_1) \cdots \hat{\pi}(t_n, c_n)}{Z},$$

where

$$\hat{\pi}(t_i, c_i) = \frac{\lambda p_{t_i}}{K_{t_i} \mu_{t_i}}$$

and Z is a normalization constant such that

$$\sum_x \pi(x) = 1.$$

It is not difficult, but it is a bit cumbersome, to prove that this is indeed the steady-state distribution using Theorem 9.8.2, but we shall not do so here.

Given $\pi(x)$, to compute the steady-state probability of n packets in the system, note that

$$\pi(n) = \sum_{t_1=1}^{\infty} \cdots \sum_{t_n=1}^{\infty} \sum_{c_1=1}^{K_{t_1}} \cdots \sum_{c_n=1}^{K_{t_n}} \pi(x),$$

where, by abusing notation, we are using $\pi(n)$ to indicate the stationary probability that there are n customers in the system. As

$$\sum_{t_i=1}^{\infty} \sum_{c_i=1}^{K_{t_i}} \frac{p_{t_i}}{K_{t_i} \mu_{t_i}} = \sum_{t_i=1}^{\infty} \frac{p_{t_i}}{\mu_{t_i}} = \frac{1}{\mu},$$

where the second equality holds due to the definition of μ, we obtain

$$\pi(n) = \frac{1}{Z} \left(\frac{\lambda}{\mu} \right)^n.$$

Note that $\pi(\emptyset)$ is the steady-state probability that the server is idle. Therefore, as the server can hold only one customer when the system is not idle, $1 - \pi(\emptyset)$ is the average queue length at the server. As $1/\mu$ is the average delay a customer experiences at the server, according to Little's law we have $1 - \pi(\emptyset) = \lambda/\mu$, which implies that

$$\pi(\emptyset) = 1 - \rho,$$

where $\rho = \lambda/\mu$, leading to

$$\pi(0) = 1 - \rho$$

and

$$\pi(n) = (1 - \rho)\rho^n.$$

9.11 Connection-level arrivals and departures in the internet

In Chapter 2, we studied resource allocation in the Internet assuming that the set of flows (also called files) in the network is fixed. In reality, flows, also known as *connections*, arrive and depart dynamically. We now use CTMCs to model and study connection-level arrivals and departures, and characterize necessary and sufficient conditions for connection-level stability.

Denote by n_r the number of flows on route r. A flow is called a type r flow if the flow is on route r. We assume that each flow of type r arrives according to a Poisson process of rate λ_r, and that the file associated with the flow has an exponentially distributed number of bits with mean $1/\mu_r$. A flow departs when all of its bits are transferred. For example, if a flow of type r and with B bits arrives at time t, it will depart at time $T > t$, where T is the smallest time such that

$$\int_t^T x_r(s)ds = B,$$

where $x_r(t)$ is the transmission rate allocated to a flow on route r.

After the flow joins the network, the transmission rate of the flow is regulated by congestion control. We assume a time-scale separation assumption such that congestion control occurs instantaneously compared to file arrivals/departures. Specifically, we assume that the network allocates x_r to a flow on route r by solving the following optimization problem (see Chapter 2 for details):

$$\sum_r n_r U_r(x_r)$$

subject to

$$\sum_{r:l\in r} n_r x_r \leq c_l, \qquad \forall l,$$

$$x_r \geq 0,$$

where $U_r(\cdot)$ is a concave function.

We assume that

$$\sum_{r:l\in r} \frac{\lambda_r}{\mu_r} < c_l \qquad \text{or} \qquad \sum_{r:l\in r} \rho_r < c_l,$$

i.e., the aggregated workload on each link is less than the link capacity. Let $n_r(t)$ denote the number of files of type r at time t. The question we seek to answer is the following: is the network stable? In other words, does $n(t) = (n_1(t), n_2(t), \ldots)$ have a well-defined stationary distribution? Some resource allocation rules may not fully utilize resources, and may lead to network instability even when the workload at each link is less than its capacity, as the following example shows.

Example 9.11.1

Consider a tandem network with three nodes and two links. Assume that there are three types of flows in the network: type-1 flows only use link 1, type-2 flows only use link 2, and type-0 flows use both links 1 and 2, as shown in Figure 9.13. Further, assume that $\mu_r = 1$ for all r and $c_l = 1$ for both links. Note that any rate allocation among the three types of flows must satisfy

$$n_0 x_0 + n_1 x_1 \leq 1,$$

$$n_0 x_0 + n_2 x_2 \leq 1.$$

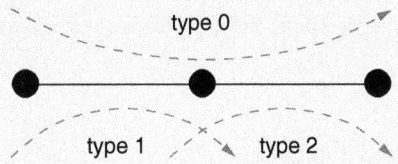

Figure 9.13 Linear network with two nodes and three links.

Consider a rate allocation scheme that gives absolute priority to flows of type 1 and 2 over type-0 flows. So, if $n_1 > 0$, the rate allocation (x_0, x_1, x_2) satisfies

$$n_0 x_0 = 0 \quad \text{and} \quad n_1 x_1 = 1.$$

Similarly, if $n_2 > 0$, the rate allocation satisfies

$$n_0 x_0 = 0 \quad \text{and} \quad n_2 x_2 = 1.$$

Clearly we need $\lambda_1 < 1$ and $\lambda_2 < 1$ for stability in this case. What about λ_0? Note that flows of type 0 can use the network only if $n_1 = n_2 = 0$ and

$$\Pr(n_1 = 0) = 1 - \lambda_1 \quad \text{and} \quad \Pr(n_2 = 0) = 1 - \lambda_2.$$

Because n_1 and n_2 are independent, we have

$$\Pr(n_1 = n_2 = 0) = (1 - \lambda_1)(1 - \lambda_2),$$

which implies that λ_0 should satisfy

$$\lambda_0 < (1 - \lambda_1)(1 - \lambda_2)$$

for stability. This region is smaller than the following region:

$$\lambda_0 + \lambda_1 < 1 \quad \text{and} \quad \lambda_0 + \lambda_2 < 1. \tag{9.13}$$

We will soon show that (9.13) is the capacity region under the utility maximization framework. In other words, the network can be stabilized under the utility maximization framework if (9.13) is satisfied, whereas the resource allocation that gives priority to type-1 and type-2 flows cannot achieve the full capacity region. □

Now we study connection-level stability under the utility maximization framework. Assume we have R types of flows in the network. Define n to be an R-vector, where n_r is the number of type-r flows in the network. Further, define an R-vector e_r such that

$$e_r = (0, 0, \ldots, 0, 1, 0, 0, \ldots, 0)^T,$$

i.e., all entries of e_r are 0 except the rth entry, which is 1.

Since file arrivals are Poisson and file sizes are exponential, n is a CTMC, and

$$n \longrightarrow n + e_r \quad \text{at rate } \lambda_r,$$

$$n \longrightarrow n - e_r, \quad \text{if } n_r > 0, \text{ at rate } \mu_r n_r x_r.$$

Assume that the utility functions are of the following form:

$$U_r(x_r) = \begin{cases} w_r x_r^{1-\alpha}/(1-\alpha), & \text{if } \alpha \neq 1, \alpha > 0, \\ w_r \log x_r, & \text{if } \alpha = 1. \end{cases}$$

Theorem 9.11.1 Suppose the network allocates transmission rates x_r for each flow type r by solving the utility maximization problem. Then, the CTMC is positive recurrent when $\sum_{r:l\in r} \rho_r < c_l$ for all l.

Proof We consider the following Lyapunov function:

$$V(n) = \sum_r \frac{\kappa_r}{1+\alpha} n_r^{1+\alpha},$$

where κ_r will be chosen later. By the Foster–Lyapunov theorem for CTMCs (Theorem 9.1.8), we need to consider the drift

$$\sum_{j:j\neq i} (V(j) - V(i)) Q_{ij},$$

where i and j are two possible values that the state n of the CTMC can take, and Q is the rate transition matrix of the CTMC.

For our model, the drift is given by

$$\sum_{j:j\neq i} (V(j) - V(i)) Q_{ij} = \sum_r \frac{\kappa_r}{1+\alpha} \left((n_r + 1)^{1+\alpha} - n_r^{1+\alpha} \right) \lambda_r$$
$$+ \sum_r \frac{\kappa_r}{1+\alpha} \left((n_r - 1)^{1+\alpha} - n_r^{1+\alpha} \right) \mu_r n_r x_r,$$

where $\mu_r n_r x_r = 0$ if $n_r = 0$. From the mean-value theorem, we have

$$\frac{1}{1+\alpha} \left((n_r + 1)^{1+\alpha} - n_r^{1+\alpha} \right) = n_r^\alpha + \frac{1}{2}\alpha \tilde{n}_r^{\alpha-1}, \qquad n_r \leq \tilde{n}_r \leq n_r + 1,$$

$$\frac{1}{1+\alpha} \left((n_r - 1)^{1+\alpha} - n_r^{1+\alpha} \right) = -n_r^\alpha + \frac{1}{2}\alpha \hat{n}_r^{\alpha-1}, \qquad n_r - 1 \leq \hat{n}_r \leq n_r.$$

After ignoring the $\hat{n}_r$ and $\tilde{n}_r$ terms (they will be small compared to n_r^α when n_r is large),[1] we obtain

$$\sum_{j:j\neq i} (V(j) - V(i)) Q_{ij} = \sum_r \kappa_r n_r^\alpha (\lambda_r - \mu_r n_r x_r)$$
$$= \sum_r \kappa_r \mu_r n_r^\alpha (\rho_r - n_r x_r).$$

Further, defining $n_r x_r = X_r$, we have

$$\sum_{j:j\neq i} (V(j) - V(i)) Q_{ij} = \sum_r \kappa_r \mu_r n_r^\alpha (\rho_r - X_r),$$

where $\{X_r\}$ solves

$$\max_{X_r \geq 0} \sum_r \frac{w_r n_r}{1-\alpha} \left(\frac{X_r}{n_r} \right)^{1-\alpha} \qquad \text{such that} \qquad \sum_{l:l\in r} X_r \leq c_l, \qquad \forall l,$$

which is equivalent to

$$\max_{X_r \geq 0} \sum_r \frac{w_r n_r^\alpha}{1-\alpha} X_r^{1-\alpha} \qquad \text{such that} \qquad \sum_{l:l\in r} X_r \leq c_l, \qquad \forall l.$$

1 This argument is straightforward but tedious. We leave the details to the reader.

Recall that, for a concave function $f(x)$ over a convex set $\mathcal{C}$,

$$f(x) + \nabla f^T(x)(y - x) \geq f(y).$$

Suppose $y = x^*$, the maximizer, then the above inequality implies that

$$\nabla f^T(x)(x^* - x) \geq f(x^*) - f(x) \geq 0, \qquad \forall x,$$

or

$$\nabla f^T(x)(x - x^*) \leq 0, \qquad \forall x.$$

Because $\sum_{r:l\in r} \rho_r < c_l$ implies that there exists $\epsilon > 0$ such that $\{\rho_r(1 + \epsilon)\}$ are feasible values, we have

$$\sum_r \frac{w_r n_r^\alpha}{(\rho_r(1 + \epsilon))^\alpha}(\rho_r(1 + \epsilon) - X_r) \leq 0,$$

which implies that

$$\sum_r \frac{w_r n_r^\alpha}{\rho_r^\alpha}(\rho_r - X_r) \leq -\epsilon \sum_r \frac{w_r n_r^\alpha}{\rho_r^\alpha}\rho_r = -\epsilon \sum_r \frac{w_r n_r^\alpha}{\rho_r^{\alpha-1}}.$$

Choosing $\kappa_r \mu_r = w_r/\rho_r^\alpha$ or $\kappa_r = w_r/\mu_r \rho_r^\alpha$, we obtain

$$\sum_{j:j\neq i}(V(j) - V(i))Q_{ij} = \sum_r \frac{w_r n_r^\alpha}{\rho_r^\alpha}(\rho_r - X_r)$$

$$\leq -\epsilon \sum_r \frac{w_r n_r^\alpha}{\rho_r^{\alpha-1}}.$$

We therefore conclude that the CTMC is positive recurrent by the Foster–Lyapunov theorem. $\square$

From Theorem 9.11.1 we have that the network is stable under the utility maximization framework if $\sum_{r:l\in r} \rho_r < c_l$ for all l. It is not difficult to see that if $\sum_{r:l\in r} \rho_r > c_l$, the network cannot be stable. Thus, the utility maximization framework achieves the largest possible connection-level throughput.

9.12   Distributed admission control

In this section, we consider control schemes to regulate the admission of inelastic sources in a communication network. Inelastic sources are those whose packet generation rate cannot be subject to congestion control without adversely affecting the quality of reception at the destinations. Examples of such sources include certain types of voice and video traffic. Thus, the only type of control that the network can exercise over such traffic is to deny admission to new sources when the network load is high, so that the sources already admitted to the network continue to enjoy good Quality of Service (QoS).

To develop a simple model of admission control for inelastic sources, in this section we assume that all sources accessing the network are inelastic and, hence, are not congestion controlled, and that each source injects packets at a fixed rate into the system. We will also assume that all sources are bottlenecked at a single node whose capacity is normalized to 1 bit per time unit. Thus, when the number of sources is fixed (not time varying), the network can be modeled as a single-server system accessed by n sources, each generating packets at rate λ_p according to a Poisson process. Suppose that the packet sizes of each source are exponentially distributed with mean $1/\mu_p$. Then, the system is an M/M/1 queue with arrival rate $n\lambda_p$ and average service time $1/\mu_p$ (per packet).

Assume that the objective of the admission control is to guarantee that

$$\Pr(q(\infty) \geq B) \leq \epsilon,$$

where $q(\infty)$ denotes the queue length in steady state. We can use the formula for the steady-state distribution of an M/M/1 queue to find $N_{\max}$, the maximum number of sources that can be admitted in the network to satisfy the QoS requirement. If we have a centralized admission control scheme that can monitor the number of sources admitted to the system, it should not allow more than $N_{\max}$ sources into the network.

Next, suppose that the number of sources in the system is time varying. Assume that sources arrive according to a Poisson process of rate λ and stay for an exponentially distributed amount of time with mean $1/\mu$. Further suppose that the time scale at which arrivals and departures of sources occur is much slower compared to the time scale at which packets are generated by each source. Then, it is still reasonable to block a newly arriving source if there are already $N_{\max}$ sources in the system. Therefore, the number of sources in the network under a centralized admission control can be modeled as an M/M/$N_{\max}$/$N_{\max}$ queue, and the blocking probability (the probability a source is not admitted when it arrives at the system) can be obtained from the Erlang-B formula. So, in some sense, centralized admission control is relatively easy.

We are interested in distributed admission control schemes that closely approximate the above centralized admission control model. In particular, we are interested in schemes under which sources need to make their own decisions about whether they should join the network. We assume that a newly arriving source probes the network by sending a few packets into the network. We also assume that congested queues in the network have the ability to *mark* packets and that the destination feeds back the number of marked packets to the source. The source then decides to join the network or not based on the number of marked probe packets. The proposed scheme is summarized in Algorithm 11.

We are interested in analyzing this distributed admission control scheme. Assume that the packet queue converges to the steady state instantaneously compared to the time scale

---

Algorithm 11 Distributed admission control

```
1: Before joining the network, a source sends m probing packets.
2: A probing packet is marked when it arrives and finds at least B̃ packets
   in the queue.
3: if one of the probing packets is marked then
4:    the source does not join the network.
5: else
6:    the source joins the network.
7: end if
```

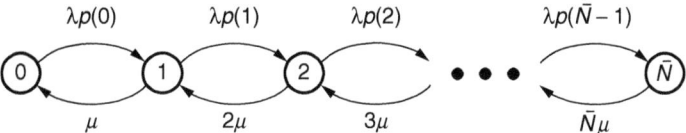

Figure 9.14 CTMC for distributed admission control.

of source arrivals/departures. Let $p_{mark}(n)$ denote the probability that a probing packet is marked when there are n sources in the network. We assume the packets sizes are exponential. So,

$$p_{mark}(n) = \Pr(q^{(n)}(\infty) \geq \tilde{B}),$$

where $q^{(n)}(\infty)$ is the steady-state queue length of the M/M/1 queue with arrival rate $n\lambda_p$ and average service time $1/\mu_p$. Assume that $p_{mark}(n) = 1$ if $n\lambda_p \geq \mu_p$ and define

$$\bar{N} = \left\lfloor \frac{\mu_p}{\lambda_p} \right\rfloor.$$

Typically, we have $N_{\max} < \bar{N}$.

Further, we denote by $p(n)$ the probability that a new source joins the network when there are n other sources in the network. Since a new source joins the network only if none of its m probing packets is marked, assuming the m probing packets are marked independently, we have

$$p(n) = (1 - p_{mark}(n))^m.$$

The network can be modeled as a CTMC, as shown in Figure 9.14. By studying this CTMC, we can compute $\Pr(n \geq N_{\max})$, and

$$p_b = \text{fraction of rejected flows.}$$

Note that, by Little's law,

$$E[n] = (1 - p_b)\lambda \frac{1}{\mu} = (1 - p_b)\rho,$$

so

$$p_b = 1 - \frac{E[n]}{\rho}.$$

The parameters m (the number of probing packets) and $\tilde{B}$ (the threshold for marking a probing packet) can be used to tune the tradeoff between $\Pr(n \geq N_{\max})$ and p_b. If p_b is increased, $\Pr(n \geq N_{\max})$ decreases, and vice versa. We would like to keep them both small.

9.13   Loss networks

Loss networks are models of circuit switched networks, traditionally used to study telephone networks. In a traditional telephone network, a telephone call requests one unit of resource (originally, it was 64 Kbps, but can be much less using data compression techniques). Each link l in the network has capacity c_l, where c_l is an integer. If $c_l = 10$, it means that the link can carry ten telephone calls at a time. Assume that a route from a source S to a destination D is fixed. We will denote a route by r, which is simply a set of links connecting S to D.

Example 9.13.1

Consider the network shown in Figure 9.15. Links $(1, 5)$ and $(5, 6)$ can be a route connecting source 1 and destination 6.

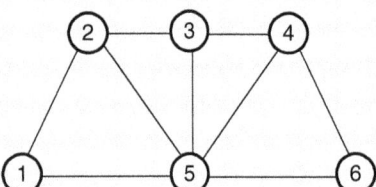

Figure 9.15 Circuit switched network.

A call on route r can be admitted to the network only if there is at least one unit of capacity on all the links on the route. Otherwise, the call is blocked and lost forever. We are interested in the following question: given traffic statistics, how can we compute the fraction of calls lost? This question will be studied in this section under the following assumptions.

- The traffic to route r is a Poisson process of rate λ_r.
- Calls stay in the network for an exponential amount of time, with mean $1/\mu$, and then they depart.
- The arrival processes and service times are independent of each other.

Under the above assumptions, the circuit switched network can be modeled as a loss network, i.e., a network generalization of the M/M/s/s loss model. The following notation will be used throughout this section:

- n_r: number of calls on route r;
- $\mathcal{R}$: set of all possible routes;
- n: system state given by $(n_1, n_2, \ldots, n_R)$, where $R = |\mathcal{R}|$;
- $\pi(n)$: steady-state distribution of state n.

In loss networks, the number of calls on a link cannot exceed the link capacity, so we have the constraints

$$\sum_{r:l \in r} n_r \le c_l \tag{9.14}$$

for all links l. Note that $l \in r$ means link l is a part of route r. The network state n is a CTMC, and we will compute the steady-state distribution of this CTMC. We write the local balance equation first:

$$\lambda_r \pi(n_1, \ldots, n_r, \ldots, n_R) = (n_r + 1)\mu \pi(n_1, \ldots, n_r + 1, \ldots, n_R).$$

Defining $\rho_r = \lambda_r/\mu$, we obtain

$$\rho_r \pi(n_1, \ldots, n_r, \ldots, n_R) = (n_r + 1)\pi(n_1, \ldots, n_r + 1, \ldots, n_R).$$

It is easy to verify that

$$\pi(n) = \frac{\prod_{r \in \mathcal{R}} \frac{\rho_r^{n_r}}{n_r!}}{Z} \tag{9.15}$$

satisfies the preceding equation for a suitable normalization constant Z, where Z is computed by noting that

$$\sum_{n: \sum_{r:l \in r} n_r \leq c_l \; \forall l} \pi(n) = 1.$$

In other words,

$$Z = \sum_{n: \sum_{r:l \in r} n_r \leq c_l \; \forall l} \prod_{r \in \mathcal{R}} \frac{\rho_r^{n_r}}{n_r!}.$$

Before we proceed further, it is useful to understand how (9.15) was derived. Suppose that $c_l = \infty$ for all l. Then, the n_r's are independent and each n_r behaves like an M/M/∞ queue. In this case, $\pi(n_r) = \rho_r^{n_r}/n_r!$ and $\pi(n) = \prod_{r \in \mathcal{R}} \rho_r^{n_r}/n_r!$. Therefore, (9.15) is simply a renormalization of the above expression to account for the constraints on n. Luckily, such a guess satisfies the local balance equation and hence is the correct steady-state distribution. This is an example of the "truncation theorem" for reversible CTMCs (see Exercise 9.10). If we can calculate Z, the steady-state probability of a call being blocked on route r can be calculated using the PASTA property (see Section 9.4) as

$$b_r = \text{Pr(at least one link on route } r \text{ is full)} = \sum_{n \in \mathcal{B}_r} \pi(n),$$

where

$$\mathcal{B}_r = \left\{ n : \sum_{s:l \in s} n_s = c_l \text{ for some } l \in r \right\}.$$

Except for small networks, this probability is difficult to compute since Z is very difficult to compute for large networks and the cardinality of the set $\mathcal{B}_r$ is very large. So we have to come up with an approximation, which we do in the following.

9.13.1 Large-system limit

Consider a regime in which the arrival rates and link speeds are all scaled by a factor k, i.e., we assume that $c_l = k\tilde{c}_l$ and $\lambda_r = k\tilde{\lambda}_r$, where we will eventually let $k \to \infty$. In other words, we want to consider a large capacity network, with correspondingly large arrival rates. We further define $x_r = n_r/k$, so the constraints (9.14) can be written as follows:

$$\sum_{r:l \in r} x_r \leq \tilde{c}_l. \tag{9.16}$$

Further, we rewrite $\pi(n)$ as

$$\pi(n) = \frac{1}{Z} \prod_{r \in \mathcal{R}} \frac{e^{n_r \log \rho_r}}{n_r!}.$$

Next, we use Stirling's formula,

$$n_r! = e^{n_r \log n_r - n_r + O(\log n_r)},$$

so that

$$\pi(n) = \frac{1}{Z} e^{\sum_{r \in \mathcal{R}} n_r \log \rho_r - n_r \log n_r + n_r - O(\log n_r)},$$

which implies that

$$\pi(x) = \frac{1}{Z} e^{\sum_{r \in \mathcal{R}} k x_r \log(k\tilde{\lambda}_r/\mu) - k x_r \log(k x_r) + k x_r - O(\log(k x_r))}.$$

Note that

$$\sum_{r \in \mathcal{R}} k x_r \log\left(k\tilde{\lambda}_r/\mu\right) - k x_r \log(k x_r) + k x_r - O(\log(k x_r))$$

$$= \sum_{r \in \mathcal{R}} k x_r \log\left(\tilde{\lambda}_r/\mu\right) - k x_r \log x_r + k x_r - O(\log(k x_r))$$

$$= k \left(\sum_{r \in \mathcal{R}} x_r \log\left(\tilde{\lambda}_r/\mu\right) - x_r \log x_r + x_r \right) - O(\log(k x_r)).$$

Given x' and x'' such that

$$\left(\sum_{r \in \mathcal{R}} x'_r \log\left(\tilde{\lambda}_r/\mu\right) - x'_r \log x'_r + x'_r \right) - \left(\sum_{r \in \mathcal{R}} x''_r \log\left(\tilde{\lambda}_r/\mu\right) - x''_r \log x''_r + x''_r \right) = \Delta > 0,$$

as $k \to \infty$, the steady-state distribution of x'' is negligible compared to that of x' because

$$\lim_{k \to \infty} \frac{\pi(x'')}{\pi(x')} = \lim_{k \to \infty} e^{-k\Delta} = 0.$$

Therefore, as $k \to \infty$, the network will converge to a state x that solves the following optimization problem:

$$\max \sum_{r \in \mathcal{R}} x_r \log \tilde{\lambda}_r/\mu - x_r \log x_r + x_r$$

subject to

$$\sum_{r:l \in r} x_r \leq \tilde{c}_l, \qquad \forall l,$$

$$x_r \geq 0.$$

Using the Lagrange multipliers, we obtain

$$L(x, p) = \sum_{r \in \mathcal{R}} x_r \log \tilde{\lambda}_r/\mu - x_r \log x_r + x_r - \sum_l p_l \left(\sum_{r:l \in r} x_r - \tilde{c}_l \right).$$

Note that

$$\frac{\partial L}{\partial x_r}(x, p) = 0$$

implies that

$$\log \tilde{\lambda}_r/\mu - \log x_r = \sum_{l \in r} p_l,$$

which further implies that

$$\log x_r = \log \tilde{\lambda}_r/\mu - \sum_{l \in r} p_l$$

and

$$\frac{x_r}{\tilde{\lambda}_r/\mu} = e^{-\sum_{l \in r} p_l}. \tag{9.17}$$

According to the definitions of x and $\tilde{\lambda}$, we have

$$\frac{x_r}{\tilde{\lambda}_r/\mu} = \frac{n_r}{\lambda_r/\mu} = \frac{n_r}{\rho_r},$$

so that

$$\frac{n_r}{\rho_r} = e^{-\sum_{l \in r} p_l}. \tag{9.18}$$

By Little's law,

$$E[n_r] = \lambda_r(1 - b_r)\frac{1}{\mu},$$

where $\lambda_r(1 - b_r)$ is the arrival rate of accepted calls on route r, which implies that

$$1 - b_r = \frac{E[n_r]}{\rho_r} = \frac{E[x_r]}{\tilde{\lambda}_r/\mu} \approx e^{-\sum_{l \in r} p_l}, \tag{9.19}$$

where the last approximation follows from the observation that the network converges to state $x_r = (\tilde{\lambda}_r/\mu)e^{-\sum_{l \in r} p_l}$ as $k \to \infty$ (equation (9.17)). So, $e^{-\sum_{l \in r} p_l}$ provides an estimate of the blocking probability if p_l can be computed.

From the KKT conditions, p_l satisfies

$$p_l\left(\sum_{s:l \in s} x_s - \tilde{c}_l\right) = 0$$

for all l; in other words,

$$p_l\left(\sum_{s:l \in s} n_s - c_l\right) = 0$$

for all l. Replacing n_s using equation (9.18), we obtain

$$p_l\left(\sum_{s:l \in s} \rho_s e^{-\sum_{j \in s} p_j} - c_l\right) = 0 \tag{9.20}$$

for all l, which provides a set of equations to compute p_l.

Further, consider a route $r = \{l\}$, i.e., a single-link route. From approximation (9.19), we have $1 - b_{\{l\}} \approx e^{-p_l}$, i.e., the probability that the calls on route $r = \{l\}$ are blocked is approximately $1 - e^{-p_l}$. Now assume that all calls that use link l experience the same blocking probability, then $b_l \triangleq 1 - e^{-p_l}$ is an approximation of the blocking probability on link l.

Hence (9.19) becomes

$$1 - b_r = \prod_{l \in r} (1 - b_l),$$

which suggests that the links behave independently in a large network. Further, (9.20) becomes

$$\sum_{s:l \in s} \rho_s \left[\prod_{j \in s} (1 - b_j) \right] \begin{cases} = c_l, & \text{if } b_l > 0, \\ \leq c_l, & \text{if } b_l = 0. \end{cases}$$

This reinforces the blocking probability interpretation for b_l because $b_l > 0$ must mean that the total number of calls on the link must be equal to c_l in the large-network limit. These statements can be made precise, but we will not do so here. In summary, the large-system limit suggests that the blocking probabilities satisfy the following equations:

$$1 - b_r = \prod_{l \in r} (1 - b_l)$$

$$\sum_{r:l \in r} \rho_r (1 - b_r) \begin{cases} = c_l, & \text{if } b_l > 0, \\ \leq c_l, & \text{if } b_l = 0. \end{cases}$$

We will use this insight to develop an algorithm to compute blocking probabilities approximately in Section 9.13.2.

9.13.2 Computing the blocking probabilities

Motivated by the large-system limit, we present the reduced-load approximation, which is a heuristic to compute blocking probabilities. Under the reduced-load approximation, we assume that the total arrival rate at a link is thinned due to blocking at other links. In other words, for each link l and for each route r containing l, we multiply the arrival rate at route r by the probability that a call is not blocked at any of the other links of this route; we assume that the arrival rate at link l is the aggregate of these quantities over all the routes containing l. Further, we assume that the blocking event at each link is independent of the blocking events at other links. Then, we use the thinned arrival rate to compute the blocking probability on link l. Thus, we get a fixed-point equation as follows: we start by assuming blocking probabilities on each link are known. Then assume that the blocking event at each link is independent of the blocking events at other links, and we compute the arrival rate at each link using the thinning heuristic. Once the arrival rates at the links are known, we compute the blocking probability on each link using the Erlang-B formula. This reduced-load heuristic is presented as Algorithm 12.

Note that step 1 is a fixed-point equation, which has to be solved iteratively as follows:

$$\tilde{b}_j(k + 1) = E_B \left(\sum_{r:j \in r} \frac{\bar{\lambda}_r(k)}{\mu}, c_j \right),$$

$$\bar{\lambda}_r(k) = \lambda_r \prod_{l:l \in r, l \neq j} \left(1 - \tilde{b}_l(k) \right).$$

Algorithm 12 The reduced-load approximation: a heuristic to compute blocking probabilities

1: Assume $\tilde{b}_l$ are known. Then, the arrival rate of route r traffic at link j (if $j \in r$) is

$$\tilde{\lambda}_r = \lambda_r \prod_{l: l \in r,\, l \neq j} \left(1 - \tilde{b}_l\right).$$

Thus, $\tilde{b}_j$ can be calculated as

$$\tilde{b}_j = E_B \left(\sum_{r: j \in r} \frac{\tilde{\lambda}_r}{\mu},\, c_j \right),$$

where $E_B(\rho, c)$ is the Erlang-B formula for a link with capacity c and load ρ.
2: The blocking probability on route r can be computed using the following equation:

$$1 - b_r = \prod_{l \in r} \left(1 - \tilde{b}_l\right).$$

Reverse the order of these steps because (i) we start by assuming that the blocking probabilities on each link are known, so step 2 is the first computation done in the course of the algorithm, and (ii) the stage $k + 1$ calculation should be performed after the stage k calculation. It has been observed that, in practice, the iterations converge, but, in general, the point to which the iterations converge may not be unique. We will comment more on this in Section 9.13.3.

9.13.3 Alternative routing

In the preceding sections on loss networks, we assumed that a call is blocked if there is no available capacity on its route. Thus, we implicitly assumed that each call is associated with only one route. In general, a call between a source and its destination can be routed on one of many paths. Often, there is a primary route that is attempted first, and, if there is no capacity available on the primary route, alternative routes are attempted according to some pre-specified rule. For simplicity, we consider a fully connected network with N nodes and with all link capacities $c_l = c$. By a fully connected network, we mean that there is an undirected link between every pair of nodes in the network. Further, we assume that there are $N(N-1)/2$ S–D pairs in this network, corresponding to every pair of nodes in the network. Assume a call tries the direct link to the destination first; if that is not available, one of the two-hop paths is chosen at random and the call is routed on this path if the capacity is available. Otherwise, the call is rejected. Computing the blocking probability exactly in such a network is difficult. In fact, one cannot even write closed-form expressions in this case. We will use the reduced-load approximation as a heuristic to calculate blocking probabilities under alternative routing. It is well known that the approximation computes the blocking probability fairly accurately in large networks. Assume $\lambda_r = \lambda$ for all r and

$\mu = 1$. Let $\tilde{b}$ be the blocking probability of a link. An alternatively routed call from direct path r arrives at link l with probability

$$\tilde{b} \times \frac{1}{N-2} \times \left(1 - \tilde{b}\right),$$

where $\tilde{b}$ is the probability that the direct link is not available, $1/(N-2)$ is the probability of choosing an alternative path that includes link l, and $1 - \tilde{b}$ is the probability the call is not blocked by the other link in the alternative path. Thus, the total arrival rate at link l is given by

$$\lambda + \lambda \frac{\tilde{b}(1 - \tilde{b})}{N-2} 2(N-2) = \lambda \left(1 + 2\tilde{b}(1 - \tilde{b})\right).$$

In the above calculation, we have used the fact that a link can be part of an alternative path for $2(N-2)$ direct paths. Thus, the fixed-point calculation becomes

$$\tilde{b} = E_B\left(\lambda(1 + 2\tilde{b}(1 - \tilde{b})), c\right).$$

For a fixed c, it is possible for $\tilde{b}$ to have two solutions for certain values of λ (such an example is explored in Exercise 9.17). Further, the two solutions for $\tilde{b}$ may be far apart. In real networks, the existence of multiple solutions for $\tilde{b}$ indicates that the network may switch from a low blocking probability regime to a high blocking probability regime quickly. The reason is as follows: when a call is accepted on an alternative route, while we have eliminated blocking for this call, we may block two calls (on the two links on the alternative path) later. So it is not always good to accept calls on alternative paths.

In practice, a call is accepted on an alternative path only if the available capacity is no less than c_t on both links for some $0 < c_t < c$. Such a scheme is called trunk reservation because it reserves some space for directly routed calls. We will see in Exercise 9.18 that such a trunk reservation scheme, with a properly chosen parameter c_t, helps to reduce blocking probabilities significantly.

9.14   Download time in BitTorrent

In Chapter 8, we discussed the BitTorrent protocol for P2P file sharing. In this section, we use a simple CTMC model to understand the performance of BitTorrent, in particular the amount of time a peer spends on downloading a file. Recall that, in BitTorrent, seeds are peers who have the complete copy of the file, and leechers are peers who have not downloaded the complete file. Assume that the arrival process of leechers is a Poisson process with rate λ and that the amount of time a peer stays in the system after it becomes a seed is exponentially distributed with parameter γ. Each node has an upload bandwidth μ and download bandwidth c. To model this system as a CTMC, we assume that the download times are exponentially distributed, where the parameters depend on the download speeds.

Let x denote the number of leechers and let y denote the number of seeds. Then (x, y) is a CTMC such that the following statements apply.

- The transition rate from state (x, y) to state $(x + 1, y)$ is λ, which occurs when a new leecher joins the network.

- The transition rate from state (x, y) to state $(x, y - 1)$ is γy, which occurs when a seed leaves the network.
- The transition rate from state (x, y) to state $(x - 1, y + 1)$ is $\min\{cx, \mu(\eta x + y)\}$, which occurs when a leecher becomes a seed. Note that $\eta \in [0, 1]$ is a parameter indicating the effectiveness of file sharing among the leechers. Note that cx is the total download capacity the leechers have and $\mu(\eta x + y)$ is the maximum upload speed to the leechers. So $\min\{cx, \mu(\eta x + y)\}$ is the overall download speed. Note that leechers have a partial file so they can upload to other leechers as well. But the upload capacity of a leecher can be used only if other leechers who need the partial file can find the leecher. Thus, the efficiency of file sharing is difficult to capture exactly; the parameter η is an approximation to capture file-sharing efficiency.
- The transition rate from state $(x - 1, y)$ to state (x, y) is λ, which occurs when a new leecher joins the network.
- The transition rate from state $(x + 1)$ to state (x, y) is $\gamma(y + 1)$, which occurs when a seed leaves the network.
- The transition rate from state $(x + 1, y - 1)$ to state (x, y) is $\min\{c(x + 1), \mu(\eta(x + 1) + (y - 1))\}$, which occurs when a leecher becomes a seed.

The transitions into and out of state (x, y) are presented in Figure 9.16.

Let π denote the steady-state distribution. Since $\pi \mathbf{Q} = 0$ if the steady-state distribution exists, the steady-state equation is

$$\pi_{x,y+1} \gamma(y + 1) + \lambda \pi_{x-1,y} + \min\{c(x + 1), \mu(\eta(x + 1) + y - 1)\}\pi_{x+1,y-1}$$
$$= \pi_{x,y} \min\{cx, \mu(\eta x + y)\} + \pi_{x,y} \gamma y + \lambda \pi_{x,y}.$$

It is difficult to obtain π in closed form, so we present a heuristic which approximately solves the steady-state equation, assuming $\lambda \to \infty$, and provides interesting insights into the operation of BitTorrent. Defining

$$\bar{x} = \frac{x}{\lambda} \qquad \text{and} \qquad \bar{y} = \frac{y}{\lambda},$$

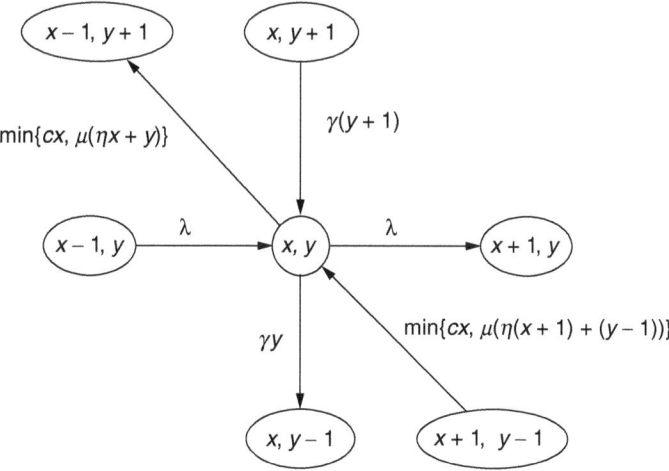

Figure 9.16 Transitions into and out of state (x, y).

and substituting them into the steady-state equation above, we have

$$\pi_{x,y+1}\gamma(\lambda\bar{y}+1) + \lambda\pi_{x-1,y} + \min\{c(\lambda\bar{x}+1), \mu(\eta(\lambda\bar{x}+1)+\lambda\bar{y}-1)\}\pi_{x+1,y-1}$$
$$= \pi_{x,y}\min\{c\lambda\bar{x}, \mu(\eta\lambda\bar{x}+\lambda\bar{y})\} + \pi_{x,y}\lambda\gamma\bar{y} + \lambda\pi_{x,y}.$$

Dividing both sides by λ, and letting $\lambda \to \infty$, we obtain

$$\pi_{x,y+1}\gamma\bar{y} + \pi_{x-1,y} + \min\{c\bar{x}, \mu(\eta\bar{x}+\bar{y})\}\pi_{x+1,y-1}$$
$$= \pi_{x,y}\min\{c\bar{x}, \mu(\eta\bar{x}+\bar{y})\} + \pi_{x,y}\gamma\bar{y} + \pi_{x,y}. \qquad (9.21)$$

Further, define

$$\tilde{\pi}_{\bar{x},\bar{y}} \triangleq \pi_{\lambda\bar{x},\lambda\bar{y}} = \pi_{x,y}.$$

Then a Taylor's series approximation yields

$$\pi_{x,y+1} = \tilde{\pi}_{\bar{x},\bar{y}(1+\frac{1}{\lambda})} \approx \tilde{\pi}_{\bar{x},\bar{y}} + \frac{1}{\lambda}\frac{\partial\tilde{\pi}_{\bar{x},\bar{y}}}{\partial\bar{y}}$$

$$\pi_{x-1,y} = \tilde{\pi}_{\bar{x}(1-\frac{1}{\lambda}),\bar{y}} \approx \tilde{\pi}_{\bar{x},\bar{y}} - \frac{1}{\lambda}\frac{\partial\tilde{\pi}_{\bar{x},\bar{y}}}{\partial\bar{x}}$$

$$\pi_{x+1,y-1} = \tilde{\pi}_{\bar{x}(1+\frac{1}{\lambda}),\bar{y}(1-\frac{1}{\lambda})} \approx \tilde{\pi}_{\bar{x},\bar{y}} + \frac{1}{\lambda}\frac{\partial\tilde{\pi}_{\bar{x},\bar{y}}}{\partial\bar{x}} - \frac{1}{\lambda}\frac{\partial\tilde{\pi}_{\bar{x},\bar{y}}}{\partial\bar{y}}.$$

Substituting into equation (9.21), we have

$$\frac{\partial\tilde{\pi}_{\bar{x},\bar{y}}}{\partial\bar{y}}\gamma\bar{y} - \frac{\partial\tilde{\pi}_{\bar{x},\bar{y}}}{\partial\bar{x}} + \min\{c\bar{x}, \mu(\eta\bar{x}+\bar{y})\}\left(\frac{\partial\tilde{\pi}_{\bar{x},\bar{y}}}{\partial\bar{x}} - \frac{\partial\tilde{\pi}_{\bar{x},\bar{y}}}{\partial\bar{y}}\right) = 0,$$

which implies

$$(\gamma\bar{y} - \min\{c\bar{x}, \mu(\eta\bar{x}+\bar{y})\})\frac{\partial\tilde{\pi}_{\bar{x},\bar{y}}}{\partial\bar{y}} = (\min\{c\bar{x}, \mu(\eta\bar{x}+\bar{y})\} - 1)\frac{\partial\tilde{\pi}_{\bar{x},\bar{y}}}{\partial\bar{x}}. \qquad (9.22)$$

One solution to equation (9.22) is to assume that $\tilde{\pi}_{\bar{x},\bar{y}}$ is a delta function at some point $(\bar{x}^*, \bar{y}^*)$, i.e., $\tilde{\pi}_{\bar{x},\bar{y}} = 0$ if $(\bar{x}, \bar{y}) \neq (\bar{x}^*, \bar{y}^*)$. This means that (9.22) has to be satisfied only at $(\bar{x}^*, \bar{y}^*)$. A possible choice of $(\bar{x}^*, \bar{y}^*)$ that satisfies (9.22) is

$$\gamma\bar{y}^* = \min\{c\bar{x}^*, \mu(\eta\bar{x}^* + \bar{y}^*)\},$$
$$1 = \min\{c\bar{x}^*, \mu(\eta\bar{x}^* + \bar{y}^*)\}, \qquad (9.23)$$

i.e.,

$$\bar{y}^* = 1/\gamma, \qquad (9.24)$$

$$\min\{c\bar{x}^*, \mu(\eta\bar{x}^* + \bar{y}^*)\} = 1. \qquad (9.25)$$

Equations (9.24) and (9.25) determine the values of $\bar{y}^*$ and $\bar{x}^*$. According to the heuristic argument above, the delta function $\pi_{x,y}$ at point $(\lambda\bar{x}^*, \lambda\bar{y}^*)$ satisfies the global balance equation when $\lambda \to \infty$, so π is the stationary distribution of the CTMC. We next compute the value of $\bar{x}^*$ by considering two cases.

• For the first case, we assume

$$c\bar{x}^* \leq \mu(\eta\bar{x}^* + \bar{y}^*), \qquad (9.26)$$

i.e.,

$$cx^* \leq \mu(\eta x^* + y^*),$$

where $x^* = \lambda \bar{x}^*$ and $y^* = \lambda \bar{y}^*$. So, in steady state, the net download bandwidth is less than or equal to the net upload bandwidth. In this case, from (9.25) we obtain

$$\bar{x}^* = \frac{1}{c}. \tag{9.27}$$

- For the second case, we assume

$$c\bar{x}^* > \mu\left(\eta\bar{x}^* + \bar{y}^*\right),$$

so the upload bandwidth becomes the bottleneck. From (9.25), we obtain

$$\bar{x}^* = \left(\frac{1}{\mu} - \frac{1}{\gamma}\right)\frac{1}{\eta}. \tag{9.28}$$

Summarizing the two cases, we have

$$\bar{x}^* = \begin{cases} 1/c, & \text{if } c\bar{x}^* \leq \mu(\eta\bar{x}^* + \bar{y}^*), \\ \left(\dfrac{1}{\mu} - \dfrac{1}{\gamma}\right)\dfrac{1}{\eta}, & \text{if } c\bar{x}^* > \mu\left(\eta\bar{x}^* + \bar{y}^*\right). \end{cases}$$

Substituting the values of $\bar{x}^*$ and $\bar{y}^*$ into the conditions, we obtain

$$\bar{x}^* = \begin{cases} 1/c, & \text{if } 1 \leq \mu\left(\dfrac{\eta}{c} + \dfrac{1}{\gamma}\right), \\ \left(\dfrac{1}{\mu} - \dfrac{1}{\gamma}\right)\dfrac{1}{\eta}, & \text{if } \dfrac{c}{\eta}\left(\dfrac{1}{\mu} - \dfrac{1}{\gamma}\right) > 1. \end{cases} \tag{9.29}$$

Note that $1 \leq \mu\left(\eta/c + 1/\gamma\right)$ is equivalent to $\left(1/\mu - 1/\gamma\right)1/\eta \leq 1/c$; and $c/\eta\left(1/\mu - 1/\gamma\right) > 1$ is equivalent to $1/\eta\left(1/\mu - 1/\gamma\right) > 1/c$. So (9.29) is equivalent to

$$\bar{x}^* = \begin{cases} 1/c, & \text{if } \dfrac{1}{\eta}\left(\dfrac{1}{\mu} - \dfrac{1}{\gamma}\right) \leq \dfrac{1}{c}, \\ \left(\dfrac{1}{\mu} - \dfrac{1}{\gamma}\right)\dfrac{1}{\eta}, & \text{if } \dfrac{1}{\eta}\left(\dfrac{1}{\mu} - \dfrac{1}{\gamma}\right) > \dfrac{1}{c}, \end{cases}$$

which can be rewritten as follows:

$$\bar{x}^* = \max\left\{\frac{1}{c}, \left(\frac{1}{\mu} - \frac{1}{\gamma}\right)\frac{1}{\eta}\right\}.$$

Thus, the number of peers in steady state (as $\lambda \to \infty$) is given by

$$x^* + y^* = \lambda(\bar{x}^* + \bar{y}^*) = \lambda\left(\max\left\{\frac{1}{c}, \left(\frac{1}{\mu} - \frac{1}{\gamma}\right)\frac{1}{\eta}\right\} + \frac{1}{\gamma}\right).$$

According to Little's law, the amount of time a peer spends on downloading a file is given by

$$\frac{x^*}{\lambda} = \frac{\lambda\bar{x}^*}{\lambda} = \max\left\{\frac{1}{c}, \frac{1}{\eta}\left(\frac{1}{\mu} - \frac{1}{\gamma}\right)\right\}.$$

The total time a peer stays in the system is

$$\frac{x^*}{\lambda} + \frac{1}{\gamma} = \max\left\{\frac{1}{c}, \frac{1}{\eta}\left(\frac{1}{\mu} - \frac{1}{\gamma}\right)\right\} + \frac{1}{\gamma}.$$

Note that both the download and waiting times in the system are independent of λ, the arrival rate. Normally, when the arrival rate increases, queueing systems become congested and therefore the download times get larger. However, in a P2P network, if the number of peers in the network increases, the total capacity available to upload files also increases proportionally. This is the reason why the download and waiting times in the system do not increase with λ in a P2P network.

9.15   Summary

- **Positive recurrence of CTMCs** Suppose CTMC $X(t)$ is irreducible and non-explosive, then $X(t)$ is positive recurrent if one of the following conditions holds.
  - There exists a vector π such that $\sum_{i \neq j} \pi_i Q_{ij} = \pi_j \sum_{i \neq j} Q_{ji}$ for any j (called the global balance equation), $\pi \geq 0$, and $\sum_i \pi_i = 1$.
  - There exists a vector π such that $\pi_i Q_{ij} = \pi_j Q_{ji}$ for all $i \neq j$ (called the local balance equation), $\pi \geq 0$, and $\sum_i \pi_i = 1$.
  - There exists a function $V : \mathcal{S} \to \mathcal{R}^+$ such that

$$\sum_{j \neq i} Q_{ij}(V(j) - V(i)) \leq -\epsilon \mathbb{I}_{i \in \mathcal{B}^c} + M \mathbb{I}_{i \in \mathcal{B}}$$

    for some $\epsilon > 0$, $M < \infty$, and a bounded set $\mathcal{B}$.

- **The M/M/1 queue** The mean queue length is $\rho/(1-\rho)$ and the mean delay is $1/(\mu - \lambda)$.

- **The M/M/s/s queue** The probability that all servers are busy is

$$\frac{\rho^s/s!}{\sum_{k=0}^s \rho^k/k!}.$$

- **The M/M/s queue** The probability that an arrival finds all servers busy is

$$\frac{\pi(0)(s\rho)^s}{s!(1-\rho)},$$

  where

$$\pi(0) = \left[\sum_{n=0}^{s-1} \frac{(s\rho)^n}{n!} + \frac{(s\rho)^s}{s!(1-\rho)}\right]^{-1}.$$

- **The M/GI/1 queue** The mean number of customers in the queue is

$$\frac{\lambda^2 E[S^2]}{2(1-\rho)},$$

  and the mean waiting time is

$$\frac{\lambda E[S^2]}{2(1-\rho)}.$$

- **Reversed chain** Let $X(t)$ be a CTMC with transition rate matrix $\mathbf{Q}$. If there exists a transition rate matrix $\mathbf{Q}^*$ and probability vector π such that

$$Q^*_{ij}\pi_i = Q_{ji}\pi_j, \qquad \forall i, j,$$

  then $\mathbf{Q}^*$ must be the transition rate matrix of the reversed chain and π must be the stationary distribution of the forward and the reversed chains.

- **Reversibility** A CTMC X is reversible if $Q^*_{ij} = Q_{ij}$. A CTMC is reversible if and only if the local balance equation is satisfied, i.e.,

$$\pi_i Q_{ij} = \pi_j Q_{ji} \qquad \text{for all } i \neq j.$$

  The M/M/1 queue and the tandem M/M/1 queue are reversible CTMCs.

- **Insensitivity to service-time distribution** Certain queueing models have stationary distributions that are insensitive to service-time distributions, i.e., the stationary distribution depends only on the mean of the service time and not on the entire distribution. An example of such a queueing model is the M/GI/1 queue with processor sharing. The stationary distribution of the M/GI/1 queue with processor sharing is $\pi(n) = \rho^n(1 - \rho)$, which is independent of the service-time distribution.

- **Connection-level arrivals and departures** A flow is called a type-r flow if the flow uses route r. We assume that each flow of type r arrives according to a Poisson process of rate λ_r, and that the file associated with the flow has an exponentially distributed number of bits with mean $1/\mu_r$. A flow departs when all of its bits are transferred. With connection-level arrivals and departures, the network is stable only if

$$\sum_{r:l\in r} \frac{\lambda_r}{\mu_r} < c_l \qquad \text{or} \qquad \sum_{r:l\in r} \rho_r < c_l,$$

  i.e., the aggregated workload on each link is less than the link capacity. If the network allocates x_r to a flow on route r by solving the network utility maximization problem $\sum_r n_r(t)U_r(x_r)$, where $n_r(t)$ is the number of flows on route r, the network is stable when the necessary conditions hold.

- **Loss networks** Loss networks are CTMC models of circuit switched networks. In a large capacity network, based on the reduced-load approximation, the blocking probabilities satisfy

$$1 - b_r = \prod_{l\in r}(1 - b_l),$$

$$\sum_{r:l\in r} \rho_r(1 - b_r) \begin{cases} = c_l, & \text{if } b_l > 0, \\ \leq c_l, & \text{if } b_l = 0, \end{cases}$$

  where b_l is the blocking probability on link l and b_r is the blocking probability on route r. Motivated by the reduced-load approximation, Algorithm 12 is a heuristic algorithm to compute the blocking probabilities in a loss network.

- **Download time in BitTorrent** A BitTorrent file-sharing system can be studied using a CTMC model. Assume that the arrival process of leechers is a Poisson process with rate λ, and that the amount of time a peer stays in the system after it becomes a seed is exponentially distributed with parameter γ. Each node has an upload bandwidth μ and download bandwidth c. From the CTMC model, we approximate the amount of time a peer spends on downloading a file as

$$\max\left\{\frac{1}{c}, \frac{1}{\eta}\left(\frac{1}{\mu} - \frac{1}{\gamma}\right)\right\},$$

and the total time a peer stays in the system as

$$\max\left\{\frac{1}{c}, \frac{1}{\eta}\left(\frac{1}{\mu} - \frac{1}{\gamma}\right)\right\} + \frac{1}{\gamma},$$

where $\eta \in [0, 1]$ is a parameter indicating the effectiveness of the file sharing among the leechers.

9.16   Exercises

Exercise 9.1 (Merging and splitting Poisson processes) Prove Result 9.2.1 (the sum of K independent Poisson processes is a Poisson process) and Result 9.2.2 (independent splitting of a Poisson process results in Poisson processes).

Exercise 9.2 (M/M/s/s queues) Consider an M/M/s/k queue with arrival rate λ and mean service time $1/\mu$. Find the steady-state distribution of this queueing system. Are there conditions on λ and μ for the steady-state distribution to exist?

Exercise 9.3 (The M/M/s/s model in heavy traffic) In this exercise, we will show a central limit theorem like result for the M/M/s/s loss model when the number of servers and the traffic intensity are both large and nearly equal to each other.

Let $B(s, \rho)$ denote the Erlang-B formula for the blocking probability in an M/M/s/s system. Let $s = \lfloor \rho + \gamma \sqrt{\rho} \rfloor$. Show that

$$\lim_{\rho \to \infty} \sqrt{\rho} B(s, \rho) = \phi(\gamma)/\Phi(\gamma),$$

where $\phi(\cdot)$ and $\Phi(\cdot)$ are the PDF and CDF of the standard normal random variable $N(0, 1)$. Use the following facts to show the above result.

- Central limit theorem for Poisson random variables: let $p(k, \rho) = e^{-\rho} \rho^k / k!$. Then,

$$\lim_{\rho \to \infty} \sum_{k=0}^{\lfloor \rho + \gamma \sqrt{\rho} \rfloor} p(k, \rho) = \Phi(\gamma).$$

- Stirling's formula:

$$\lim_{n \to \infty} n! = \sqrt{2\pi n}(n/e)^n.$$

- $\lim_{\rho \to \infty} e^{\gamma \sqrt{\rho}}(1 + \gamma/\sqrt{\rho})^{-\rho} = e^{\gamma^2/2}$. (Prove this fact. You don't have to prove the previous two facts.)

Exercise 9.4 (P-K formula) Consider an M/GI/1 queue. Recall

$$A(z) = M(\lambda(z - 1)),$$

$$Q(z) = \frac{\pi(0)A(z)(z - 1)}{z - A(z)}.$$

Compute the closed-form expression of $Q(z)$ and then derive the P-K formula.

Exercise 9.5 (M/GI/1 queues with two classes of customers) Consider an M/GI/1 queue with two classes of customers, with the class i arrival rate, the mean service time, and the variance of the service time being λ_i, $1/\mu_i$, and σ_i^2, respectively. Let us assume that the service discipline is non-pre-emptive priority with higher priority to class 1 customers; i.e., when the server becomes free, the first customer in the class 1 queue begins service. However, a class 2 customer undergoing service cannot be interrupted even if a class 1 customer arrives (hence, the terminology "non-pre-emptive priority"). Show that the average waiting time in the queue for a class i customer is given by

$$W_q^i = \frac{\sum_{i=1}^2 \lambda_i \left(1/\mu_i^2 + \sigma_i^2\right)}{2(1 - \rho_{i-1})(1 - \sum_{j=1}^i \rho_j)},$$

where $\rho_0 = 0$ and $\rho_i = \lambda_i/\mu_i$, $i = 1, 2$. Hint: Note that a class 2 customer has to wait for the service completion of all class 2 and class 1 customers that arrived prior to it, and for the service completion of all class 1 customers that arrived while the class 2 customer is waiting in the queue.

Exercise 9.6 (The Kingman bound for GI/GI/1 queues) For a GI/GI/1 queue with $\lambda < \mu$, show that

$$W_q \le \frac{\lambda(\sigma_a^2 + \sigma_s^2)}{2(1 - \rho)},$$

where σ_a^2 and σ_s^2 are the inter-arrival-time and service-time variances, respectively, and $\rho = \lambda/\mu$. This result is called the Kingman bound, and it is the continuous-time analog of the result in Section 3.4.4 for discrete-time queues. Hint: Let w_k be the waiting time of the kth packet. Write w_{k+1} as

$$w_{k+1} = w_k + S_k - A_{k+1} + u_k,$$

where u_k ensures that w_{k+1} does not become negative. Show that $-E[u_k] = E[S_k - A_{k+1}]$ in steady state. Finally, use the Lyapunov function w_k^2 and assume that the system is in steady state to derive the result.

Exercise 9.7 (Resource pooling)

(1) Consider an M/M/2 queue with arrival rate 2λ and mean service time $1/\mu$. Find the expected waiting time in steady state of a packet in this queue.

(2) Now consider a system consisting of two independent M/M/1 queues with mean service time equal to $1/\mu$ in each queue. Assume that packets arrive to this system according to a Poisson process of rate 2λ. When a packet arrives, it joins queue 1 with probability $1/2$ and queue 2 with probability $1/2$. Compute the expected steady-state waiting time in this system.

Note: The mean waiting time in part (1) is smaller because it has a single queue so that no server is idle when there is work to be done.

Exercise 9.8 (The M/GI/s/s loss model) Consider the M/GI/s/s loss model, where the service-time distribution is Erlang with K stages, each with mean $1/(K\mu)$. Show that the blocking probability depends only on the mean of the service-time distribution $1/\mu$ and not on the number of stages. Hint: Proceed as in the insensitivity proof for the M/GI/1-PS queue. Note: this result can be extended to mixtures of Erlang distributions, and, more generally, to any service-time distribution, to prove the insensitivity of the model to the distribution of service time beyond the mean.

Exercise 9.9 (Insensitivity in M/GI/∞ queues) Consider the M/M/∞ queueing system, with arrival rate λ and mean service time $1/\mu$.

(1) Show that the steady-state distribution is the Poisson distribution with mean $\rho = \lambda\mu$.

(2) Now consider an M/GI/∞ model with the following service-time distribution: the service time is $1/\mu_i$ with probability p_i, where $\sum_{i=1}^{K} p_i = 1$ and $\sum_{i=1}^{K} p_i/\mu_i = 1/\mu$. In other words, the service time consists of a mixture of K deterministic service times. Let $(n_1, n_2, \ldots, n_K)$ be the state of this system, where n_i is the number of customers with service time $1/\mu_i$ in the system. Find the steady-state distribution of $(n_1, n_2, \ldots, n_K)$. Hint: Use Result 9.2.1.

(3) In part (2) above, let $n = \sum_i n_i$. Show that π_n, the steady-state distribution, only depends on μ, and not on K, p_i, or μ_i explicitly. Hint: Use Result 9.2.2.

(4) Is the departure process of the model in part (2) Poisson? Clearly explain your answer.

Exercise 9.10 (Truncation theorem) Let X be a CTMC over a state space $\mathcal{S}$ and suppose that X is time reversible in steady state. Now consider a CTMC Y that is a restriction of X to the state space $\mathcal{A}$, where $\mathcal{A} \subset \mathcal{S}$. By restriction, we mean that Y takes values only in $\mathcal{A}$, has the same transition rates as X, but transitions out of $\mathcal{A}$ are not allowed. Show that Y is also time reversible in steady state and that its steady-state distribution is the same as the steady-state distribution of X restricted to $\mathcal{A}$ but rescaled to add up to 1. This result is called the truncation theorem.

Exercise 9.11 (Trunk reservation in the M/M/s/s loss model) Consider the following variation of the M/M/s/s loss model. Consider a single link of capacity 50 with two call classes. Each call from each class requests one unit of capacity on arrival. Assume both call classes have holding times that are exponentially distributed with unit mean. The call arrival processes are independent Poisson processes, with a mean arrival rate of 40 for class 1 calls and a mean arrival rate of 20 for class 2 calls. Class 2 calls are blocked from entering (and lost) if the available capacity is less than or equal to 5. Class 1 calls are blocked and lost only if there is no available capacity. Thus, this is a loss model with some priority given to Class 1 calls. This is an example of the trunk reservation introduced in Section 9.13.

(1) Compute the blocking probabilities of class 1 and class 2 calls in this system. Hint: Consider the Markov chain describing the total number of calls in the system.

(2) What are the expected numbers of class 1 and class 2 calls in the system in steady state? Hint: Use Little's law.

Exercise 9.12 (Aggregated arrival rates in a Jackson network) Consider the Jackson network in Section 9.9.1 with K nodes. Recall that the external arrivals into node i comprise a Poisson process with rate r_i. A customer moves from node i to node j with probability P_{ij} after the service is completed, and leaves the network with probability $1 - \sum_j P_{ij} = P_{i0}$.

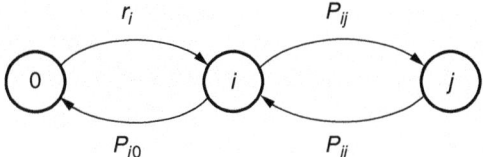

Figure 9.17 CTMC related to the Jackson network.

Therefore, the aggregated arrival rate to node i is

$$\lambda_i = r_i + \sum_{j=1}^{K} \lambda_j P_{ji}. \tag{9.30}$$

Assume that, for every node j, there exists a node i such that $P_{i0} > 0$ and a route $(r_1, r_2, \ldots, r_k)$ such that $P_{jr_1} P_{r_1 r_2} \cdots P_{r_{k-1} r_k} P_{r_k i} > 0$, i.e., every customer will eventually leave the network with probability 1. Prove that there exists a unique solution λ to (9.30). Hint: Divide (9.30) by $(1 + \sum_k \lambda_k)$ and argue that the resulting equation is simply the equation used to obtain the stationary distribution of the CTMC in Figure 9.17. Note that only three states of the CTMC are shown in Figure 9.17.

Exercise 9.13 (Congestion control with multiple ON–OFF sources) Consider S ON–OFF sources that switch between an active (ON) state and an inactive (OFF) state. Let us suppose that the sources access a bottleneck node with transmission rate C bps, and that each source accesses this node via an access link whose transmission rate is 1 bps. In other words, when the number of ON sources in the system is less than or equal to C, each source is served at rate 1. Assume $C < S$. When the number of ON sources exceeds C, there is some form of resource allocation mechanism that divides the available capacity C to all the sources equally. In other words, if there are $N(t) \geq C$ active sources in the system at time t, each source that is ON is served at rate $C/N(t)$ bps. We assume that the OFF times are exponential with mean $1/\lambda$. Once a source enters the ON state, it remains in this state till it receives an exponentially distributed amount of service whose mean is $1/\mu$ bits.

(1) Let $N(t)$ be the number of active sources in the system at time t. Draw the state transition diagram of the Markov chain that describes the evolution of $N(t)$.

(2) Let π_i denote the steady-state probability that there are i active sources in the system. Write down the equations needed to compute $\{\pi_i\}$.

(3) Derive an expression for the average amount of time spent in the ON state by a source as a function of $\{\pi_i\}$.

Exercise 9.14 (Q-CSMA in continuous time) Consider an ad hoc network whose scheduling algorithm operates in continuous time. Assume that if a link l interferes with the transmission of link j, then j also interferes with l. Associate with each link l in the network a weight w_l. Let $S_1, S_2, \ldots, S_K$ be the independent sets in the interference graph of the network, i.e., each S_k represents a set of links that can be scheduled simultaneously without interference. We include the empty set in the set of independent sets. Consider the following Medium Access Control (MAC) algorithm: each link can be either active or inactive at each time instant. At time 0, assume that the set of all active links is an independent set. Further, at time 0, each link starts an exponentially distributed timer with mean 1. When a link's timer expires (call this link l), it checks to see if any other link in its neighborhood (where the neighborhood is the set of other links with which it interferes) is active. If there is an active link in its neighborhood, link l maintains its current state (which must be inactive), starts another timer, and repeats the process. If no link in its neighborhood is active, the link chooses to be active with probability $e^{w_l}/(1+e^{w_l})$ and chooses to be inactive with probability $1/(1+e^{w_l})$. Then, it starts a new timer and repeats the process. Note that the state of a link does not change between two of its consecutive timer expiration moments.

(1) Let $X(t)$ be the set of links active at time t. Note that the set of active links must be an independent set in the interference graph, and also note that $X(t)$ is a CTMC. Find the transition rate matrix of $X(t)$.

(2) Show that the steady-state distribution of $X(t)$ is given by

$$\pi_{S_k} = \frac{\prod_{l \in S_k} e^{w_l}}{Z} = \frac{e^{\sum_{l \in S_k} w_l}}{Z},$$

where Z is a normalization constant and $\pi_{\emptyset} = 1/Z$.

Exercise 9.15 (Loss model with two types of calls) Consider a loss model with two types of calls. Calls of type 1 arrive according to a Poisson process of rate $\lambda_1 = 50$ calls per second. Calls of type 2 arrive at rate $\lambda_2 = \infty$. Both call types have holding times that are exponentially distributed with mean 1. Calls of type 2 are admitted into the system only if the total number of calls (of both types) already in the system is less than 45. Type 1 calls are admitted if the total number of calls is less than 50. If $n(t)$ denotes the total number of calls in the system, $n(t)$ is a CTMC whose transition diagram looks like that shown in Figure 9.18. Find the expected number of type 1 and type 2 calls in the system in steady state.

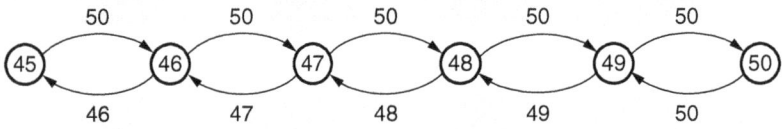

Figure 9.18 Markov chain for the system in Exercise 9.15.

Hints:

(1) $E[\text{type 2 calls}]$ in steady state is *not* equal to 45.

(2) $E[\text{type 1 + type 2 calls}]$ is $E[n]$, which can be calculated after solving the steady-state distribution of $n(t)$.

(3) Compute the blocking probability of type 1 calls and then use it to compute $E[\text{type 1 calls}]$.

Exercise 9.16 (Asymptotic expression for blocking probability) Recall the Erlang-B blocking probability formula for the M/M/c/c loss model with arrival rate λ and mean holding time μ:

$$p_b = \frac{\rho^c/c!}{\displaystyle\sum_{n=1}^{c} \rho^n/n!},$$

where $\rho = \lambda/\mu$. Now suppose that $\rho = \alpha c$ and let $c \to \infty$. Show that p_b converges to

$$p_b = \left(\frac{\alpha - 1}{\alpha}\right)^{+}.$$

Note: This suggests the following approximation for the blocking probability when c and ρ are large:

$$p_b \approx \left(\frac{\rho - c}{c}\right)^{+}.$$

Exercise 9.17 (Alternative routing without trunk reservation) Recall the reduced-load approximation for alternative routing without trunk reservation in a fully connected, symmetric network in Section 9.13.3:

$$\tilde{b} = E_B\left[\lambda(1 + 2\tilde{b}(1 - \tilde{b})), c\right],$$

where $\tilde{b}$ is the blocking probability on a link and $E_B(\rho, C)$ is the Erlang-B formula for a link with capacity c and traffic intensity ρ. Also, assuming that the link blocking events are independent, the end-to-end blocking probability is given by

$$b = \tilde{b}(1 - (1 - \tilde{b})^2).$$

Let $c = 1000$ and $\lambda = 950$. Compute b by first solving the fixed-point equation for $\tilde{b}$ iteratively starting from some initial guess. Note: There will be two values for $\tilde{b}$, depending on the initial guess.

Exercise 9.18 (Alternative routing with trunk reservation) Consider a symmetric loss network with alternative routing and trunk reservation, as in Section 9.13.3. Let $c = 1000$, $\lambda = 950$, and $c_t = 50$. Consider a single link in the network, and let ν be the arrival rate of alternatively routed calls to this link. Assume ν is given for now, and assume that the arrival process of alternatively routed calls on the link is Poisson and independent of the external arrival process of rate λ. The calls stay in the network for an exponential amount of time, with mean 1. The CTMC describing the number of calls on this link is shown in Figure 9.19. Let $\pi(i)$ be the stationary probability that there are i calls on the link. If $\tilde{b}_d$ and $\tilde{b}_a$, respectively, denote the blocking probabilities of direct and alternatively routed calls on the link, $\tilde{b}_d = \pi(c)$ and $\tilde{b}_a = \sum_{i=c-c_t}^{c} \pi(i)$.

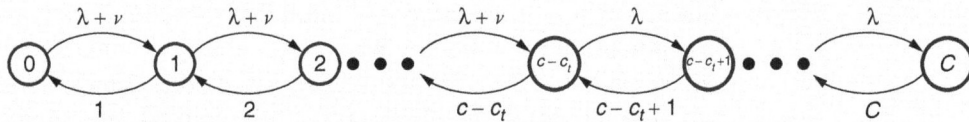

Figure 9.19 CTMC describing the number of calls on a link.

(1) Using the idea behind the reduced-load approximation, write down an expression for v in terms of $\tilde{b}_d$ and $\tilde{b}_a$.

(2) Solve the fixed-point equations from part (1) to obtain $\tilde{b}_d$ and $\tilde{b}_a$.

(3) Using $\tilde{b}_d$ and $\tilde{b}_a$, compute the probability that a call is blocked.

Exercise 9.19 (Performance of BitTorrent) In the analysis of BitTorrent, we assume the system is stable (i.e., the CTMC is in steady state). The evolution of the number of leechers and seeds can be described using the following fluid model:

$$\dot{x}(t) = \lambda - \min\{cx(t), \mu(\eta x(t) + y(t))\},$$

$$\dot{y}(t) = \min\{cx(t), \mu(\eta x(t) + y(t))\} - \gamma y(t).$$

Linearize the system and show that the linear system reaches its equilibrium starting from any initial condition when $1/c \neq (1/\eta)(1/\mu - 1/\gamma)$. Hint: Consider two separate cases: (i) $\frac{1}{c} < \frac{1}{\eta}\left(\frac{1}{\mu} - \frac{1}{\gamma}\right)$ and (ii) $\frac{1}{c} > \frac{1}{\eta}\left(\frac{1}{\mu} - \frac{1}{\gamma}\right)$.

Exercise 9.20 (Continuous-time gossip model) Consider the following model, called a gossip model, for disseminating a single packet in a network with N nodes. Initially, node 1 is the only node with the packet. Packet dissemination then proceeds as follows: each node that has the packet waits for an amount of time that is exponentially distributed with mean 1, and then selects one other node in the network, uniformly at random, and gives the packet to the selected node (irrespective of whether the node already has the packet or not). Each node with the packet repeats this process until all the nodes in the network have received a packet. (A packet transmission from one node to another is assumed to be instantaneous.) Let $X(t)$ denote the number of nodes that have the packet at time t. Note that $X(t)$ is a CTMC with $X(0) = 1$. Show that the expected amount of time that it takes for the packet to reach all the nodes is of the order of $\log N$.

Exercise 9.21 (Join the shortest queue routing) Consider two independent exponential server queues, with mean service times $1/\mu_1$ and $1/\mu_2$. A common stream of packets arrives at the two queues according to a Poisson process of rate λ. An arriving packet is routed to the server with the smaller queue, with ties broken uniformly at random. Let (q_1, q_2) be the CTMC representing this system, where q_i denotes the queue length at server i. Show that this Markov chain is stable (positive recurrent) if $\lambda < \mu_1 + \mu_2$ using the continuous-time Foster's theorem.

Exercise 9.22 (The G/M/1/1 loss model) Consider the following G/M/1/1 loss model: the source generating packets is in one of two states, ON or OFF. The state of the source is a Markov chain, with q_{ON} being the rate of transition from OFF to ON, and q_{OFF} being the rate of transition from ON to OFF. When the source is in the ON state, packets are generated according to a Poisson process of rate 1, and no packets are generated by the source if it is in the OFF state. In other words, conditioned on the source being ON, the probability that one packet arrives in a time interval of duration δ is $\lambda\delta$, with $\lambda = 1$, independent of past arrivals. If there is already a packet in the queue, new arrivals are blocked and lost. Once a packet enters the queue, it takes an exponentially distributed amount of time, with mean 1, to process. Assume $q_{OFF} = q_{ON} = 1$. Find (i) the probability that there is one packet in the queue, and (ii) the probability that an arriving packet is blocked.

9.17   Notes

Continuous-time queueing theory is the subject of many books (see, for example, [6, 175]), and its applications to communication networks have also been treated in many texts (see, for example, [10, 112, 117, 167]). Reversible queueing networks and their many applications have been presented in [68]. The heavy-traffic approximation to the Erlang-B formula was derived in [174].

A simple model of connections generated by ON–OFF sources was studied in [55]. More general models of connection arrivals and departures in the Internet were studied in [12, 33, 77, 145, 155, 156]. The approach presented here is based on the analysis in [155, 156]. The result has been extended to the case of general file-size distributions in [113] for proportionally fair resource allocation. A fluid model motivated by general file-size distributions operating under α-fair allocation schemes has been proved to be stable in [137]. The connection-level model here assumes a time-scale separation between the connection arrival and departure processes and the congestion control algorithm. It has been shown in [101] that the time-scale separation assumption is not necessary to prove connection-level stability. The time-scale separation assumption has been justified in [168] for a slightly different model, by making an interesting connection between reversible queueing networks and connection-level models with congestion control.

Distributed admission control was studied in [76]. Loss networks have been studied extensively as models of telephone networks; [71, 117] present surveys that also contain an exhaustive list of references. The Markov chain model of BitTorrent was developed in [177], and the fluid limit in the large-arrival-rate regime was presented in [141]. More detailed models of P2P networks can be found in [115, 179]. CTMC models of book-ahead systems are considered in [48, 157].

10 Asymptotic analysis of queues

Stationary distributions of queueing systems with general arrivals and service-time distributions are difficult to compute. In this chapter, we will develop techniques to understand the behavior of such systems in certain asymptotic regimes. We consider two such regimes: the heavy-traffic regime and the large-deviations regime. The analysis in the heavy-traffic regime can be thought of as the analog of the central limit theorem for random variables, and the analysis of the large-deviations regime can be thought of as the analog of the Chernoff bound. Traditionally, heavy-traffic analysis is performed by scaling arrival and service processes so that they can be approximated by Brownian motions, which are the stochastic process analogs of Gaussian random variables. Here, we take a different approach: we will show that heavy-traffic results can be obtained by extending the ideas behind the discrete-time Kingman bound presented in Chapter 3. The analysis of the large-deviations regime in this chapter refines the probability of overflow estimates obtained using the Chernoff bound, also in Chapter 3.

Roughly speaking, heavy traffic refers to a regime in which the mean arrival rate to a queueing system is close to the boundary of the capacity region. We will first use a Lyapunov argument to derive bounds on the moments of the scaled queue length $\epsilon q(t)$ for the discrete-time G/G/1 queue; where $\epsilon = \mu - \lambda$; μ is the service rate and λ is the arrival rate. The reason for considering this scaling is that, based on elementary queueing models, we now know that the mean queue length is proportional to $1/\epsilon$, and therefore it makes sense that we should study ϵq to get meaningful results when $\epsilon \to 0$. We will see that the moments of the scaled queue length depend only on the variances of the arrival and service processes, thus justifying our earlier statement that the heavy-traffic regime is the analog of the central limit theorem. Then, we will introduce the concept of *state-space collapse*, and use a Lyapunov-drift analysis to study the heavy-traffic performance of a multi-server queueing system under the Join the Shortest Queue (JSQ) routing algorithm. The following questions will be answered in this chapter.

- *What do the moments of $\epsilon q(t)$ converge to as $\epsilon \to 0$?*
- *What is state-space collapse?*
- *Is JSQ heavy-traffic optimal?*

We then, introduce the large-deviations regime, which is used to study the probabilities of rare events, e.g., the probability that the sum of n i.i.d. random variables deviates from its mean by $O(n)$. We will show that the upper bound on the queue overflow probability obtained in Chapter 3 is tight in a large deviations sense, and that the effective bandwidth is the minimum required bandwidth to support a source with QoS constraint θ. We will first introduce the concept of large deviations, and the Cramer–Chernoff theorem. Then, we will look at two specific large-deviations regimes: queues with large buffers (large buffer, large deviations) and queues with many data sources (many sources, large deviations). The following large-deviations questions will be addressed in this chapter.

- *What is the large-deviations regime, and how does it differ from the central limit theorem?*
- *What is a large-deviations rate function?*
- *What are the rate functions of a queue with a large buffer, and of a queue with a large buffer and large link capacity, shared by many sources?*

10.1 Heavy-traffic analysis of the discrete-time G/G/1 queue

We consider a discrete-time G/G/1 queue (see Figure 10.1), where the number of arrivals in each time slot is i.i.d., the number of bits that can be served in each time slot is i.i.d., and the arrival and service processes are independent of each other. Let $a(t)$ be the number of bits arriving to the queue at the beginning of time slot t, $s(t)$ be the maximum number of bits that can be drained from the queue during time slot t, and $u(t)$ be the unused capacity at time slot t due to the lack of enough bits in the queue. Therefore, the queue dynamics can be described as

$$q(t+1) = q(t) + a(t) - s(t) + u(t).$$

We assume that $s(t) \leq c_{\max}$ for all t, i.e., the number of bits that can be drained from the queue is upper bounded by $c_{\max}$.

We will use a Lyapunov argument to derive bounds on the moments of $q(t)$. Consider the Lyapunov function $V(t) = q^n(t)$, where n is a positive integer. The drift of the Lyapunov function is given by

$$q^n(t+1) - q^n(t) = (q(t) + a(t) - s(t) + u(t))^n - q^n(t).$$

Note that, if $u(t) > 0$, $q(t) + a(t) - s(t) + u(t) = 0$, so

$$(q(t) + a(t) - s(t) + u(t))^n = (q(t) + a(t) - s(t))^n - (-u(t))^n.$$

Therefore,

$$
\begin{aligned}
V(t+1) - V(t) &= q^n(t+1) - q^n(t) \\
&= (q(t) + a(t) - s(t))^n - q^n(t) - (-u(t))^n \\
&= \sum_{i=0}^{n-1} \binom{n}{i} q^i(t) (a(t) - s(t))^{n-i} - (-u(t))^n \\
&= \sum_{i=0}^{n-2} \binom{n}{i} q^i(t) (a(t) - s(t))^{n-i} + nq^{n-1}(t)(a(t) - s(t)) - (-u(t))^n.
\end{aligned}
$$

$$(10.1)$$

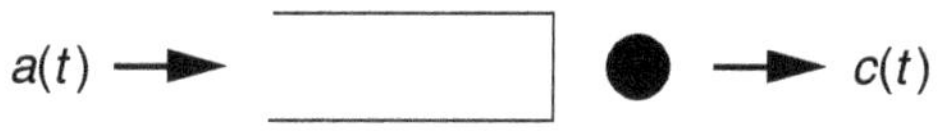

$a(t) \longrightarrow$ [ ] $\bullet \longrightarrow c(t)$

Figure 10.1 Single-server queue.

Define $\lambda = E[a(t)]$ and $\mu = E[s(t)]$, and assume that $\lambda < \mu$. For convenience, we omit the time variable t in the following analysis. Taking expectations on both sides of equation (10.1), and assuming that the system is in steady state, we obtain

$$0 = \sum_{i=0}^{n-2} \binom{n}{i} E[q^i] E\left[(a-s)^{n-i}\right] - nE[q^{n-1}](\mu - \lambda) - E[(-u)^n].$$

Moving $nE[q^{n-1}](\mu - \lambda)$ to the left-hand side and dividing both sides by $n(\mu - \lambda)$, we have

$$E[q^{n-1}] = \frac{1}{n(\mu - \lambda)} \sum_{i=0}^{n-2} \binom{n}{i} E[q^i] E\left[(a-s)^{n-i}\right] - \frac{E[(-u)^n]}{n(\mu - \lambda)}.$$

Multiplying both sides of the equation by $(\mu - \lambda)^{n-1}$, we obtain

$$(\mu - \lambda)^{n-1} E[q^{n-1}]$$

$$= \frac{(\mu - \lambda)^{n-2}}{n} \frac{n(n-1)}{2} E[q^{n-2}] E[(a-s)^2] - \frac{(\mu - \lambda)^{n-2}}{n} E[(-u)^n]$$

$$+ \frac{(\mu - \lambda)^{n-2}}{n} \sum_{i=0}^{n-3} \binom{n}{i} E[q^i] E\left[(a-s)^{n-i}\right]. \tag{10.2}$$

We now derive bounds on the moments of q in the heavy-traffic regime. In particular, we consider a sequence of queueing systems indexed by ϵ such that the arrival process $a^{(\epsilon)}(t)$ satisfies

$$\lambda^{(\epsilon)} = E[a^{(\epsilon)}] = \mu - \epsilon,$$

then derive bounds on $\epsilon^n E\left[\left(q^{(\epsilon)}\right)^n\right]$ as $\epsilon \to 0$ under the assumption that

$$\lim_{\epsilon \to 0} E\left[\left(a^{(\epsilon)}(t) - s(t)\right)^2\right] = \sigma^2, \tag{10.3}$$

where $q^{(\epsilon)}$ is the steady-state queue length of the ϵth queueing system. We now present an example to show that the assumption (10.3) is reasonable.

Example 10.1.1

If the arrival process is Poisson and the service process is deterministic,

$$E\left[\left(a^{(\epsilon)} - s\right)^2\right] = E\left[\left(a^{(\epsilon)}\right)^2\right] - 2E[a^{(\epsilon)}]\mu + \mu^2$$

$$= \left(\lambda^{(\epsilon)}\right)^2 + \lambda^{(\epsilon)} - 2\lambda^{(\epsilon)}\mu + \mu^2$$

$$= (\mu - \epsilon)^2 + \mu - \epsilon - 2(\mu - \epsilon)\mu + \mu^2,$$

which converges to μ as $\epsilon \to 0$. $\qquad \square$

We continue with the heavy-traffic analysis of the moments of the queue lengths. Recall that $u^{(\epsilon)}$ is the unused capacity, so $E\left[u^{(\epsilon)}\right] = \mu - \lambda^{(\epsilon)} = \epsilon$, and

$$\epsilon^{n-2} E\left[\left|u^{(\epsilon)}\right|^n\right] \le \epsilon^{n-2} c_{\max}^{n-1} E\left[u^{(\epsilon)}\right] = \epsilon^{n-1} c_{\max}^{n-1}.$$

So $\epsilon^{n-2} E\left[\left|u^{(\epsilon)}\right|^n\right] \to 0$ as $\epsilon \to 0$ for any $n \ge 2$. We first consider the case $n = 2$. From (10.2), we obtain

$$\epsilon E[q] = \frac{1}{2}\left(E\left[(a-s)^2\right] - E\left[u^2\right]\right),$$

which implies that

$$\lim_{\epsilon \to 0} \epsilon E[q] = \lim_{\epsilon \to 0} \frac{1}{2}\left(E\left[(a-s)^2\right] - E\left[u^2\right]\right) = \frac{\sigma^2}{2}. \qquad (10.4)$$

Now consider the case $n = 3$. From (10.2), we have

$$\epsilon^2 E\left[\left(q^{(\epsilon)}\right)^2\right] = \epsilon E\left[q^{(\epsilon)}\right] E\left[\left(a^{(\epsilon)} - s\right)^2\right] - \frac{\epsilon}{3} E\left[\left(-u^{(\epsilon)}\right)^3\right] + \frac{\epsilon}{3} E\left[\left(a^{(\epsilon)} - s\right)^3\right].$$

Note that $\lim_{\epsilon \to 0} \epsilon E\left[\left|u^{(\epsilon)}\right|^3\right] = 0$ and

$$\lim_{\epsilon \to 0} \epsilon E\left[\left|a^{(\epsilon)} - s\right|^3\right] \le \lim_{\epsilon \to 0} \epsilon c_{\max} E\left[\left(a^{(\epsilon)} - s\right)^2\right] = 0.$$

So we have

$$\lim_{\epsilon \to 0} \epsilon^2 E\left[\left(q^{(\epsilon)}\right)^2\right] = \frac{\sigma^2}{2} \times \sigma^2 = 2\left(\frac{\sigma^2}{2}\right)^2. \qquad (10.5)$$

In Exercise 10.1 you will be asked to prove that

$$\lim_{\epsilon \to 0} \epsilon^n E\left[\left(q^{(\epsilon)}\right)^n\right] = n!\left(\frac{\sigma^2}{2}\right)^n.$$

The following theorem summarizes the result on the moment bounds of q.

Theorem 10.1.1 Assume (i) both the arrival process $a^{(\epsilon)}(t)$ and the service process $s(t)$ are stationary and independent of $q^{(\epsilon)}(t)$; (ii) $s(t) \le c_{\max}$ for all t; and (iii) $\lim_{\epsilon \to 0} E\left[\left(a^{(\epsilon)} - s\right)^2\right] = \sigma^2$. Then, the following limit holds for any $n \ge 1$:

$$\lim_{\epsilon \to 0} \epsilon^n E\left[\left(q^{(\epsilon)}\right)^n\right] = n!\left(\frac{\sigma^2}{2}\right)^n. \qquad (10.6)$$

$\square$

There are two interesting observations about the moment bounds in the heavy-traffic regime.

(i) The nth moment of the scaled queue length $\epsilon q^{(\epsilon)}$ depends only on σ^2, i.e., the variances of the arrival and service processes. Therefore, the behavior of $\epsilon q^{(\epsilon)}$ in the heavy-traffic limit (i.e., $\epsilon \to 0$) depends only on the first and second moments of the arrival and service processes.

(ii) The second observation is that the moments of $\epsilon q^{(\epsilon)}$ in steady state as $\epsilon \to 0$ are identical to those of the exponential distribution with mean $\sigma^2/2$. This suggests that the distribution of $\epsilon q^{(\epsilon)}$, as $\epsilon \to 0$, is an exponential distribution with mean $\sigma^2/2$.

10.2   Heavy-traffic optimality of JSQ

In Section 10.1, we used a Lyapunov argument to derive bounds on the nth moments of a discrete-time G/G/1 queue. In this section, we will demonstrate that a Lyapunov-drift argument can also be used to study the heavy-traffic performance of more complex queueing systems. In particular, we will study a system with parallel queues, each being served by a separate server. This system models a number of applications. For example, it can model a supermarket checkout, in which each checkout counter is a server and the customers waiting at a counter form the queue. In the context of communication networks, parallel queues can be used to model multiple paths from a source to its destination in the following manner. Assume that there is a single congested link on each path, and that the paths do not share any congested links. Then, packets routed on each path will form queues at their corresponding congested links. Thus, the system as a whole can be viewed as a collection of parallel queues. In both of these applications, there is a common arrival stream to all the queues, and each arrival has to select a queue to join. Thus, the resource allocation problem here is a routing problem: each arrival has to be routed to a queue to achieve some objective, such as minimizing the mean sum-queue length in steady state.

To study the above routing problem, we consider a two-server system served by a common arrival stream, as shown in Figure 10.2. We assume that two service processes are Bernoulli with means μ_1 and μ_2, respectively, and the arrival process is Bernoulli with mean λ. The results in this section can be extended to more general arrival and service processes, as outlined in the exercises at the end of the chapter. We assume $\lambda < \mu_1 + \mu_2 \leq 1$. The arrivals can be routed to either of the two servers. We consider a routing policy, called Join the Shortest Queue (JSQ), under which an arrival joins the shorter of the two queues. If both queues have equal length, one of two queues is picked at random. Let $q_i(t)$ denote the queue length at server i at the beginning of time slot t, and Let $u_i(t)$ denote the unused service in time slot t. The queue evolution is as follows:

$$q_i(t+1) = q_i(t) + a_i(t) - s_i(t) + u_i(t),$$

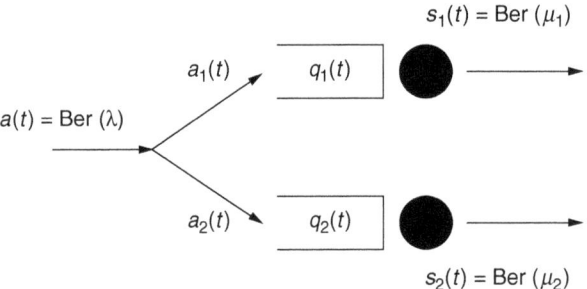

Figure 10.2 Two parallel queues served by a common arrival stream.

where $a_i(t)$ is the arrival routed to server i at the beginning of time slot t and $a_1(t) + a_2(t) = a(t)$.

We will show that the JSQ policy minimizes the expected sum-queue length in the heavy-traffic regime. Here, by the heavy-traffic regime, we mean a regime where the arrival rate λ approaches the total system capacity $\mu_1 + \mu_2$. The heavy-traffic analysis of the JSQ policy will consist of three major steps as follows.

(1) **Derive a lower bound on the expected sum-queue length, by relating the two-queue system to a single-server queue** It should be intuitively clear that a single-server queue, whose service capacity is equal to the sum of the capacities of the two servers (see Figure 10.3), will have a smaller expected queue length than the expected sum-queue length of the two-server system. We will prove in Theorem 10.2.2 that the queue length of the single-server system is a lower bound (in a stochastic sense) on the total queue length of the two-server system under *any routing policy*. In particular, we will show that

$$E[q_1(\infty) + q_2(\infty)] \geq \frac{E[(a - s_1 - s_2)^2]}{2(\mu_1 + \mu_2 - \lambda)}, \tag{10.7}$$

where $q_i(\infty)$ is the queue in steady-state.

(2) **Derive an upper bound on the expected steady-state queue lengths** For the JSQ policy, by analyzing the drift of the Lyapunov function $V(t) = (q_1(t) + q_2(t))^2$, we will establish an upper bound on the sum-queue length in Theorem 10.2.3 such that

$$E[q_1(\infty) + q_2(\infty)] \leq \frac{E[(a - s_1 - s_2)^2]}{2(\mu_1 + \mu_2 - \lambda)} + \frac{1}{\mu_1 + \mu_2 - \lambda}$$
$$\times \sqrt{E[(q_1 - q_2)^2]}\left(\sqrt{E[u_1^2]} + \sqrt{E[u_2^2]}\right). \tag{10.8}$$

(3) **Prove heavy-traffic optimality of JSQ using state-space collapse** Let $\epsilon = \mu_1 + \mu_2 - \lambda$ be the difference between the total service rate and the arrival rate. We will show that a phenomenon called *state-space collapse* occurs when $\epsilon \to 0$. Specifically, we will prove $E[(q_1 - q_2)^2] \leq K$ for some constant K independent of ϵ. Note that $E[q_1 + q_2] = O(1/\epsilon)$ according to the lower bound (10.7), so $E[(q_1 - q_2)^2] \leq K$ implies that $q_1(t) \approx q_2(t)$ as ϵ approaches zero. Hence, the two-dimensional $q(t)$ becomes essentially a one-dimensional system (the line $q_1 = q_2$), as shown in Figure 10.4. This state-space collapse occurs because JSQ routes each arrival to the shorter queue to "force" the two queues to be equal. Because of this state-space collapse, in the heavy-traffic regime the sum queue can be approximated by the queue of the single-server system used in the lower bound, so JSQ is heavy-traffic optimal.

Mathematically, note that the first term in the upper bound (10.8) is the same as the lower bound (10.7). Given $E[(q_1 - q_2)^2] \leq K$ and the fact that $E[u_i^2] \leq \epsilon$, the first term of the upper bound (10.8) dominates the other two terms and the upper bound asymptotically coincides with the lower bound as $\epsilon \to 0$.

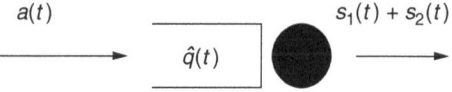

Figure 10.3 The queue length of this single-server system is a lower bound on the total queue length of the two-server system.

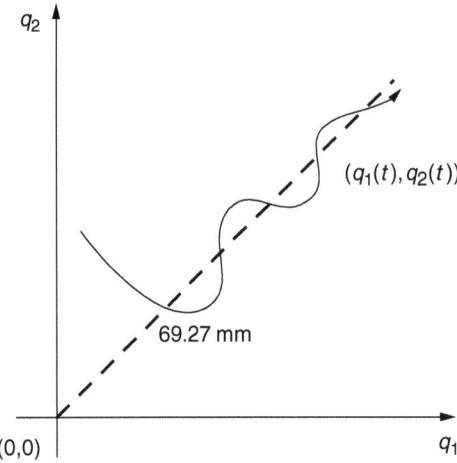

Figure 10.4 The two queue lengths (whose evolution is shown as the solid curve) are close to each other under JSQ.

We now formally establish the results mentioned above. We first prove that the two-server queueing system under JSQ is a positive recurrent Markov chain, so a unique stationary distribution exists.

Theorem 10.2.1 Consider the two-server system in Figure 10.2 and assume $\lambda < \mu_1 + \mu_2$. We have that $q(t) = (q_1(t), q_2(t))$ is a positive recurrent Markov chain under JSQ.

Proof Consider the Lyapunov function

$$V(q(t)) = q_1^2(t) + q_2^2(t).$$

We have

$$
\begin{aligned}
V(q(t+1)) &- V(q(t)) \\
&= \left((q_1 + a_1 - s_1 + u_1)^2 - q_1^2\right) + \left((q_2 + a_2 - s_2 + u_2)^2 - q_2^2\right) \\
&\leq \left((q_1 + a_1 - s_1)^2 - q_1^2\right) + \left((q_2 + a_2 - s_2)^2 - q_2^2\right) \\
&= (a_1 - s_1)^2 + 2q_1(a_1 - s_1) + (a_2 - s_2)^2 + 2q_2(a_2 - s_2).
\end{aligned}
$$

Note that $(a_i(t) - s_i(t))^2 \leq 1$ because both $a_i(t)$ and $s_i(t)$ are Bernoulli random variables, so

$$
\begin{aligned}
E\left[V(t+1) - V(t)|q(t) = q\right] \\
&\leq 2E[q_1 a_1 + q_2 a_2 | q(t) = q] - 2(q_1 \mu_1 + q_2 \mu_2) + 2 \\
&=_{(a)} 2 \min\{q_1, q_2\} E[a(t)] - 2(q_1 \mu_1 + q_2 \mu_2) + 2 \\
&= 2 \min\{q_1, q_2\} \lambda - 2(q_1 \mu_1 + q_2 \mu_2) + 2
\end{aligned}
$$

$$\leq 2 \left(q_1 \frac{\mu_1 - \epsilon/2}{\lambda} + q_2 \frac{\mu_2 - \epsilon/2}{\lambda} \right) \lambda - 2(q_1\mu_1 + q_2\mu_2) + 2$$

$$= -\epsilon(q_1 + q_2) + 2,$$

where inequality (a) holds due to the JSQ routing, and the last equality holds because there exists $\epsilon > 0$ such that $\mu_1 + \mu_2 - \lambda = \epsilon$. $\qquad\square$

Using $q(\infty)$ to denote informally the queues in steady state, we next establish a lower bound on $E[q_1(\infty) + q_2(\infty)]$ using a coupling argument.

Theorem 10.2.2 Consider the two-server system in Figure 10.2 and assume $\lambda < \mu_1 + \mu_2$. Under any routing policy, the following lower bound holds:

$$(\mu_1 + \mu_2 - \lambda)E[q_1(\infty) + q_2(\infty)] \geq \frac{E[(a - s_1 - s_2)^2]}{2} - (\mu_1 + \mu_2 - \lambda).$$

Proof Consider a queue with arrival process $a(t)$ and departure process $s(t) = s_1(t) + s_2(t)$, as shown in Figure 10.3. Denote the queue length in this system by $\hat{q}(t)$. We couple this system, called System 2, with the two-server system, called System 1, as follows: the arrival to System 2 at the beginning of each time slot is the same as the arrival to System 1. In other words, when there is an arrival to System 1, an arrival also occurs to System 2. Similarly, the service available in System 2 in each time slot is equal to the sum of the services available in the two servers in System 1. Note that some available service could be wasted in System 1 when it is not wasted in System 2; for example, in System 1, if there is a packet in queue 1 and no packet in queue 2, but the service capacity of queue 2 is 1, this service capacity cannot be used in this time slot. However, in System 2, since there is a common queue, any available service capacity can be used to serve the packets in the queue. This is the reason why we expect the queue length in System 2 to be a lower bound on the sum of the queue lengths in System 1. It is easy to make this argument precise using induction.

Suppose that $\hat{q}(0) = q_1(0) + q_2(0)$. Now, if $q_1(k) + q_2(k) \geq \hat{q}(k)$, it is not difficult to see that

$$q_1(k+1) + q_2(k+1) \geq \hat{q}(k+1)$$

because $s(k) = s_1(k) + s_2(k)$. Therefore, by induction, we have

$$q_1(t) + q_2(t) \geq \hat{q}(t)$$

for all t. Using the discrete-time Kingman bound in Exercise 3.13 on $\hat{q}(\infty)$, we obtain

$$E[\hat{q}(\infty)] \geq \frac{E[(a - s_1 - s_2)^2] - E[u^2]}{2(\mu_1 + \mu_2 - \lambda)},$$

which implies

$$(\mu_1 + \mu_2 - \lambda)E[q_1(\infty) + q_2(\infty)] \geq \frac{E[(a - s_1 - s_2)^2]}{2} - (\mu_1 + \mu_2 - \lambda).$$

Note $E[u^2] \leq 2E[u] = 2(\mu_1 + \mu_2 - \lambda)$. $\qquad\square$

In the following theorem, we will establish an upper bound on $E[q_1(\infty) + q_2(\infty)]$.

Theorem 10.2.3 Consider the two-server system in Figure 10.2 with the JSQ routing. Assume $\lambda < \mu_1 + \mu_2$ and that the system is in steady state, then the following upper bound holds:

$$E[q_1(\infty) + q_2(\infty)] \le \frac{E\left[(a - s_1 - s_2)^2\right]}{2(\mu_1 + \mu_2 - \lambda)} + \frac{1}{\mu_1 + \mu_2 - \lambda}$$
$$\times \sqrt{E[(q_1 - q_2)^2]} \left(\sqrt{E[u_1^2]} + \sqrt{E[u_1^2]}\right).$$

Proof We consider the Lyapunov function

$$V(q(t)) = (q_1(t) + q_2(t))^2.$$

The drift of the Lyapunov function is given by

$$E[V(t+1) - V(t)|q(t) = q]$$
$$= (q_1 + a_1 - s_1 + u_1 + q_2 + a_2 - s_2 + u_2)^2 - (q_1 + q_2)^2$$
$$= (q_1 + q_2 + a - (s_1 + s_2) + (u_1 + u_2))^2 - (q_1 + q_2)^2. \tag{10.9}$$

Note that the above expression is nearly identical to the expression we would get if we computed the Lyapunov drift for the lower bounding system in Theorem 10.2.2, except for the term $u_1 + u_2$, which would be replaced by a single common unused service term. Expanding further the first term in (10.9) yields

$$E[V(t+1) - V(t)|q(t) = q]$$
$$= (q_1 + q_2 + a - (s_1 + s_2))^2 + (u_1 + u_2)^2 + 2(q_1 + q_2 + a - s_1 - s_2)(u_1 + u_2)$$
$$\quad - (q_1 + q_2)^2$$
$$= (a - (s_1 + s_2))^2 + 2(q_1 + q_2)(a - (s_1 + s_2))$$
$$\quad + 2(q_1 + a_1 - s_1 + u_1 + q_2 + a_2 - s_2 + u_2)(u_1 + u_2) - (u_1 + u_2)^2$$
$$= (a - (s_1 + s_2))^2 + 2(q_1 + q_2)(a - (s_1 + s_2)) + 2(q_1(t+1) + q_2(t+1))(u_1 + u_2)$$
$$\quad - (u_1 + u_2)^2.$$

Note that $u_i(t) > 0$ only if $q_i(t) + a_i(t) < s_i(t)$, so $u_i(t) > 0$ implies $q_i(t+1) = 0$, and we have $q_i(t+1)u_i(t) = 0$. From this fact, we obtain

$$E[V(t+1) - V(t)|q(t) = q]$$
$$= (a - (s_1 + s_2))^2 + 2(q_1 + q_2)(a - (s_1 + s_2)) + 2q_1(t+1)u_2 + 2q_2(t+1)u_1$$
$$\quad - (u_1 + u_2)^2$$
$$= (a - (s_1 + s_2))^2 + 2(q_1 + q_2)(a - (s_1 + s_2))$$
$$\quad + 2(q_1(t+1) - q_2(t+1))u_2 + 2(q_2(t+1) - q_1(t+1))u_1 - (u_1 + u_2)^2.$$

As the system is in steady state, we have $E[V(t+1)-V(t)] = 0$ and $E[q_i(t)] = E[q_i(t+1)]$. Taking the expectation on both sides of the above equation yields

$$E[q_1(\infty) + q_2(\infty)]$$

$$= \frac{E[(a-s)^2]}{2(\mu_1 + \mu_2 - \lambda)} + \frac{1}{\mu_1 + \mu_2 - \lambda} E[(q_1(\infty) - q_2(\infty))u_2]$$

$$+ \frac{1}{\mu_1 + \mu_2 - \lambda} E[(q_2(\infty) - q_1(\infty))u_1] - E\left[(u_1^2 + u_2^2)\right]$$

$$\leq \frac{E\left[(a-s)^2\right]}{2(\mu_1 + \mu_2 - \lambda)} + \frac{1}{\mu_1 + \mu_2 - \lambda} \sqrt{E[(q_1(\infty) - q_2(\infty))^2]E[u_2^2]}$$

$$+ \frac{1}{\mu_1 + \mu_2 - \lambda} \sqrt{E[(q_1(\infty) - q_2(\infty))^2]E[u_1^2]},$$

where the last step follows from the Cauchy–Schwartz inequality, which states that, for two random variables X and Y,

$$|E(XY)| \leq \sqrt{E(|X|^2)E(|Y|^2)}. \qquad \square$$

In the following lemma, we will show that $E[u_i^2] \leq \epsilon$.

Lemma 10.2.4 Consider the two-server system in Figure 10.2 with the JSQ routing. Assume $\lambda < \mu_1 + \mu_2$ and that the system is in steady state, then we have $E[u_i^2] \leq \epsilon$.

Proof As the service process is Bernoulli, $u_i \in \{0, 1\}$, which implies that $u_i^2 = u_i$. In steady state,

$$E[q_1(t+1) + q_2(t+1) - (q_1(t) + q_2(t))] = 0,$$

so

$$E[a_1 - s_1 + u_1 + a_2 - s_2 + u_2] = 0,$$

which implies

$$\lambda - (\mu_1 + \mu_2) + E[u_1 + u_2] = 0$$

and

$$E[u_1 + u_2] = \mu_1 + \mu_2 - \lambda = \epsilon.$$

Therefore, we conclude

$$E[u_1^2] = E[u_1] \leq \epsilon \qquad \text{and} \qquad E[u_2^2] = E[u_2] \leq \epsilon. \qquad \square$$

In the following theorem, we will complete the main result of this section by proving the state-space collapse mentioned earlier, i.e., $E[(q_1 - q_2)^2] \leq K$ for some K independent of ϵ. If we have $E[(q_1 - q_2)^2] \leq K$, then

$$\sqrt{E[(q_1(\infty) - q_2(\infty))^2]E[u_i^2]} \leq \sqrt{K\epsilon},$$

which converges to zero as $\epsilon \to 0$. Further, the numerator in the first term of (10.8) becomes a constant in the limit $\epsilon \to 0$:

$$
\begin{aligned}
E[(a - s_1 - s_2)^2] &= \lambda + \mu_1 + \mu_2 - 2\lambda\mu_1 - 2\lambda\mu_2 + 2\mu_1\mu_2 \\
&= 2\mu_1 + 2\mu_2 - \epsilon - 2(\mu_1 + \mu_2)(\mu_1 + \mu_2 - \epsilon) + 2\mu_1\mu_2 \\
&\geq 2\mu_1\mu_2 - \epsilon,
\end{aligned}
$$

where the last inequality holds because $\mu_1 + \mu_2 \leq 1$. So, $\lim_{\epsilon \to 0} E[(a - s_1 - s_2)^2] \geq 2\mu_1\mu_2$. Therefore, the first term of (10.8) dominates the other two terms when $\epsilon \to 0$. The first term matches the lower bound in Theorem 10.2.2, proving the heavy-traffic optimality of JSQ. The result is presented in the following theorem.

Theorem 10.2.5 Consider the two-server system in Figure 10.2 with the JSQ routing. Assume $\lambda < \mu_1 + \mu_2$ and that the system is in steady state. Then JSQ is heavy-traffic optimal, i.e., it minimizes $\lim_{\epsilon \to 0} \epsilon E\left[q_1^{(\epsilon)}(\infty) + q_2^{(\epsilon)}(\infty)\right]$ among all possible routing policies, where $q^{(\epsilon)}(\infty)$ is the steady-state queue length of the two-server system with arrival rate $\lambda^{(\epsilon)} = \mu_1 + \mu_2 - \epsilon$.

Proof We consider the following Lyapunov function:

$$
V(q(t)) = |q_1(t) - q_2(t)|^3.
$$

We are interested in computing the Lyapunov drift,

$$
E[V(t+1) - V(t)|q(t) = q],
$$

under JSQ. Note that q_1 and q_2 can each decrease or increase by at most 1 in each time slot because arrival and service processes are Bernoulli. We consider the following three cases.

(1) **Case 1:** $|q_1 - q_2| \leq 2$. In this case, we have

$$
\begin{aligned}
&E[V(t+1) - V(t)|q(t) = q] \\
&= |q_1 - q_2 + (a_1 - s_1 + u_1 - (a_2 - s_2 + u_2))|^3 - |q_1 - q_2|^3 \leq 64
\end{aligned}
$$

because $|a_1 - s_1 + u_1 - (a_2 - s_2 + u_2)| \leq 2$.

(2) **Case 2:** $q_1 - q_2 \geq 3$. In this case, $q_1(t+1) - q_2(t+1) > 0$ because $q_1 - q_2$ can decrease by at most 2. Further, $q_1 - q_2 \geq 3$ implies that $a_2 = a$ and $a_1 = 0$ due to JSQ, and $u_1 = 0$. So we have

$$
\begin{aligned}
q_1(t+1) - q_2(t+1) &= q_1 - s_1 - (q_2 + a - s_2 + u_2) \\
&= (q_1 - q_2) - (a + s_1 - s_2 + u_2)
\end{aligned}
$$

and

$$
\begin{aligned}
&|q_1(t+1) - q_2(t+1)|^3 - |q_1(t) - q_2(t)|^3 \\
&= -3(q_1 - q_2)^2(a + s_1 - s_2 + u_2) + 3(q_1 - q_2)(a + s_1 - s_2 + u_2)^2 \\
&\quad - (a + s_1 - s_2 + u_2)^3.
\end{aligned}
$$

Taking the conditional expectation on both sides yields

$$E[V(t+1) - V(t)|q(t) = q] \le -3(q_1 - q_2)^2(\lambda + \mu_1 - \mu_2) + 3(q_1 - q_2)K_1 + K_2,$$

where K_1 and K_2 are constants and $K_1 > 0$. Without loss of generality, we can assume

$$\lambda > \mu_2 - \mu_1 \quad \text{and} \quad \lambda > \mu_1 - \mu_2$$

because we are interested in the heavy-traffic regime, and in that regime $\lambda \to \mu_1 + \mu_2$. Thus the above drift is less than zero for sufficiently large $q_1 - q_2$.

(3) **Case 3:** $q_2 - q_1 \ge 3$. Similar to Case 2, we have

$$E[V(t+1) - V(t)|q(t) = q] \le -3(q_1 - q_2)^2(\lambda + \mu_2 - \mu_1) + 3(q_2 - q_1)K_3 + K_4$$

for some constants $K_3 > 0$ and K_4.

Let π denote the steady-state queue length distribution. In the steady state, we have

$$0 = E[V(t+1) - V(t)]$$

$$= \sum_{q:|q_1-q_2|\le 2} E[V(t+1) - V(t)|q(t) = q]\pi(q)$$

$$+ \sum_{q:|q_1-q_2|\ge 3} E[V(t+1) - V(t)|q(t) = q]\pi(q)$$

$$\le 64 \sum_{q:|q_1-q_2|\le 2} \pi(q) + \sum_{q:|q_1-q_2|\ge 3} (-3|q_1 - q_2|^2\delta + 3|q_1 - q_2|C_1 + C_2)\pi(q)$$

$$\le \sum_{q:|q_1-q_2|\le 2} \left(64 - (-3|q_1 - q_2|^2\delta + 3|q_1 - q_2|C_1 + C_2)\right)\pi(q)$$

$$- 3E[|q_1 - q_2|^2]\delta + 3E[|q_1 - q_2|]C_1 + C_2,$$

where

$$\delta = \min(\lambda - \mu_1 + \mu_2, \lambda - \mu_2 + \mu_1) > 0$$
$$C_1 = \max(K_1, K_3) > 0$$
$$C_2 = \max(K_2, K_4).$$

Thus, there exists a positive constant K_5, independent of ϵ, such that

$$0 \le K_5 - 3\delta E\left[|q_1 - q_2|^2\right] + 3C_1 E[|q_1 - q_2|] + C_2,$$

which implies that

$$\delta E[|q_1 - q_2|^2] - C_1 E[|q_1 - q_2|] \le \frac{K_5 + C_2}{3},$$

where all the above constants are independent of ϵ.

Starting with the Lyapunov function

$$V(t) = (q_1(t) - q_2(t))^2,$$

we can also show that $E[|q_1 - q_2|]$ is bounded by a positive constant independent of ϵ using a similar argument. Thus, both $E\left[|q_1 - q_2|^2\right]$ and $E[|q_1 - q_2|]$ are upper bounded by some positive constants independent of ϵ.

Recall that $q^{(\epsilon)}(\infty)$ is the steady-state queue length of the two-server system with arrival rate $\lambda^{(\epsilon)} = \mu_1 + \mu_2 - \epsilon$. Since $E[u_i^2] \leq \epsilon$ (Lemma 10.2.4) and $E\left[|q_1 - q_2|^2\right]$ is upper bounded by some positive constants independent of ϵ, we conclude that, under JSQ,

$$\lim_{\epsilon \to 0} \sqrt{E\left[(q_1^{(\epsilon)}(\infty) - q_2^{(\epsilon)}(\infty))^2\right] E\left[\left(u_i^{(\epsilon)}\right)^2\right]} = 0$$

and

$$\lim_{\epsilon \to 0} \epsilon E\left[q_1^{(\epsilon)} + q_2^{(\epsilon)}\right] = \frac{E[(a - s_1 - s_2)^2]}{2},$$

which coincides with the lower bound in Theorem 10.2.2 when $\epsilon \to 0$. □

10.3 Large deviations of i.i.d. random variables: the Cramer–Chernoff theorem

In the preceding sections, we presented techniques to analyze queueing systems that are heavily loaded, i.e., the arrival rate is close to capacity. In such a regime, we showed that the queue lengths were large (of the order of $1/\epsilon$, where ϵ is the difference between the service and arrival rates), and presented techniques to estimate the rate at which they grow as a function of $1/\epsilon$. In the rest of the chapter, we will consider traffic regimes in which large queue length events are rare. We will be interested in the probabilities of these rare events. This regime is called the *large-deviations regime*. Both the heavy-traffic and large-deviations regimes are important to understand in practice: they represent two extremes of communication network operation. Before we present techniques to analyze queues in the large-deviations regime, in this section we first present techniques to estimate the probability that the empirical average of a large number of independent, identically distributed random variables deviates from their common mean. These techniques will then be used in the subsequent sections to study large deviations in queues.

Consider a sequence of i.i.d. random variables $\{X_i\}$. The Central Limit Theorem (CLT) provides an estimate of the probability,

$$\Pr\left(\frac{\sum_{i=1}^{n} X_i - n\mu}{\sigma\sqrt{n}} \geq x\right),$$

where $\mu = E[X_1]$ and $\sigma^2 = Var(X_1)$. Thus, the CLT estimates the probability of $O(\sqrt{n})$ deviations from the mean of the sum of the random variables $\{X_i\}_{i=1}^{n}$. These deviations are small compared to the mean of $\sum_{i=1}^{n} X_i$, which is an $O(n)$ quantity. On the other hand, *large deviations* of the order of the mean itself, i.e., $O(n)$ deviations, form the subject of the rest of this chapter.

The simplest large-deviations result is the Chernoff bound, which states that, for any constant x,

$$\Pr\left(\sum_{i=1}^{n} X_i \geq nx\right) \leq e^{-n \sup_{\theta \geq 0}(\theta x - \log M(\theta))}, \tag{10.10}$$

where $M(\theta) = E(e^{\theta X_1})$ is the *moment generating function* of X_1. The following theorem quantifies the tightness of the Chernoff bound.

Theorem 10.3.1 (Cramer–Chernoff theorem) Let $X_1, X_2, \ldots$ be i.i.d. random variables with $E[X_1] = \mu$, and suppose that their common moment generating function $M(\theta) < \infty$ for all θ in some neighborhood $\mathcal{B}_0$ of $\theta = 0$. Further suppose that the supremum in the following definition of the *rate function* $I(x)$ is obtained at some interior point in this neighborhood:

$$I(x) = \sup_{\theta} \theta x - \Lambda(\theta), \tag{10.11}$$

where $\Lambda(\theta) \triangleq \log M(\theta)$ is called the *log moment generating function* or the *cumulant generating function*. In other words, we assume that there exists $\theta^* \in int(\mathcal{B}_0)$ such that

$$I(x) = \theta^* x - \Lambda(\theta^*).$$

Fix any $x > E[X_1]$. Then, for each $\epsilon > 0$, there exists N such that, for all $n \geq N$,

$$e^{-n(I(x)+\epsilon)} \leq \Pr\left(\sum_{i=1}^{n} X_i \geq nx\right) \leq e^{-nI(x)}. \tag{10.12}$$

In other words,

$$\lim_{n \to \infty} \frac{1}{n} \log \Pr\left(\sum_{i=1}^{n} X_i \geq nx\right) = -I(x). \tag{10.13}$$

Proof We first prove the upper bound and then the lower bound.

Upper bound in (10.12): This follows from the Chernoff bound if we show that the value of the supremum in (10.10) does not change if we relax the condition $\theta \geq 0$ and allow θ to take negative values. Since $e^{\theta x}$ is a convex function of x, by Jensen's inequality we have

$$M(\theta) = E[e^{\theta X_1}] \geq e^{\theta \mu}.$$

If $\theta < 0$, because $x - \mu > 0$, $e^{-\theta(x-\mu)} > 1$. Thus,

$$M(\theta)e^{-\theta(x-\mu)} \geq e^{\theta \mu}.$$

Taking the logarithm of both sides yields

$$\theta x - \Lambda(\theta) \leq 0.$$

Noting that $\theta x - \Lambda(\theta) = 0$ when $\theta = 0$, we have

$$\sup_{\theta} \theta x - \Lambda(\theta) = \sup_{\theta \geq 0} \theta x - \Lambda(\theta).$$

Lower bound in (10.12): Let $p(x)$ be the probability density function (PDF) of X_1. We assume that such a PDF exists, but the result holds more generally and can be proved in the same manner as below with slightly more complicated notation. Then, for any $\delta > 0$,

$$
\Pr\left(\sum_{i=1}^n X_i \geq nx\right) = \int_{\sum_{i=1}^n x_i \geq nx} \prod_{i=1}^n p(x_i)dx_i
$$

$$
\geq \int_{nx \leq \sum_{i=1}^n x_i \leq n(x+\delta)} \prod_{i=1}^n p(x_i)dx_i
$$

$$
= \frac{M^n(\theta^*)}{e^{n(x+\delta)\theta^*}} \int_{nx \leq \sum_{i=1}^n x_i \leq n(x+\delta)} \frac{e^{n\theta^*(x+\delta)}}{M^n(\theta^*)} \prod_{i=1}^n p(x_i)dx_i
$$

$$
\geq \frac{M^n(\theta^*)}{e^{n\theta*(x+\delta)}} \int_{nx \leq \sum_{i=1}^n x_i \leq n(x+\delta)} \prod_{i=1}^n \frac{e^{\theta^* x_i} p(x_i)}{M(\theta^*)} dx_i
$$

$$
= \frac{M^n(\theta^*)}{e^{n\theta*(x+\delta)}} \int_{nx \leq \sum_{i=1}^n x_i \leq n(x+\delta)} \prod_{i=1}^n q(x_i)dx_i, \qquad (10.14)
$$

where

$$
q(y) = \frac{e^{\theta^* y} p(y)}{M(\theta^*)}.
$$

Note that $\int_{-\infty}^{\infty} q(y)dy = 1$, so $q(y)$ is a PDF. Let Y be a random variable with $q(y)$ as its PDF. The moment generating function of Y is given by

$$
M_Y(\theta) = \int_{-\infty}^{\infty} e^{\theta y} q(y)dy = \frac{M(\theta + \theta^*)}{M(\theta^*)}.
$$

Thus,

$$
E[Y] = \frac{dM_Y(\theta)}{d\theta}\Big|_{\theta=0} = \frac{M'(\theta^*)}{M(\theta^*)}.
$$

From the assumptions of the theorem, θ^* achieves the supremum in (10.11). Thus,

$$
\frac{d}{d\theta}(\theta x - \log M(\theta))\Big|_{\theta=\theta^*} = 0.
$$

The above equations assume that $M(\theta)$ is differentiable, but this can be guaranteed by the assumptions of the theorem. From the above equation, we obtain $x = (M'(\theta^*))/(M(\theta^*))$. Therefore, $E[Y] = x$. In other words, the PDF $q(y)$ defines a set of i.i.d. random variables Y_i, each with mean x. Thus, from (10.14), the probability of a large deviation of the sum $\sum_{i=1}^n X_i$ can be lower bounded by probability that $\sum_{i=1}^n Y_i$ is near its mean nx as follows:

$$
\Pr\left(\sum_{i=1}^n X_i \geq nx\right) \geq e^{-nI(x) - n\delta\theta^*} \Pr\left(nx \leq \sum_{i=1}^n Y_i \leq n(x+\delta)\right).
$$

By the CLT,

$$\Pr\left(nx \le \sum_{i=1}^{n} Y_i \le n(x+\delta)\right) = \Pr\left(0 \le \frac{\sum_{i=1}^{n}(Y_i - x)}{\sqrt{n}} \le \sqrt{n}\,\delta\right) \xrightarrow{n\to\infty} \frac{1}{2}.$$

Given $\epsilon > 0$, fix $\delta < \epsilon$, and choose a sufficiently large n (dependent on δ) such that

$$\Pr\left(nx \le \sum_{i=1}^{n} Y_i \le n(x+\delta)\right) \ge \frac{1}{4}$$

and

$$e^{-n\delta\theta^*}\frac{1}{4} \ge e^{-n\epsilon}.$$

We have

$$\Pr\left(\sum_{i=1}^{n} X_i \ge nx\right) \ge e^{-nI(x)-n\epsilon}.$$

Thus, the theorem is proved.   □

The key idea in the proof of Theorem 10.3.1 is the definition of a new PDF $q(y)$ under which the random variable has a mean at x instead of at μ. This changes the nature of the deviation from the mean to a small deviation, instead of a large deviation, and thus allows the use of the CLT to complete the proof. Changing the PDF from $p(y)$ to $q(y)$ is called a *change of measure*. The new distribution, $\int_{-\infty}^{x} q(y)dy$, is called the *twisted* distribution or *exponentially tilted* distribution.

Note that the theorem is also applicable when $x < E[X_1]$. To see this, define $Z_i = -X_i$ and consider

$$\Pr\left(\sum_{i=1}^{n} Z_i \ge -nx\right).$$

Note that $M(-\theta)$ is the moment generating function of Z, and

$$I(x) = \sup_{\theta} \theta x - \Lambda(\theta) = \sup_{\theta} -\theta x - \Lambda(-\theta).$$

Thus, the rate function is the same for Z. Since $-x > E(Z_1)$, the theorem applies to $\{Z_i\}$.

It should also be noted that the proof of the theorem can be easily modified to yield the following result: for any $\delta > 0$,

$$\lim_{n\to\infty} \frac{1}{n} \log \Pr\left(nx \le \sum_{i=1}^{n} X_i < n(x+\delta)\right) = -I(x).$$

Noting that, for small δ, $\Pr\left(nx \le \sum_{i=1}^{n} X_i < n(x+\delta)\right)$ can be interpreted as $\Pr\left(\sum_{i=1}^{n} X_i \approx nx\right)$, this result states that the probability that the sum of the random variables exceeds nx is approximately equal (up to logarithmic equivalence) to the probability that the sum is "equal" to nx.

The rate function $I(\cdot)$ determines the decay rate of the tail probability. We present in the following several properties of the rate function.

Lemma 10.3.2 The rate function $I(x)$ is a convex function.

Proof Given α such that $0 \le \alpha \le 1$, we have

$$I(\alpha x_1 + (1-\alpha)x_2) = \sup_\theta \theta(\alpha x_1 + (1-\alpha)x_2) - \Lambda(\theta)$$

$$= \sup_\theta \theta\alpha x_1 - \alpha\Lambda(\theta) + \theta(1-\alpha)x_2 - (1-\alpha)\Lambda(\theta)$$

$$\le \sup_\theta \theta\alpha x_1 - \alpha\Lambda(\theta) + \sup_\theta \theta(1-\alpha)x_2 - (1-\alpha)\Lambda(\theta)$$

$$= \alpha I(x_1) + (1-\alpha)I(x_2). \qquad \square$$

Lemma 10.3.3 Let $I(x)$ be the rate function of a random variable X with mean μ. Then,

$$I(x) \ge I(\mu) = 0.$$

Proof Recall that $I(x) = \sup_\theta \theta x - \Lambda(\theta)$, so

$$I(x) \ge 0 \cdot x - \Lambda(0) = 0.$$

By Jensen's inequality,

$$M(\theta) = E[e^{\theta X}] \ge e^{\theta\mu}.$$

So we have

$$\Lambda(\theta) = \log M(\theta) \ge \theta\mu,$$

which implies that

$$\theta\mu - \Lambda(\theta) \le 0$$

for any θ, and

$$I(\mu) = \sup_\theta \theta\mu - \Lambda(\theta) \le 0.$$

We have already shown that $I(x) \ge 0$, $\forall x$, so we have the desired result. $\qquad \square$

Lemma 10.3.4 The log moment generating function $\Lambda(\theta)$ and the rate function $I(x)$ satisfy

$$\Lambda(\theta) = \sup_x \theta x - I(x).$$

Proof We will prove this under the assumption that all functions of interest are differentiable, which is the assumption throughout this chapter. From the definition of $I(x)$, we have

$$I(x) \geq \theta x - \Lambda(\theta), \qquad \forall x, \theta,$$

i.e.,

$$\Lambda(\theta) \geq \theta x - I(x), \qquad \forall x, \theta,$$

which implies that

$$\Lambda(\theta) \geq \sup_{x} \theta x - I(x).$$

To prove the lemma, it is enough to show that, for each θ, there exists an x^* such that

$$\Lambda(\theta) = \theta x^* - I(x^*). \tag{10.15}$$

We claim that such an x^* is given by $x^* = \Lambda'(\theta)$.

To see this, we note that, according to the definition of $I(x)$ and the convexity of $\Lambda(\theta)$, we have

$$I(x) = \theta^* x - \Lambda(\theta^*),$$

where θ^* solves

$$\Lambda'(\theta^*) = x.$$

Therefore, for $x^* = \Lambda'(\theta)$, we have

$$I(x^*) = \theta x^* - \Lambda(\theta),$$

which verifies (10.15). $\qquad \square$

10.4 Large-buffer large deviations

In Chapter 3, we studied a single-server queue with n data sources. Denote by $a_i(t)$ the number of bits arriving from source i at the beginning of time slot t, which is assumed to be i.i.d. across time slots and sources. Denote by c the number of bits that can be drained from the queue in one time slot. Note that c is a constant, thus we assume that the link capacity is deterministic, which is typically the case in a wireline link, although most of the results in this section can be generalized to link capacities that are i.i.d. across time slots. For any $\theta > 0$ such that $(\Lambda(\theta))/(\theta) < (c/n)$, we have shown in Chapter 3 that

$$\Pr(q(t) \geq B) \leq \frac{e^{(n\Lambda(\theta) - \theta c)}}{1 - e^{(n\Lambda(\theta) - \theta c)}} e^{-\theta B}, \qquad \forall t \geq 0, \tag{10.16}$$

where $\Lambda(\theta)$ is the moment generating function of $a_i(t)$.

In this section, we will revisit this upper bound for a single-server, single-source ($n = 1$) queue, as shown in Figure 10.5. We will show that this bound is tight in steady state, in the large-deviations sense. Since there is only one source, we drop the index i for the source

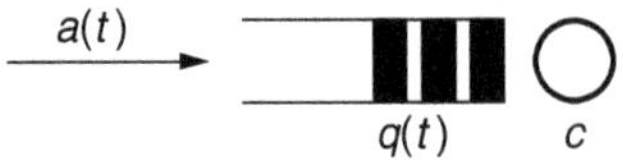

Figure 10.5 Single server queue with a single source.

in this section. We assume $E[a(t)] = \lambda < c$, and that the arrival process $a(t)$ is such that $\{q(t)\}$ is an aperiodic, irreducible, positive recurrent discrete-time Markov chain. We are interested in the steady-state probability $\Pr(q(\infty) \geq B)$. By our assumption on the Markov chain $q(t)$, because the system reaches a steady state starting from any initial condition, we will assume that $q(0) = 0$.

According to inequality (10.16), for $n = 1$ and any $\theta > 0$ such that $\Lambda_a(\theta)/\theta < c$, we have

$$\lim_{B \to \infty} \frac{1}{B} \log \Pr(q(\infty) \geq B) \leq -\theta,$$

where the subscript a in $\Lambda_a(\theta)$ indicates the associated random variable. Therefore, defining

$$\theta^* = \sup \{\theta > 0 : \Lambda_a(\theta) < c\theta\},$$

we obtain

$$\lim_{B \to \infty} \frac{1}{B} \log \Pr(q(\infty) \geq B) \leq -\theta^*.$$

In the following theorem, we will prove that this upper bound is tight.

Theorem 10.4.1 Assume that $\Lambda_a(\theta) < \infty$ for all $\theta < \infty$. Then,

$$\lim_{B \to \infty} \frac{1}{B} \log \Pr(q(\infty) \geq B) = -\theta^*.$$

Further, θ^* can also be written as

$$\theta^* = \inf_x \frac{I(x)}{x} = \inf_x \frac{I_a(x + c)}{x}.$$

To prove this theorem, we first establish a lower bound on $\lim_{B \to \infty} \frac{1}{B} \log \Pr(q(\infty) \geq B)$ in the following lemma.

Lemma 10.4.2 For any $t > 0$,

$$\lim_{B \to \infty} \frac{1}{B} \log \Pr(q(\infty) \geq B) \geq -tI\left(\frac{1}{t}\right),$$

where $I(\cdot)$ is the rate function of $a(t) - c$ (not the rate function of $a(t)$).

Proof We first recall that

$$q(t) = \max \left\{ \sup_{1 \leq k \leq t} \sum_{s=1}^{k} a(t - s) - kc, 0 \right\}. \tag{10.17}$$

Since $a(s)$ are i.i.d., note that

$$\sum_{s=1}^{k} a(t - s) \overset{d}{=} \sum_{s=1}^{k} a(s),$$

where $\overset{d}{=}$ means equality in distribution. Thus,

$$\Pr(q(\infty) > B) = \Pr\left(\sup_{k \geq 1} \sum_{t=1}^{k} a(t) - kc > B\right). \qquad (10.18)$$

Defining $A(k) = \sum_{t=1}^{k} a(t)$, we have, for any $k \geq 1$,

$$\Pr(q(\infty) \geq B) \geq \Pr\left(\frac{A(k) - kc}{B} \geq 1\right).$$

To apply the Cramer–Chernoff theorem, choose $k = \lceil tB \rceil$ for some $t > 0$. Substituting k into the above inequality, we obtain

$$\Pr(q(\infty) \geq B) \geq \Pr\left(\frac{A(\lceil tB \rceil) - \lceil tB \rceil c}{\lceil tB \rceil} \geq \frac{B}{\lceil tB \rceil}\right)$$

$$\geq \Pr\left(\frac{A(\lceil tB \rceil) - \lceil tB \rceil c}{\lceil tB \rceil} \geq \frac{1}{t}\right).$$

According to the Cramer–Chernoff theorem, given any $\epsilon > 0$, there exists a constant b_ϵ such that, $\forall B \geq b_\epsilon$,

$$\Pr\left(\frac{A(\lceil tB \rceil) - \lceil tB \rceil c}{\lceil tB \rceil} \geq \frac{1}{t}\right) \geq e^{-\lceil tB \rceil \left(I\left(\frac{1}{t}\right) + \epsilon\right)},$$

where $I(\cdot)$ is the rate function of $a(t) - c$.

Therefore, we can conclude that

$$\frac{1}{B} \log \Pr(q(\infty) \geq B) \geq -\frac{\lceil tB \rceil}{tB}\left(tI\left(\frac{1}{t}\right) + t\epsilon\right)$$

and

$$\lim_{B \to \infty} \frac{1}{B} \log \Pr(q(\infty) \geq B) \geq -\left(tI\left(\frac{1}{t}\right) + t\epsilon\right).$$

As $\epsilon > 0$ can be arbitrarily small, we obtain the desired result. □

Now define $\delta = \inf_{t > 0} tI(1/t)$. From Lemma 10.4.2, we have

$$\lim_{B \to \infty} \frac{1}{B} \log \Pr(q(0) \geq B) \geq -\delta.$$

Note that we already proved

$$\lim_{B \to \infty} \frac{1}{B} \log \Pr(q(0) \geq B) \leq -\theta^*,$$

so the theorem holds if $-\delta \geq -\theta^*$, which we will prove next.

Proof The proof consists of three steps.

Step 1 We first prove that θ^* must satisfy

$$\Lambda_a(\theta^*) = c\theta^*.$$

Note that $\Lambda_a(\theta)$ is convex because, according to Holder's inequality, for any $0 \leq \alpha \leq 1$,

$$E\left[e^{(\alpha\theta_1+(1-\alpha)\theta_2)a}\right] = E\left[e^{\alpha\theta_1 a}e^{(1-\alpha)\theta_2 a}\right] \leq \left(E\left[e^{\theta_1 a}\right]\right)^{\alpha}\left(E\left[e^{(1-\alpha)\theta_2 a}\right]\right)^{1-\alpha},$$

which implies that

$$\Lambda_a\left(\alpha\theta_1 + (1-\alpha)\theta_2\right) \leq \alpha\Lambda_a(\theta_1) + (1-\alpha)\Lambda_a(\theta_2).$$

We further note the following three facts.

(i) Because $\Lambda_a(0) = \log E\left[e^{0\times a}\right] = 0$, we have

$$\Lambda_a(\theta) - \theta c = 0, \qquad \text{when } \theta = 0. \tag{10.19}$$

(ii) Because $E[a] < c$,

$$\left.\frac{d}{d\theta}\left(\Lambda_a(\theta) - \theta c\right)\right|_{\theta=0} = E[a] - c < 0. \tag{10.20}$$

(iii) The buffer cannot overflow unless a takes some value, say A, such that $A > c$ and $\Pr(a(t) \geq A) > 0$. Note

$$E\left[e^{\theta a}\right] \geq E\left[e^{\theta a}\mathbb{I}_{a\geq A}\right] \geq e^{\theta A}\Pr(a \geq A),$$

so we have

$$\Lambda_a(\theta) - \theta c \geq A\theta + \log\Pr(a \geq A) - \theta c \tag{10.21}$$

$$= \theta(A - c) + \log\Pr(a \geq A) \tag{10.22}$$

$$\to \infty \quad \text{as} \quad \theta \to \infty. \tag{10.23}$$

Based on the observations above, and the fact that $\Lambda_a(\theta)$ is convex, $\Lambda_a(\theta) - c\theta$ must look like Figure 10.6. Therefore, we conclude that θ^* satisfies

$$\Lambda_a(\theta^*) = c\theta^*.$$

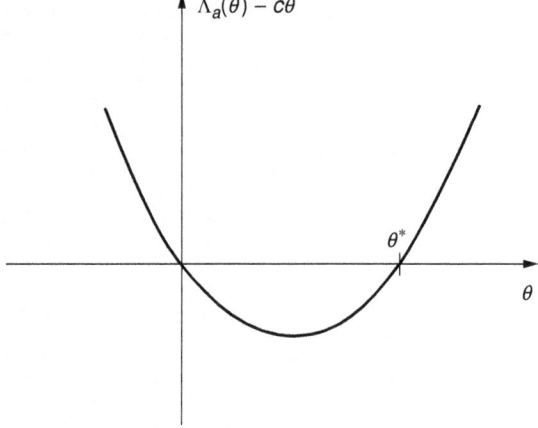

Figure 10.6 The plot of $\Lambda_a(\theta) - c\theta$.

Step 2 By Lemma 10.3.4, the moment generating function of $a(t) - c$, denoted by $\Lambda(\theta)$, satisfies

$$\Lambda(\theta) = \Lambda_a(\theta) - c\theta = \sup_x \theta x - I(x).$$

So $\Lambda_a(\theta^*) = c\theta^*$ implies

$$\sup_x \theta^* x - I(x) = 0.$$

Thus, θ^* is the largest slope of the line such that the line is below $I(x)$.

Step 3 Note that

$$\delta = \inf_{t>0} tI\left(\frac{1}{t}\right) = \inf_{x>0} \frac{I(x)}{x},$$

so δ is the largest number such that $\delta \leq I(x)/x$ for all $x > 0$ (or $\delta x - I(x) \leq 0$ for all $x > 0$). Note that $I(x) \geq 0$, so $\delta x - I(x) \leq 0$ for $x \leq 0$. Therefore, δ is the largest number such that

$$\delta x - I(x) \leq 0, \qquad \forall x.$$

Pictorially, δ is the slope of the line such that δx is tangential to $I(x)$, as shown in Figure 10.7. Therefore, we have $\theta^* = \delta$, according to the figure. To draw the figure, we used the following properties shown in the notes after the Cramer–Chernoff theorem: (a) $I(x) \geq I(\lambda - c) = 0$, and (b) $I(x)$ is convex.

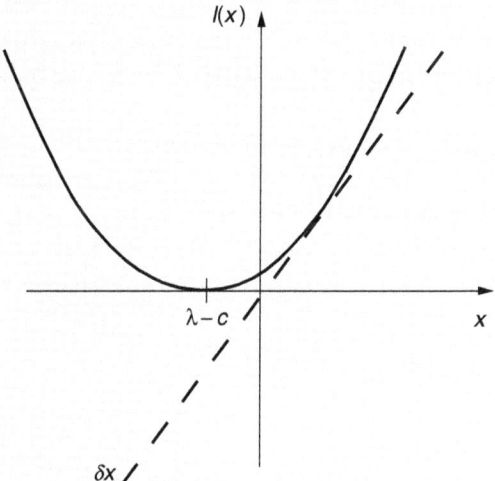

Figure 10.7 The figure depicts δx and $I(x)$.

We can also rigorously prove $\theta^* \geq \delta$ by contradiction. Suppose that $\theta^* < \delta$. In other words, suppose that there exists $\epsilon > 0$ such that $\theta^* + \epsilon = \delta$. Then, according to the definition of δ, we have

$$(\theta^* + \epsilon)x - I(x) \leq 0, \qquad \forall x,$$

which implies that

$$\theta^* x - I(x) \leq -\epsilon x, \qquad \forall x.$$

Recall from Figure 10.6 that

$$\sup_x \theta^* x - I(x) = 0.$$

Therefore, (i) the supremum is achieved at $x = 0$ and (ii)

$$I(0) = \sup_\theta \theta c - \Lambda_a(\theta) = 0,$$

which contradicts the facts that $\Lambda_a(0) = 0$ and

$$\frac{d}{d\theta} \theta c - (\Lambda_a(\theta)) \bigg|_{\theta=0} = c - E[a] > 0;$$

see Figure 10.6.

Therefore, we conclude that $\theta^* \geq \delta = \inf(I(x)/x)$, and the theorem holds. $\qquad \square$

10.5   Many-sources large deviations

Roughly speaking, the large-buffer large-deviations result says that the probability of overflow can be approximated (up to logarithmic equivalence) by $e^{-\theta^* B}$ for some $\theta^* > 0$. This estimate of the overflow probability approaches 1 as B approaches zero. However, in a real network, the probability that overflow occurs is not equal to 1 when the buffer size is very small.

To see this, we consider a single-server queue with n sources. Let $a_i(t)$ be the number of bits generated by source i at time t. Assume that the sources are stationary and i.i.d. Denote by $\tilde{c}$ the capacity of the link accessed by these sources, and assume $\tilde{c}$ scales with n such that $\tilde{c} = nc$ for some constant c and $c > E[a_i(t)]$. Then, from the Cramer–Chernoff theorem, the steady-state probability of overflow when there is *no buffer* is given by

$$\Pr\left(\sum_{i=1}^{n} a_i(t) \geq nc\right) \approx e^{-nI_a(c)},$$

where

$$I_a(c) = \sup_\theta \theta c - \log E\left[e^{\theta a_i(t)}\right].$$

Thus, the probability of overflow even when there does no buffer does not equal 1, but it is rather small due to the fact that there are many sources accessing the link. Thus, the chance of the sum of the arrival rates exceeding their mean is a rare event. The large-buffer large-deviations result captures the fact that overflows are rare due to the presence of a buffer that absorbs (short-lived) fluctuations in the arrival process, which results in the arrival rate exceeding the capacity of the link. However, it does not capture the fact that, if the capacity is larger than the sum of the mean arrival rates of the sources, this capacity itself could lead

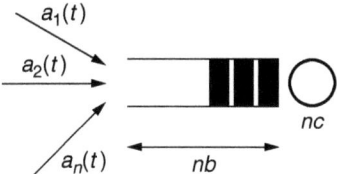

Figure 10.8 Single-server queue shared by n sources.

to small overflow probabilities, as we have seen in Chapter 3. In this section, we develop large-deviations results to capture the effects of both the zero-buffer *statistical multiplexing* gain and the presence of a large buffer, i.e., we consider the case in which both the number of sources and the buffer size are large.

Consider a single-server queue shared by n sources, as shown in Figure 10.8, along with a buffer of size nb. From the earlier discussion, the probability of overflow is given by

$$\Pr(q(\infty) \geq nb) = \Pr\left(\max_{k \geq 0} A(k) - nck \geq nb\right),$$

where

$$A(k) = \sum_{t=1}^{k} \sum_{i=1}^{n} a_i(t).$$

The following theorem characterizes the probability of overflow when $n \to \infty$.

Theorem 10.5.1

$$\lim_{n \to \infty} \frac{1}{n} \log \Pr(q(\infty) \geqslant nb) = -\min_{k \geq 1} k I_a\left(\frac{b + kc}{k}\right),$$

where $I_a(\cdot)$ is the rate function associated with random variable $a_i(t)$.

Proof Note that, for any $k \geq 1$, we have

$$\Pr(q(\infty) \geq nb) \geq \Pr(A(k) - nkc \geqslant nb)$$

$$= \Pr(A(k) \geqslant n(b + kc))$$

$$= \Pr\left(\frac{A(k)}{n} \geqslant b + kc\right).$$

Define

$$z_i(k) = \sum_{t=1}^{k} a_i(t),$$

then

$$A(k) = \sum_{i=1}^{n} z_i(k).$$

Defining $\Lambda_k(\theta)$ to be the log moment generating function of $z_i(k)$,

$$\Lambda_k(\theta) = \log E\left[e^{\theta \sum_{t=1}^{k} a_1(t)}\right]$$

$$=_{(a)} \log \left(E\left[e^{\theta a_1(t)}\right]\right)^k$$

$$= k \log E\left[e^{\theta a_1(t)}\right]$$

$$= k \Lambda_a(\theta),$$

where equality (a) holds because $a_1(t)$ are i.i.d. across time. Therefore, we have

$$I_k(x) = \sup_\theta \theta x - k \Lambda_a(\theta)$$

$$= k \sup_\theta \theta \left(\frac{x}{k}\right) - \Lambda_a(\theta)$$

$$= k I_a\left(\frac{x}{k}\right).$$

Then, by the Cramer–Chernoff theorem,

$$\Pr(q(\infty) \geq nb) \geq \Pr\left(\frac{A(k)}{n} \geqslant b + kc\right) \geq e^{-n(I_k(b+kc)+\epsilon)} = e^{-n\left(kI_a\left(\frac{b+kc}{k}\right)+\epsilon\right)}$$

for some large n (given $\epsilon > 0$). Thus,

$$\lim_{n\to\infty} \frac{1}{n} \log \Pr(q(\infty) \geq nb) \geq -kI_a\left(\frac{b+kc}{k}\right).$$

The inequality holds because the lower bound holds for any $k > 0$. Note that we have used min instead of inf because the inf is achievable. To see this, note that

$$\lim_{k\to\infty} kI_a\left(\frac{b}{k}+c\right) \approx \lim_{k\to\infty} kI_a(c) \to \infty.$$

Thus, the infimum is achieved at some finite k.

Now we prove this bound is tight. Based on the union bound and the Chernoff bound, we have

$$\Pr(q(\infty) \geq nb) \leq \sum_{k=1}^{\infty} \Pr\left(\frac{\sum_{i=1}^{n} z_i(k)}{n} \geq b + kc\right)$$

$$\leq \sum_{k=1}^{\infty} e^{-nI_k(b+kc)}$$

$$\leq \sum_{k=1}^{\infty} e^{-nkI_a\left(\frac{b}{k}+c\right)}.$$

We consider the rate function $I_k(\cdot)$. From the definition of the rate function, for any θ,

$$I_k(b+kc) \geq \theta(b+kc) - \Lambda_k(\theta)$$

$$= \theta b + \theta k c - k \Lambda_a(\theta)$$
$$= \theta b + k(\theta c - \Lambda_a(\theta)).$$

For any $\theta > 0$ such that $\theta c > \Lambda_a(\theta)$ and $k_1 > 0$, based on the above, inequality we can obtain

$$\Pr(q(\infty) \geq nb) \leq \sum_{k=1}^{k_1-1} e^{-nkI_a\left(\frac{b}{k}+c\right)} + \sum_{k=k_1}^{\infty} e^{-n\theta b} e^{-kn(\theta c - \Lambda_a(\theta))}$$

$$= \sum_{k=1}^{k_1-1} e^{-nkI_a\left(\frac{b}{k}+c\right)} + \frac{e^{-n\theta b} e^{-k_1 n(\theta c - \Lambda_a(\theta))}}{1 - e^{-n(\theta c - \Lambda_a(\theta))}}$$

$$\leq \sum_{k=1}^{k_1-1} e^{-nkI_a\left(\frac{b}{k}+c\right)} + e^{-n\theta b} e^{-k_1 n(\theta c - \Lambda_a(\theta))}. \qquad (10.24)$$

Note that we have argued that the infimum of $kI_a\,(b/k + c)$ is achieved at some finite k. Therefore, given θ such that $\theta c > \Lambda_a(\theta)$, there exists k_1 such that

$$\theta b + k_1(\theta c - \Lambda_a(\theta)) > \min_{k \geq 1} kI_a\left(\frac{b}{k} + c\right).$$

Then, from inequality (10.24), we have

$$\Pr(q(\infty) \geq nb) \leq k_1 e^{-\min_{k \geq 1} kI_a\left(\frac{b}{k}+c\right)},$$

which implies

$$\lim_{n \to \infty} \frac{1}{n} \log \Pr(q(0) \geq nb) \leq -\min_{k \geq 1} kI_a\left(\frac{b}{k} + c\right). \qquad \square$$

We now define

$$J(b, c) \triangleq \min_{k \geq 1} kI_a\left(\frac{b}{k} + c\right). \qquad (10.25)$$

In general, $J(b, c)$ could be difficult to compute. The following approximation is often used:

$$J(b, c) \approx I_a(c) + b \inf_{x \geq 0} \frac{I_a(x + c)}{x}. \qquad (10.26)$$

Note that

$$I_a(x) = \sup_{\theta} \theta x - \Lambda_a(\theta)$$

is the rate function associated with random variable $a_i(t)$. Thus, $I_a(c)$ is the zero-buffer large-deviations exponent, and $\inf_{x \geq 0}(I_a(x+c)/x)$ is the large-buffer large-deviations exponent. The above approximation corrects for the large-buffer approximation by adding the exponent due to the zero-buffer statistical multiplexing gain. This approximation is justified in the following two lemmas.

Lemma 10.5.2 In the zero-buffer case,

$$J(0, c) = I_a(c).$$

Proof This follows from the definition of $J(b, c)$. □

Lemma 10.5.3 When the buffer size is large, we have

$$\lim_{b \to \infty} \frac{J(b, c)}{b} = \inf_{x \geq 0} \frac{I_a(x + c)}{x}.$$

Proof

$$\frac{J(b, c)}{b} = \min_{k \geq 1} \frac{k}{b} I_a \left(\frac{b}{k} + c \right)$$

$$= \min_{y = b, \frac{b}{2}, \frac{b}{3}, \ldots} \frac{I_a(y + c)}{y},$$

where $y = b/k$. So we have

$$\frac{J(b, c)}{b} \geq \inf_{x \geq 0} \frac{I_a(x + c)}{x}.$$

Next, we will prove that

$$\frac{J(b, c)}{b} \leq \theta^*, \tag{10.27}$$

where

$$\theta^* = \sup \{\theta > 0 : \Lambda_a(\theta) < c\theta\}.$$

Note that it has been proved in Section 10.4 that $\theta^* = \inf_{x \geq 0}(I_a(x + c)/x)$, so the proof is complete after (10.27) is established.

Given any b, let

$$k_b \triangleq \left\lceil \frac{b}{\Lambda'_a(\theta^*) - c} \right\rceil \qquad \text{and} \qquad \tilde{b} \triangleq k_b(\Lambda'_a(\theta^*) - c).$$

Note that $\Lambda'_a(\theta^*) - c > 0$ from Figure 10.6. According to the definition of $J(b, c)$, we have

$$\frac{J(b, c)}{b} \leq \frac{k_b}{b} I_a \left(\frac{b}{k_b} + c \right)$$

$$= \frac{k_b}{b} \sup_\theta \left(\frac{b}{k_b} \theta + c\theta - \Lambda_a(\theta) \right)$$

$$\leq \frac{k_b}{b} \sup_\theta \left(\frac{\tilde{b}}{k_b} \theta + c\theta - \Lambda_a(\theta) \right)$$

$$\leq_{(a)} \frac{k_b}{b} \sup_{\theta} \left(\frac{\tilde{b}}{k_b}\theta + c\theta - \Lambda_a(\theta^*) - (\theta - \theta^*)\Lambda'(\theta^*) \right)$$

$$= \frac{\tilde{b}}{b} \sup_{\theta} \left(\theta + \frac{k_b}{\tilde{b}} \left(c\theta - \Lambda_a(\theta^*) - (\theta - \theta^*)\Lambda'(\theta^*) \right) \right),$$

where inequality (a) holds because $\Lambda_a(\theta)$ is convex. According to the definition of $\tilde{b}$ and the fact that $\Lambda_a(\theta^*) = c\theta^*$, we have

$$\theta + \frac{k_b}{\tilde{b}} \left(c\theta - \Lambda_a(\theta^*) - (\theta - \theta^*)\Lambda'(\theta^*) \right) = \theta^*,$$

which implies

$$\frac{J(b,c)}{b} \leq \frac{\tilde{b}}{b}\theta^*.$$

From the definition of $\tilde{b}$, $\tilde{b}/b \to 1$ as $b \to \infty$, so

$$\frac{J(b,c)}{b} \leq \theta^*,$$

and the lemma holds. $\qquad\qquad\qquad\qquad\qquad\qquad\qquad\qquad\qquad\qquad \square$

10.6   Summary

- **nth moment bounds in heavy traffic** Consider a single-server queue. Denote by $a(t)$ the number of bits arriving to the queue at the beginning of time slot t and by $c(t)$ the maximum number of bits that can be drained from the queue during time slot t. Let $\epsilon = E[c] - E[a]$. Under the assumptions that: (i) both the arrival process $a(t)$ and the service process $c(t)$ are stationary, and are independent of $q(t)$; (ii) $c(t) \leq c_{max}$ for all t; and (iii) $\lim_{\epsilon \to 0} E[(a - c)^2] = \sigma^2$, the following limit holds for any $n \geq 1$:

$$\lim_{\epsilon \to 0} \epsilon^n E\left[\left(q^{(\epsilon)} \right)^n \right] = n! \left(\frac{\sigma^2}{2} \right)^n.$$

- **Heavy-traffic optimality of JSQ** Consider a two-server queueing system with Bernoulli arrival and service processes, as in Figure 10.2. The JSQ policy routes an arrival to the shorter of the two queues. Let $\epsilon = \mu_1 + \mu_2 - \lambda$. JSQ minimizes $\lim_{\epsilon \to 0} \epsilon E\left[q_1^{(\epsilon)} + q_1^{(\epsilon)} \right]$ among all routing policies, and, under JSQ,

$$\lim_{\epsilon \to 0} \epsilon E\left[q_1^{(\epsilon)} + q_1^{(\epsilon)} \right] = \frac{E[(a - s_1 - s_2)^2]}{2}.$$

- **Cramer–Chernoff theorem** Let $X_1, X_2, \ldots$ be i.i.d. random variables with $E[X_1] = \mu$, and suppose that their common moment generating function $M(\theta) < \infty$ for all θ in some neighborhood $\mathcal{B}_0$ of $\theta = 0$. Further suppose that the supremum in the following definition of the *rate function* $I(x)$ is obtained at some interior point in this neighborhood:

$$I(x) = \sup_{\theta} \theta x - \Lambda(\theta).$$

For each $x > \mu$,

$$\lim_{n \to \infty} \frac{1}{n} \log \Pr \left(\sum_{i=1}^{n} X_i \geq nx \right) = -I(x). \tag{10.28}$$

- **Large-buffer large-deviations** Consider a single-source, single-server queue. The buffer overflow probability $\Pr(q(0) > B)$ satisfies

$$\lim_{B \to \infty} \frac{1}{B} \log \Pr(q(0) \geq B) = -\theta^*,$$

where

$$\theta^* = \sup \{\theta > 0 : \Lambda_a(\theta) < c\theta\}.$$

Suppose we impose a QoS requirement that $\theta^* \geq \theta$, then a necessary and sufficient condition on the capacity to meet this QoS requirement is

$$\frac{\Lambda_a(\theta)}{\theta} < c.$$

Because the minimum bandwidth needed to meet the QoS requirement of the source is $\Lambda_a(\theta)/\theta$, this is called the effective bandwidth of the source.

- **Many-sources large deviations** Consider a single-server queue with n sources. Let $a_i(t)$ be the number of bits generated by source i at time t. Assume that the sources are stationary and i.i.d. Denote by $\tilde{c}$ the capacity of the link accessed by these sources, and assume $\tilde{c}$ scales with n such that $\tilde{c} = nc$ for some constant c, and $c > E[a_i(t)]$. The following result holds:

$$\lim_{n \to \infty} \frac{1}{n} \log \Pr(q(0) \geq nb) = -\min_{k>0} kI_a \left(\frac{b + kc}{k} \right),$$

where $I_a(\cdot)$ is the rate function associated with random variable $a_i(t)$.

Define

$$J(b,c) \triangleq \min_{k \geq 1} kI_a \left(\frac{b}{k} + c \right).$$

The following approximation is often used:

$$J(b,c) \approx I_a(c) + b \inf_{x \geq 0} \frac{I_a(x + c)}{x}.$$

Thus, $I_a(c)$ is the zero-buffer large-deviations exponent and $\inf_{x \geq 0}(I_a(x + c)/x)$ is the large-buffer large-deviations exponent.

10.7   Exercises

Exercise 10.1 (The nth moment of the GI/GI/1 queue) Consider the GI/GI/1 queue studied in Section 10.1. Prove that, in the heavy-traffic regime, the nth moment of $q(t)$ satisfies

$$\lim_{\epsilon \to 0} \epsilon^n E \left[\left(q^{(\epsilon)} \right)^n \right] = n! \left(\frac{\sigma^2}{2} \right)^n.$$

Exercise 10.2 (State-space collapse with Bernoulli arrivals and service I) Consider the two-server queue in Section 10.2 and JSQ routing. Prove that $E[|q_2 - q_1|]$ is bounded by a positive constant independent of ϵ using the Lyapunov function $V(t) = (q_1(t) - q_2(t))^2$.

Exercise 10.3 (Moment bounds from drift inequalities) Let $Y_0, Y_1, Y_2, Y_3, \ldots$ be a sequence of random vectors, each of which takes values in a countable set $\mathcal{S}$. Let $V(y) \geq 0$ be a function such that

$$E[V(Y_{k+1}) - V(Y_k)|\mathcal{F}_k] \leq -\delta, \qquad \text{if } V(Y_k) \geq C,$$

for some $C < \infty$, where $\mathcal{F}_k = \{Y_1, Y_2, \ldots, Y_k\}$. Further suppose that there exists a constant M such that

$$|V(Y_{k+1}) - V(Y_k)| \leq M, \qquad \forall k.$$

More precisely,

$$\Pr\left(|V(Y_{k+1}) - V(Y_k)| \leq M, \qquad \forall k\right) = 1.$$

(1) Show that, given $\theta \in (0,1)$, we have

$$E\left[e^{\theta(V(Y_{k+1})-V(Y_k))}\mathbb{I}_{V(Y_k)\geq C}\,\Big|\,\mathcal{F}_k\right] \leq 1 - \theta\delta + \theta^2(e^M - 1 - M).$$

(2) Show that there exists $\theta_0 > 0$ such that, for any $\theta \in (0, \theta_0]$,

$$\limsup_{k\to\infty} E\left[e^{\theta V(Y_k)}\right] \leq \frac{e^{\theta(M+C)}}{1 - \eta_\theta},$$

where $\eta_\theta = 1 - \theta\delta + \theta^2(e^M - 1 - M)$. Hint: Consider

$$E\left[e^{\theta V(Y_k)}\,\Big|\,\mathcal{F}_k\right] = E\left[e^{\theta(V(Y_{k+1})-V(Y_k))+\theta V(Y_k)}\,\Big|\,\mathcal{F}_k\right].$$

Exercise 10.4 (State space collapse with Bernoulli arrivals and service II) Consider the two-server queue model with JSQ routing in Section 10.2. Let

$$Y_k = \begin{pmatrix} q_1(k) \\ q_2(k) \end{pmatrix} \qquad \text{and} \qquad V(Y_k) = |q_1(k) - q_2(k)|.$$

Use the result from Exercise 10.3 to show that there exist positive constants ϵ_0, θ, and M such that

$$\limsup_{k\to\infty} E\left[e^{\theta V(Y_k)}\right] \leq M < \infty, \qquad \forall \epsilon < \epsilon_0,$$

where θ and M do not depend on ϵ.

Exercise 10.5 (State-space collapse with general arrivals and departures) Consider the two-server queue model with JSQ routing in Section 10.2, but modifying the arrival and service processes as follows. Arrivals have a mean λ and variance σ_a^2, and the number of packets that queue i can serve in one time slot is a random variable with mean μ_i and variance σ_i^2. Further, assume that the maximum number of arrivals in a time slot is upper bounded by $A_{\max}$, and that the number of packets that can be served by each queue is upper bounded by $S_{\max}$. The rest of the model is identical to Section 10.2: arrivals and departures are independent of each other and independent across time slots, and the arrival and service processes are identical across time slots. Show that there exist positive constants ϵ_0, θ, and M such that

$$\lim_{k\to\infty} E\left[e^{\theta|q_1(k)-q_2(k)|}\right] \leq M < \infty, \qquad \forall \epsilon \leq \epsilon_0,$$

where θ and M do not depend on ϵ.

Exercise 10.6 (Lower bound with general arrivals and departures) Consider the model in Exercise 10.5. Let $q_1^{(\epsilon)}(k)$ and $q_2^{(\epsilon)}(k)$ be the queue lengths at queues 1 and 2, respectively, in time slot k. Show that

$$E\left[q_1^{(\epsilon)}(\infty) + q_2^{(\epsilon)}(\infty)\right] \geq \frac{\sigma^2}{2(\mu_1 + \mu_2 - \lambda)} - s_{\max},$$

where $q_i^{(\epsilon)}(\infty)$ denotes the steady-state queue length at queue i, and $\sigma^2 = E\left[(a(k) - s_1(k) - s_2(k))^2\right]$. Hint: The second question of Exercise 3.13 suggests a lower bound derivation for a single-server queue.

Exercise 10.7 (Heavy-traffic optimality with general arrivals and departures) Consider the model in Exercise 10.5. Show that JSQ is heavy-traffic optimal in the sense that it minimizes

$$\lim_{\epsilon \to 0} \epsilon E\left[q_1^{(\epsilon)}(\infty) + q_2^{(\epsilon)}(\infty)\right].$$

Exercise 10.8 (The nth moment in the heavy-traffic regime) Prove that, under JSQ,

$$\lim_{\epsilon \to 0} \epsilon^n E\left[\left(q_1^{(\epsilon)}(\infty)\right)^n + \left(q_2^{(\epsilon)}(\infty)\right)^n\right] = \left(\frac{\sigma^2}{2}\right)^n n!.$$

Exercise 10.9 (Rate functions) Let X be a random variable. Recall that the rate function of X is

$$I(x) := \sup_s sx - \log M(s),$$

where $M(s) := E[e^{sX}]$. Compute the rate function of each of the following random variables:

(1) X is Gaussian with mean μ and variance σ^2;
(2) X is exponential with mean $1/\lambda$;
(3) X is Poisson with mean λ;
(4) $X = 1$ with probability p and $X = 0$ with probability $1 - p$.

Exercise 10.10 (Burstiness and the Chernoff bound) Suppose that there are N i.i.d. sources accessing a bufferless router with capacity Nc. Let X_i denote the rate at which source i transmits data. Assume that the peak rate of each source is equal to M, and that the mean rate is less than or equal to ρ. Show that the distribution of X_i that maximizes the Chernoff bound estimate of the overflow probability is given by

$$X_i = \begin{cases} M, & \text{with probability} \quad q, \\ 0, & \text{with probability} \quad 1 - q, \end{cases}$$

where $q = \rho/M$. Hint: Maximizing the Chernoff bound estimate of the overflow probability is the same as minimizing the rate function $I_X(c)$ over all distributions such that $X \leq M$ with probability 1 and $E[X] \leq \rho$.

Exercise 10.11 (Many-sources large-deviations example) Consider the many-sources large-deviations model in Section 10.5. Assume that each source generates one packet with probability 0.7 in each time slot. Assume the packet arrival processes are i.i.d. across time slots and sources. Let the link capacity per source $c = 0.9$. Numerically compute the many-sources large-deviations exponent $J(b, c)$ given in (10.25), and plot it as a function of b. On the same graph, plot the approximation to $J(b, c)$ given in (10.26).

Exercise 10.12 (Effective bandwidth) The definition of effective bandwidth for discrete-time sources can be extended to cover continuous-time sources. Let $a(t)$ be a stationary process describing the rate (in bps) at which a source is generating data at time t. The log moment generating function of this source is defined as

$$\Lambda_a(\theta) = \lim_{T \to \infty} \frac{1}{T} \log E\left[e^{\theta \int_0^T a(t)dt}\right].$$

The effective bandwidth is given by $\Lambda_a(\theta)/\theta$.

Consider a continuous-time source that switches between ON and OFF states; it generates data at rate 1 when ON, and does not generate any data when OFF. The ON and OFF times are independent, exponentially distributed, with an average ON time of $1/q_0$ and an average OFF time of $1/q_1$. Show that the effective bandwidth of this source is given by

$$\frac{1}{2\theta}\left(\theta - q_1 - q_0 + \sqrt{(\theta + q_1 - q_0)^2 + 4q_1q_0}\right).$$

Hint: Define

$$M_t^u(\theta) := E\left[e^{\theta A(0,t)}|x(0) = \text{ON}\right] \quad \text{and} \quad M_t^d(\theta) := E\left[e^{\theta A(0,t)}|x(0) = \text{OFF}\right],$$

where $A(0,t) = \int_0^t a(s)ds$. Show that

$$\frac{dM_t^d}{dt} = -q_1 M_t^d(\theta) + q_1 M_t^u(\theta)$$

and

$$\frac{dM_t^u}{dt} = (-q_0 + \theta)M_t^u(\theta) + q_0 M_t^d(\theta).$$

Exercise 10.13 (Link with time-varying capacity) Consider calls accessing a link with time-varying capacity. The call holding times are independent and exponentially distributed with mean $1/\mu$, and each call requires one unit of bandwidth from the link. The link has an available capacity of nc_0 in the time interval $[0, \tau_1]$, an available capacity of nc_1 in $[\tau_1, \tau_2]$, and an available capacity of nc_2 in $[\tau_2, \infty)$, where $\tau_2 > \tau_1 > 0$. We assume that $nc_0 > nc_1 > nc_2$ and $n > 0$. Suppose that there are $n\alpha$ calls in progress at time $t = 0$ (assume $\alpha \in (0, c_0)$ is an integer), and that no further calls ever arrive at the system. Let $N(t)$ denote the number of calls in progress at time t. Thus, the bandwidth requested by the calls at time t is $N(t)$. We are interested in estimating the probability that the required bandwidth ever exceeds the available capacity, up to a logarithmic equivalence. To this end, prove the following fact for appropriate functions $I_i(x)$:

$$\lim_{n \to \infty} \frac{1}{n} \log \Pr\left(\text{there exists a } t \in [0, \infty) \text{ such that } N(t) \geq c(t)\right) = - \min_{1,2} I_i(c_i),$$

where $c(t)$ is the available capacity at time t. Explicitly compute the functions $I_i(x)$.

10.8   Notes

Heavy-traffic performance analysis of queues was originated in [87] using the Lyapunov-drift approach and using a Brownian approximation in [86]. The Lyapunov approach presented here follows [39], which also provides a comprehensive survey of prior work on establishing heavy-traffic optimality of various control policies for queueing networks. The bound on the exponential moments in steady state from the Lyapunov-drift condition was obtained in [52]. Although we did not study the diffusion-limit approach to heavy-traffic analysis, the interested reader can find a good introduction to this approach in [23].

Large deviations is the subject of many books; see [20, 43, 148]. The large-deviations analysis of queues presented in this chapter follows the ideas in many papers on effective bandwidths mentioned in Chapter 3. In particular, our treatment is influenced by the presentation in [27, 34].

11 Geometric random graph models of wireless networks

In earlier chapters, we considered distributed resource allocation algorithms for ad hoc wireless networks. We showed that there exist algorithms, called throughput-optimal algorithms, that maximize the network throughput. However, we do not yet have a simple expression for the total amount of information that can be transferred in an ad hoc wireless network. The problem of computing such an expression is challenging due to interference caused by simultaneous wireless transmissions. This chapter focuses on understanding the limitations on network throughput imposed by wireless interference. We will introduce geometric random graph models of wireless networks, in which n wireless nodes are distributed over a geographical area. Two nodes can communicate if the distance between them is smaller than a threshold and if no nodes near the receiver are transmitting at the same time. The capacity region of such a network is difficult to characterize exactly. Therefore, we will study the network capacity when the number of wireless nodes becomes large. We will show that, in such an asymptotic regime, one can obtain expressions for the network capacity that are asymptotically correct in a sense that will be made precise. A key insight of this chapter is that the throughput per source diminishes as the number of wireless nodes (in a fixed geographical area) increases because of wireless interference. The following questions will be addressed in this chapter.

(1) *What is the maximum achievable throughput in a network of n wireless nodes distributed over a geographical area, and how does the maximum throughput scale with the number of wireless nodes?*
(2) *What algorithms should be used to achieve the maximum throughput?*

11.1 Mathematical background: the Hoeffding bound

Recall the Chernoff bound for i.i.d. random variables $\{X_i\}$:

$$\Pr\left(\frac{1}{n}\sum_{i=1}^{n} X_i \geq x\right) \leq e^{-nI(x)},$$

where $I(\cdot)$ is the rate function of X_i. Often, $I(\cdot)$ is in a form that is not tractable for further computation. The Hoeffding bound instead provides a simple upper bound of the form e^{-ncx^2} for some constant c under the additional assumption that the X_i's are bounded. In fact, we will derive the Hoeffding bound for random variables that are independent but not identically distributed.

Theorem 11.1.1 (Hoeffding bound) Consider a sequence of n independent random variables $\{X_i\}$. Further, random variable X_i takes values in $[a_i, b_i]$. Then the following inequality holds:

$$\Pr\left(\left|\frac{\sum_i X_i - E\left[\sum_i X_i\right]}{n}\right| \geq x\right) \leq 2\exp\left(-\frac{2n^2 x^2}{\sum_{i=1}^{n}(b_i - a_i)^2}\right).$$

Proof We will prove the Hoeffding bound for the case when the random variables are i.i.d. Bernoulli. We leave the proof of the Hoeffding bound for general random variables as an exercise (see Exercise 11.3).

Let $\{X_i\}$ be i.i.d. Bernoulli random variables with parameter p. First, according to the Markov inequality, we have, for $\theta > 0$,

$$\Pr\left(\frac{1}{n}\sum_{i=1}^{n}X_i \geq p + x\right) \leq \frac{E\left[e^{\theta \sum_{i=1}^{n}X_i}\right]}{e^{\theta(p+x)n}}$$

$$= \left(E\left[e^{\theta X_1}\right]\right)^n e^{-n\theta(p+x)}$$

$$= \left(e^\theta p + (1-p)\right)^n e^{-n\theta(p+x)}$$

$$= e^{n\log(pe^\theta + 1 - p)} e^{-n\theta(p+x)}.$$

The basic idea behind the Hoeffding bound is to bound $\log\left(pe^\theta + 1 - p\right)$ using Taylor's theorem. Let

$$f(\theta) = \log\left(pe^\theta + 1 - p\right).$$

According to Taylor's theorem, we have

$$f(\theta) = f(0) + f'(0)\theta + \frac{1}{2}f''(u)\theta^2$$

for some $u \in [0, \theta]$.

Note that $f(0) = 0$. Also

$$f'(\theta) = \frac{pe^\theta}{pe^\theta + 1 - p},$$

so $f'(0) = p$. Furthermore,

$$f''(\theta) = \frac{\left(pe^\theta + 1 - p\right)pe^\theta - \left(pe^\theta\right)^2}{\left(pe^\theta + 1 - p\right)^2}$$

$$= \frac{pe^\theta(1-p)}{\left(pe^\theta + 1 - p\right)^2}$$

$$= \left(\frac{pe^\theta}{pe^\theta + 1 - p}\right)\left(\frac{1-p}{pe^\theta + 1 - p}\right),$$

which is of the form $y(1-y)$ for $y \in [0,1]$. Since $y(1-y) \leq 1/4$ for $y \in [0,1]$, we have

$$f'' \leq \frac{1}{4}.$$

Thus, we obtain that

$$f(\theta) \leq p\theta + \frac{1}{8}\theta^2,$$

which implies that

$$\Pr\left(\frac{1}{n}\sum_{i=1}^{n} X_i \geq p + x\right) \leq e^{-n(p+x)\theta} e^{n\left(p\theta + \frac{\theta^2}{8}\right)}$$

$$= e^{-nx\theta + \frac{1}{8}n\theta^2}.$$

The tightest bound is obtained by minimizing the exponent on the right-hand side as a function of θ, i.e.,

$$\Pr\left(\frac{1}{n}\sum_{i=1}^{n} X_i \geq p + x\right) \leq e^{\min_{\theta > 0}\left(-nx\theta + \frac{1}{8}n\theta^2\right)}.$$

The minimization can be carried out using the first-order condition for a minimum, to yield

$$-nx + \frac{2n\theta}{8} = 0,$$

which results in $\theta = 4x$. By substituting $\theta = 4x$, we obtain

$$\Pr\left(\frac{1}{n}\sum_{i=1}^{n} X_i \geq p + x\right) \leq e^{-n4x^2 + \frac{1}{8}n(4x)^2} = e^{-2nx^2}.$$

Similarly, we can obtain that

$$\Pr\left(\frac{1}{n}\sum_{i=1}^{n} X_i \leq p - x\right) \leq e^{-2nx^2}.$$

Thus,

$$\Pr\left(\left|\frac{1}{n}\sum_{i=1}^{n} X_i - p\right| \geq x\right) \leq 2e^{-2nx^2}. \qquad \square$$

11.2   Nodes arbitrarily distributed in a unit square

Consider n wireless nodes on a unit square, as shown in Figure 11.1. Each node has a radius of communication r, i.e., node a can send data to node b if node b is within r meters of node a. We will assume a very simple interference model.

Definition 11.2.1 (Interference model) Node b can successfully receive data from node a if there are no other transmitters within a distance $r(1 + \Delta)$ of node b, as shown in Figure 11.2.

We will assume that r is a parameter that can be controlled (by controlling the transmit power) to maximize network throughput. Note that, if the radius r is large, a node can communicate with a node farther away. However, by our interference assumption, this also means that a larger area around the receiver must be free of other transmitters for reception to be successful.

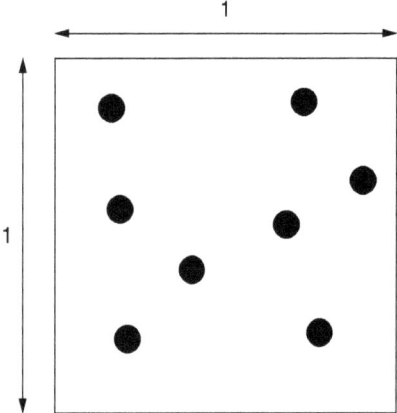

Figure 11.1 n wireless nodes are distributed on a unit square.

We further assume the following communication requirements in the network. For simplicity, assume n is an even number, and let $n/2$ nodes be sources of data and $n/2$ nodes be destinations of data. For convenience, we will name the sources $1, 2, \ldots, n/2$ and the destinations $n/2 + 1, n/2 + 2, \ldots, n$, and associate source i with destination $n/2 + i$. Data transfer from a source to its destination can occur in a multi-hop fashion: each packet can go through multiple intermediate nodes, called relays, to reach its destination. All nodes in the network can act as relays. Assume that time is slotted and that one packet can be transmitted from a transmitter to a receiver in each time slot if the interference condition stated earlier is satisfied. Each source wants to transmit λ packets/slot on average. Our first goal is to compute an upper bound on λ.

Theorem 11.2.1 Let L_i denote the distance between source i and its destination, and let $L = (2/n) \sum_{i=1}^{n/2} L_i$ be the average distance between source-destination pairs. The per-node throughput λ satisfies

$$\lambda \leq \frac{4\sqrt{2}}{\sqrt{\pi}\,\Delta L} \frac{1}{\sqrt{n}}. \tag{11.1}$$

Proof To obtain the upper bound (11.1), we first derive an upper bound on the number of transmissions that can be simultaneously supported in the network. Consider two simultaneous transmissions: node i is transmitting to node j and node a is transmitting to node b. Note that, according to the interference model, node a, which is a transmitter, must be at least $r(1 + \Delta)$ meters from node j; similarly, node i must be at least $r(1 + \Delta)$ meters from node b. Thus, any two receivers must be Δr meters from each other, as shown in Figure 11.3.

exclusion region around receiver b

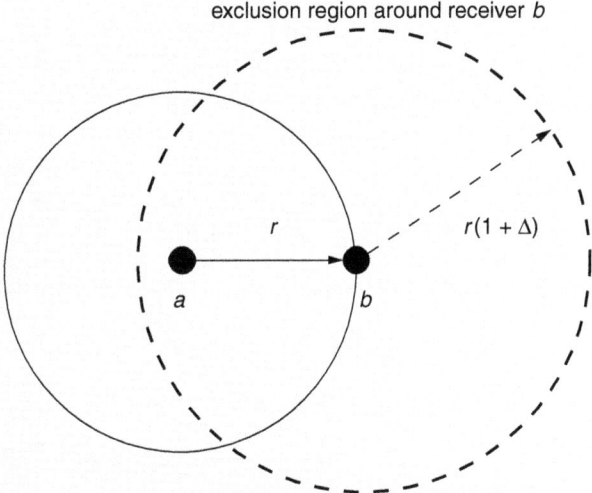

Figure 11.2 Node b can successfully receive data from node a if the two nodes are within r meters of each other and there are no other transmitters within a distance $r(1 + \Delta)$ meters of node b.

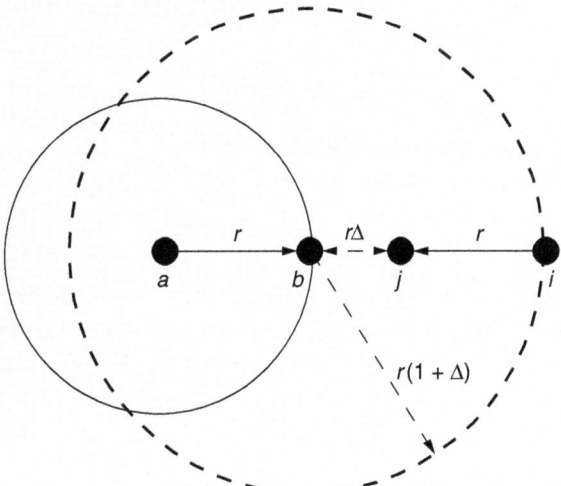

Figure 11.3 The receivers of two simultaneous transmissions must be Δr meters apart.

Each receiver can be thought of as having a *footprint* of radius $\Delta r/2$ such that no two receivers' footprints overlap. The area of the footprint must be at least $\frac{\pi}{4} (\Delta r/2)^2$, even if the receiver is at a corner of the unit square. Let $N(t)$ be the number of simultaneous transmissions in time slot t. Then $N(t)$ satisfies

$$N(t)\frac{\pi}{4}\left(\frac{\Delta r}{2}\right)^2 \leq 1,$$

which is equivalent to

$$r \leq \sqrt{\frac{16}{\pi \Delta^2 N(t)}}. \tag{11.2}$$

If each source-destination pair can communicate λ packets/slot, then, in T time slots, the number of packets to be delivered to their destinations is $\lambda nT/2$ packets. Each packet traverses L meters on average, so the traffic load, in packet-meters, is $\lambda nTL/2$. Because each transmission can cover at most r meters, a necessary condition for λ to be supportable is

$$\lambda \frac{nTL}{2} \leq \sum_{t=1}^{T} rN(t) \leq \frac{4}{\sqrt{\pi}\Delta}\sum_{t=1}^{T}\sqrt{N(t)},$$

where the second inequality arises from inequality (11.2). Dividing both sides by $nTL/2$, we obtain

$$\lambda \leq \frac{8}{\sqrt{\pi}\Delta nL}\frac{1}{T}\sum_{t=1}^{T}\sqrt{N(t)}$$

$$\leq \frac{8}{\sqrt{\pi}\Delta nL}\sqrt{\frac{1}{T}\sum_{t=1}^{T}N(t)} \tag{11.3}$$

$$\leq \frac{8}{\sqrt{\pi}\Delta nL}\sqrt{\frac{n}{2}} \tag{11.4}$$

$$= \frac{4\sqrt{2}}{\sqrt{\pi}\Delta L}\frac{1}{\sqrt{n}},$$

where Jensen's inequality is applied to the function $\sqrt{N}$ in (11.3), and (11.4) holds since there can be at most $n/2$ transmissions/slot because a node can either receive or transmit, but not both.   □

Thus, we have shown that $\lambda = O\left(1/\sqrt{n}\right)$, independently of how nodes are placed in a unit square, as long as the average distance L between the source-destination pairs is independent of n. Next, we will show that this throughput can be nearly achieved when nodes are placed randomly in the unit square.

11.3   Random node placement

Consider n nodes randomly deployed on a unit square. Each node's position is picked uniformly on the unit square and is independent of the positions of the other nodes. As before, we will assume that $n/2$ of the nodes are sources and the other $n/2$ are destinations, with each destination node assigned to a unique source and each source associated with a unique destination.

For this model, we will first prove that, with high probability (w.h.p.), L is a constant independent of n, so the throughput per node is $O(1/\sqrt{n})$ according to Theorem 11.1. Then we will present a strategy that achieves a per-node throughput of the order of $1/\sqrt{n \log n}$, which is only a factor of $1/\sqrt{\log n}$ different from the upper bound.

Theorem 11.3.1 In the random node placement model, the average distance between the source-destination pairs satisfies

$$\Pr(|L - K| \geq \epsilon) \leq 2e^{-\frac{n\epsilon^2}{2}},$$

where

$$K = \int_0^1 \int_0^1 \int_0^1 \int_0^1 \sqrt{(x_1 - x_2)^2 + (y_1 - y_2)^2} \; dx_1 \, dx_2 \, dy_1 \, dy_2$$

is a constant independent of n. Therefore, w.h.p., we have the following upper bound on the throughput per source:

$$\frac{4\sqrt{2}}{\sqrt{\pi} \Delta (K - \epsilon)} \frac{1}{\sqrt{n}}. \tag{11.5}$$

Proof Recall that L_i denotes the distance between source i and its destination and that $L = (2/n) \sum_{i=1}^{n/2} L_i$. We will first show that $L \sim O(1)$ in an appropriate probabilistic sense. Let (X_i, Y_i) be the position of node i. Note that X_i and Y_i are independent and uniform on $[0, 1]$. The distance between source i and its destination is given by

$$L_i = \sqrt{\left(X_i - X_{i+\frac{n}{2}}\right)^2 + \left(Y_i - Y_{i+\frac{n}{2}}\right)^2},$$

so the expected distance is given by

$$E[L_i] = \int_0^1 \int_0^1 \int_0^1 \int_0^1 \sqrt{(x_1 - x_2)^2 + (y_1 - y_2)^2} \; dx_1 \, dx_2 \, dy_1 \, dy_2,$$
$$= K,$$

where K is a constant independent of n. Since $E[L_i]$ is the same for all i, we have

$$E[L] = K,$$

which is also independent of n.

Note that $L_i \in [0, \sqrt{2}]$, so, by the Hoeffding bound, we obtain

$$\Pr\left(\left|2/n \sum_{i=1}^{n/2} L_i - K\right| \geq \epsilon\right) \leq 2 \exp\left(-\frac{2 (n/2)^2 \epsilon^2}{(n/2) \times 2}\right)$$

$$= 2 \exp\left(-\frac{n\epsilon^2}{2}\right).$$

The theorem then follows from the upper bound derived in Theorem 11.1. $\square$

Next, we establish an achievable throughput that is within a factor of $\sqrt{\log n}$ of the throughput upper bound, also w.h.p. First, we outline the resource allocation strategy for proving that a throughput of $\Omega\left(1/\sqrt{n \log n}\right)$ is achievable w.h.p.

(i) We divide the unit square into small squares (which we will call squarelets) of equal size, with each squarelet having an area $c \log n/n$ for $c > 1$, as shown in Figure 11.4. (Throughout, we will ignore ceiling and floor functions for ease of exposition.) Thus, there are roughly $n/c \log n$ squarelets, each containing roughly $c \log n$ nodes on average. We will prove in Lemma 11.3.2 that each squarelet contains at least one node w.h.p.

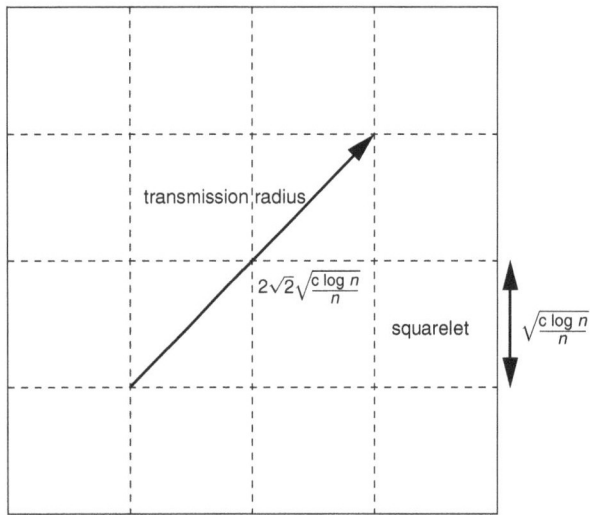

Figure 11.4 The unit square is divided into squarelets with side length $\sqrt{c \log n/n}$. The transmission radius is chosen to be $2\sqrt{2}\sqrt{c \log n/n}$ so that any two nodes in neighboring squarelets can communicate with each other.

(ii) The radius of communication will be chosen such that $r = 2\sqrt{2}\sqrt{c \log n/n}$; then, any node in a square can talk to any other node in a neighboring squarelet, as shown in Figure 11.4. So, if each squarelet contains at least one node w.h.p., the network is connected w.h.p. if $r = 2\sqrt{2}\sqrt{c \log n/n}$.

(iii) All packets will be routed using a simple routing scheme from their sources to their destinations. A packet will hop from squarelet to squarelet, first vertically and then horizontally, as shown in Figure 11.5.

(iv) We will show in Lemma 11.3.3 that the number of routes going through a squarelet is of the order of $\sqrt{n \log n}$. Thus, the offered load to a squarelet is of the order of $\lambda \sqrt{n \log n}$ w.h.p.

(v) Assuming $\Delta = 0.5$, we will show in Lemma 11.3.4 that one out of 81 squarelets can be scheduled in each time slot without interference, i.e., every one in 81 squarelets is chosen and a node in each of the squarelets is allowed to transmit. Thus, the average number of packets transmitted from a squarelet is 1/81. Thus, ignoring constants, we can support a data rate of λ packets/slot/node if $\lambda \sqrt{n \log n} \le 1/81$. In other words, a throughput per node of the order of $1/\sqrt{n \log n}$ is supportable by the network.

In the following, we will prove the claims mentioned in each step, and then summarize the result as a theorem.

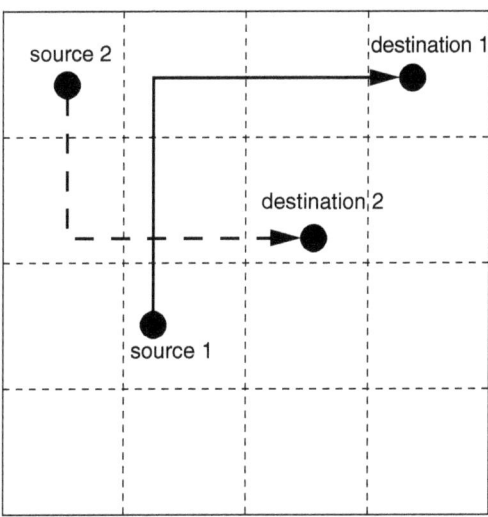

Figure 11.5 A packet is transmitted first vertically and then horizontally.

Lemma 11.3.2 The probability that there exists an empty squarelet is upper bounded by

$$\frac{n^{1-c}}{c \log n},$$

which goes to zero as $n \to \infty$ if $c \geq 1$.

Proof Recall that the area of each squarelet is $c \log n / n$. Denote by X_{ij} the event that node i is in squarelet j. Then,

$$Y_j = \sum_{i=1}^{n} X_{ij}$$

is the number of nodes in squarelet j. We are interested in computing the probability that at least one squarelet is empty, i.e.,

$$\Pr\left(\min_j Y_j = 0\right) = \Pr\left(Y_1 = 0 \text{ or } Y_2 = 0 \text{ or } \ldots \text{ or } Y_{n/c \log n} = 0\right)$$

$$\leq \frac{n}{c \log n} \Pr(Y_1 = 0) \tag{11.6}$$

$$= \frac{n}{c \log n} \Pr\left(\sum_{i=1}^{n} X_{i1} = 0\right)$$

$$= \frac{n}{c \log n} \left(1 - \frac{c \log n}{n}\right)^n \tag{11.7}$$

$$\leq \frac{n}{c \log n} e^{-c \log n} \tag{11.8}$$

$$\leq \frac{n^{1-c}}{c \log n},$$

where (11.6) is by the union bound and symmetry, (11.7) holds because

$$X_{ij} = \begin{cases} 1, & \text{with probability } \frac{c\log n}{n}, \\ 0, & \text{with probability } 1 - \frac{c\log n}{n}, \end{cases}$$

and (11.8) follows from the fact that $(1-x) \leq e^{-x}$.   □

Lemma 11.3.3 Under the simple routing scheme described in step (iii) of the resource allocation strategy, for any $\alpha \geq 2\sqrt{c} + 1/\sqrt{2}$, the following holds:

$$\Pr\left(\text{there exists a squarelet with at least } \alpha\sqrt{n\log n} \text{ routes going through it}\right) \leq \frac{2}{c\log n},$$

which goes to zero as $n \to \infty$.

Proof The number of routes crossing a squarelet is upper bounded by the number of sources and destinations in the same column and row, respectively, as the squarelet under consideration (see Figure 11.6). This is equal to the number of nodes in the same row and column.

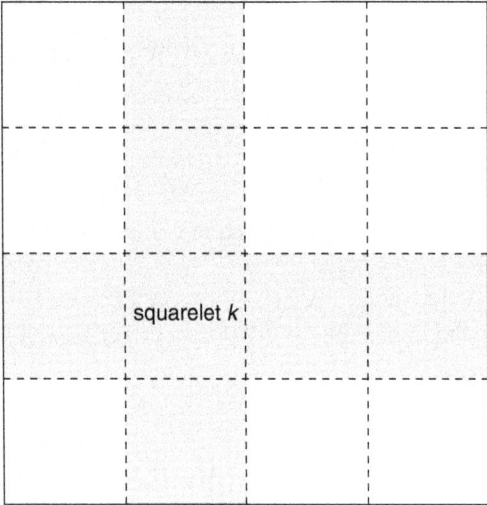

Figure 11.6 The number of routes crossing a squarelet is upper bounded by the number of nodes in the squarelets in the same row and column, i.e., the shaded squarelets.

Since each squarelet has side length $\sqrt{c\log n/n}$, the number of squarelets in the same row (or column) is $\sqrt{n/c\log n}$. Thus, the total number of squarelets in the same row or column as a given squarelet is

$$2\sqrt{n/(c\log n)} - 1.$$

The expected number of nodes in these squarelets is given by

$$n \times \text{(total area of these squarelets)} = n \left(2\sqrt{\frac{n}{c \log n}} - 1 \right) \frac{c \log n}{n}$$

$$= 2\sqrt{cn \log n} - c \log n.$$

Let $\mathcal{S}_k$ be the set of squarelets in the same row or column as squarelet k. We next compute the probability that there are at least $\alpha \sqrt{n \log n}$ nodes in $\mathcal{S}_k$.

We define a random variable Z_{ik} such that $Z_{ik} = 1$ if node i is in $\mathcal{S}_k$ and $Z_{ik} = 0$ otherwise, so $\sum_i Z_{ik}$ is the number of nodes in $\mathcal{S}_k$, and

$$\Pr\left(\sum_{i=1}^{n} Z_{ik} \geq \alpha\sqrt{n \log n} \right) = \Pr\left(\sum_{i=1}^{n} Z_{ik} - 2\sqrt{cn \log n} + c \log n \right.$$

$$\geq (\alpha - 2\sqrt{c})\sqrt{n \log n} + c \log n \Bigg)$$

$$\leq \Pr\left(\sum_{i=1}^{n} Z_{ik} - (2\sqrt{cn \log n} - c \log n) \geq (\alpha - 2\sqrt{c})\sqrt{n \log n} \right)$$

$$\leq 2 \exp\left(-2(\alpha - 2\sqrt{c})^2 \log n \right),$$

where the last inequality results from the Hoeffding bound. Using the union bound, we further obtain

$$\Pr(\text{at least } \alpha\sqrt{n \log n} \text{ routes go through squarelet } k \text{ for some } k)$$

$$\leq \frac{n}{c \log n} \times 2 \exp\left(-2(\alpha - 2\sqrt{c})^2 \log n \right)$$

$$= \frac{n}{c \log n} \times \frac{2}{n^{2(\alpha - 2\sqrt{c})^2}}.$$

When $2(\alpha - 2\sqrt{c})^2 \geq 1$, i.e., $\alpha \geq 2\sqrt{c} + 1/\sqrt{2}$, the upper bound becomes

$$\Pr(\text{at least } \alpha\sqrt{n \log n} \text{ routes go through squarelet } k \text{ for some } k) \leq \frac{2}{c \log n},$$

which goes to 0 as $n \to \infty$. □

Lemma 11.3.4 Assume $\Delta = 0.5$. There exists a squarelet scheduling algorithm such that, in each time slot, every one in 81 squarelets can be chosen, and one node in each of the chosen squarelets is allowed to transmit without interference.

Proof Recall that any two receivers of simultaneous transmissions must be Δr meters from each other. A sufficient condition to guarantee this is that the distance between any two transmitters is at least $(2 + \Delta)r$ meters from each other; see Figure 11.3. Assume $\Delta = 0.5$ and recall that $r = 2\sqrt{2}\sqrt{c \log n/n}$, then a sufficient condition to guarantee that two receivers are Δr apart is to require the distance between two transmitters to be at least

$$5\sqrt{2}\sqrt{\frac{c\log n}{n}} < 8\sqrt{\frac{c\log n}{n}},$$

i.e., a distance of eight squarelets.

So, we can group 9×9 squarelets into a cell and index the squarelets from 0 to 80, as shown in Figure 11.7. Then, in each time slot t, the squarelets with index $(t \bmod 81)$ is chosen, and one wireless node in each of the chosen squarelet is allowed to transmit. It should be clear that this can be achieved without any transmissions suffering from interference. Thus, one in every 81 squarelets is scheduled in a time slot. □

Note that, for other values of Δ, a similar result holds with different constants.

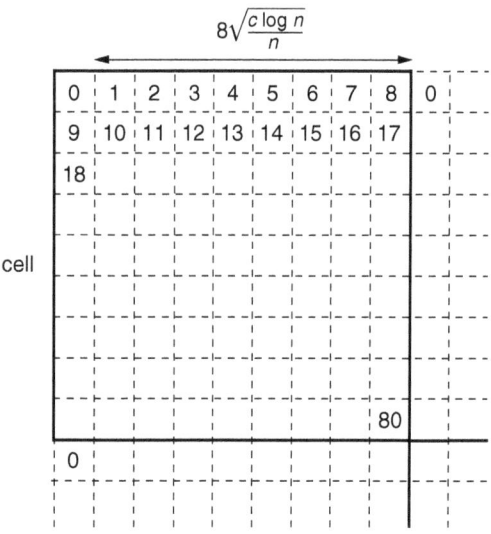

Figure 11.7 Each cell consists of 9×9 squarelets, which are indexed from 0 to 80. In each time slot t, the squarelets with index $(t \bmod 81)$ is chosen. For example, at time slot 0, all squarelets with index 0 are chosen. Note that two nodes in two squarelets with the same index are at least $8\sqrt{c\log n/n}$ meters from each other.

Theorem 11.3.5 Under random placement, using the resource allocation strategy mentioned earlier with $r = 2\sqrt{2}\sqrt{4\log n/n}$,

$$\Pr\left(\lambda \le \frac{1}{648\sqrt{n\log n}} \text{ is supportable by the network}\right) \to 1$$

as $n \to \infty$.

Proof From the statement of the theorem, $c = 4$. From Lemma 11.3.2, we have

$$\Pr(\text{there is an empty squarelet in the network}) \to 0 \text{ as } n \to \infty.$$

Given that the side length of a squarelet is $\sqrt{c\log/n}$, a transmission radius $r = 4\sqrt{2}\sqrt{\log n/n}$ guarantees that any node in a squarelet can communicate with a node in neighboring squarelets. So the network is connected w.h.p.

Further, according to Lemma 11.3.3,

$$\Pr\left(\text{number of routes through a squarelet} \geq 8\sqrt{n \log n} \text{ for some squarelet}\right)$$

$$\to 0 \text{ as } n \to \infty.$$

So the load on each of the squarelets is no more than $\lambda 8\sqrt{n \log n}$ w.h.p. According to Lemma 11.3.4, a squarelet can be scheduled once every 81 time slots, so the load can be supported if

$$\lambda 8\sqrt{n \log n} \leq \frac{1}{81},$$

i.e.,

$$\lambda \leq \frac{1}{81}\left(\frac{1}{8\sqrt{n \log n}}\right) = \frac{1}{648\sqrt{n \log n}}.$$

Note that a per-source throughput of λ is supportable in the network if (i) the network is connected and (ii) the traffic load on each squarelet is less than $1/81$. As both conditions hold w.h.p., we conclude that $\lambda \leq 1/648\sqrt{n \log n}$ is supportable by the network w.h.p. $\quad\square$

Theorem 11.3.5 suggests that the radius of communication should be chosen to be small (note that $r \to 0$ as $n \to \infty$, but is large enough to ensure network connectivity) so that multiple simultaneous transmissions are possible in the network. Since the radius is small, it will take multiple hops for a packet to travel from a source to its destination. Therefore, the theorem indicates that multi-hop transmissions are optimal from a throughput-maximization perspective.

11.4 Summary

- **The Hoeffding bound** Consider a sequence of n independent random variables $\{X_i\}$. Further, random variable X_i takes values in $[a_i, b_i]$. Then the following inequality holds:

$$\Pr\left(\left|\frac{\sum_i X_i - E\left[\sum_i X_i\right]}{n}\right| \geq x\right) \leq 2\exp\left(-\frac{2n^2 x^2}{\sum_{i=1}^{n}(b_i - a_i)^2}\right).$$

- **Capacity scaling of ad hoc wireless networks** Consider n wireless nodes on a unit square. Each node has a radius of communication r, i.e., node a can send data to node b if node b is within r meters of node a.

Let L_i denote the distance between source i and its destination, and let $L = (2/n)\sum_{i=1}^{n/2} L_i$ be the average distance between source-destination pairs. The per-node throughput λ satisfies

$$\lambda \leq \frac{4\sqrt{2}}{\sqrt{\pi}\Delta L}\frac{1}{\sqrt{n}}.$$

If nodes are randomly deployed, and each node's position is picked uniformly on the unit square and independent of the positions of the other nodes, then

$$\Pr\left(\lambda \le \frac{1}{648\sqrt{n\log n}} \text{ is supportable by the network}\right) \to 1,$$

i.e., the per-node throughput scales as $1/\sqrt{n\log n}$, as $n \to \infty$.

11.5 Exercises

Exercise 11.1 (Upper bound on the moment generating function I) Consider a random variable X_i that takes values in $[a_i, b_i]$ and has zero mean. Use the fact that an exponential function is convex, i.e., for any $a_i \le x \le b_i$,

$$e^{\theta x} \le \frac{x - a_i}{b_i - a_i}e^{\theta b_i} + \frac{b_i - x}{b_i - a_i}e^{\theta a_i},$$

to show that

$$E\left[e^{\theta X_i}\right] \le e^{\alpha u + \log(\alpha e^{-u} + (1-\alpha))},$$

where $\alpha = b_i/(b_i - a_i)$ and $u = \theta(b_i - a_i)$.

Exercise 11.2 (Upper bound on the moment generating function II) Consider a random variable X_i that takes values in $[a_i, b_i]$ and has zero mean. Use the result of Exercise 11.1 to show that the moment generating function of X_i satisfies

$$E\left[e^{\theta X_i}\right] \le \exp\left(\frac{\theta^2(b_i - a_i)^2}{8}\right).$$

Hint: Define $f(u) = \alpha u + \log(\alpha e^{-u} + (1-\alpha))$. Use Taylor's series, i.e., $f(u) = f(0) + f'(0)u + f''(v)u^2/2$, for some $v \in [0, u]$, to show that $f(u) \le u^2/8$.

Exercise 11.3 (The Hoeffding bound) Consider a sequence of n independent random variables $\{X_i\}$. Assume that each random variable X_i takes values in $[a_i, b_i]$ and $E[X_i] = 0$. Use the result in Exercise 11.2 to prove that the following inequality holds:

$$\Pr\left(\left|\frac{\sum_i X_i}{n}\right| \ge x\right) \le 2\exp\left(-\frac{2n^2 x^2}{\sum_{i=1}^n (b_i - a_i)^2}\right).$$

Exercise 11.4 (Connectivity under a spatial Poisson process I) Consider a model in which the number of nodes in the unit square is a Poisson random variable with mean n. These $Poi(n)$ nodes are randomly deployed. Each node's position is picked uniformly on the unit square and independent of the positions of the other nodes. We divide the unit square into cells with side length $3d$, and each cell is further divided into nine squarelets with side length d, as shown in Figure 11.8. Similar to Result 9.2.2, the number of nodes in each squarelet is a Poisson random variable with mean $d^2 n$; the numbers of nodes in different squarelets are independent.

(1) For each cell, compute the probability that all squarelets except the one at the center are empty.

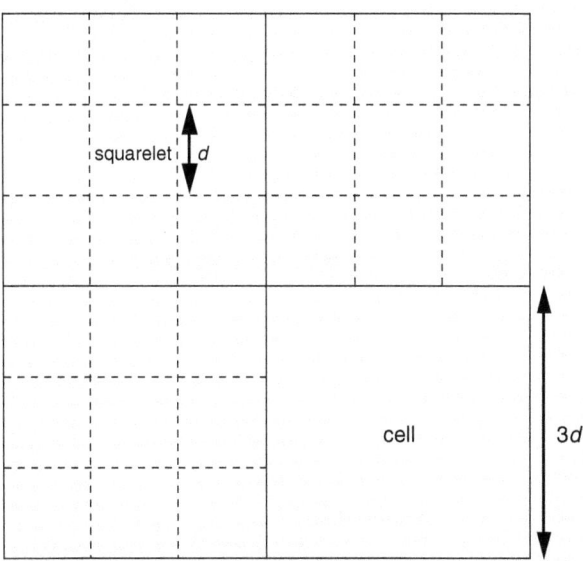

Figure 11.8 A unit square is divided into cells with side length $3d$. Each cell is further divided into squarelets with side length d.

(2) Compute the probability that a cell described in part (1) exists in the network.

(3) Assume each node has a radius of communication r. Based on the result of part (2), show that the network is disconnected (i.e., there exists an isolated node in the network) with probability 1 as $n \to \infty$ if

$$r \leq \sqrt{\frac{(1 - \epsilon) \log n}{8n}}$$

for some $\epsilon > 0$. Hint: Assume

$$d = \sqrt{\frac{(1 - \epsilon) \log n}{8n}},$$

then a node can only communicate with other nodes in the same squarelet or in one of the eight neighboring squarelets when

$$r \leq \sqrt{\frac{(1 - \epsilon) \log n}{8n}}.$$

Note that $(1 - x) \leq e^{-x}$.

Exercise 11.5 (Connectivity under a spatial Poisson process II) Consider the two-dimensional spatial process defined in Exercise 11.4.

(1) Compute the probability that none of the squarelets in the network is empty.

(2) Based on the result of part (1), show that the network is connected with probability 1 as $n \to \infty$ if $r \geq \sqrt{8 \log n / n}$. Hint: Assume $d = \sqrt{\log n / n}$, then a node can communicate with any node in its neighboring squarelets when $r \geq \sqrt{8 \log n / n}$.

Exercise 11.6 (Minimum distance between neighbors in one dimension I) Consider points distributed on a real line according to a Poisson process, i.e., the distances between consecutive points are mutually independent and identically distributed according to an exponential distribution of mean $1/n$. Consider $(n+1)$ consecutive points and label the distances between the points $X_1, X_2, \ldots, X_n$, as shown in Figure 11.9. Let

$$D_{\min} = \min\{X_1, X_2, \ldots, X_n\}.$$

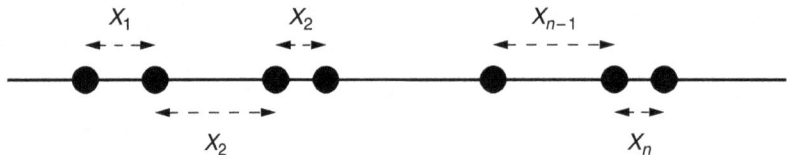

Figure 11.9 $n+1$ points distributed on a real line.

(1) Find the PDF of $D_{\min}$.

(2) Show that $D_{\min}$ concentrates in the interval

$$\left[\frac{1}{n^2 h(n)}, \frac{g(n)}{n^2}\right]$$

w.h.p., where $g(n), h(n) \to \infty$ as $n \to \infty$.

Exercise 11.7 (Minimum distance between neighbors in one dimension II) Consider dropping $(n+1)$ points uniformly at random in the interval $[0,1]$ on the real line. Let the distance between consecutive points be labeled $X_1, \ldots, X_n$, and define

$$D_{\min}(n) = \min\{X_1, \ldots, X_n\}.$$

Show that

$$\Pr\left(D_{\min} \geq \frac{1}{n^2 h(n)}\right) \to 1$$

as $n \to \infty$, where $h(n) \to \infty$ as $n \to \infty$. Hint: Assume that the $n+1$ points are dropped one by one. Let $\mathcal{E}_m$ denote the event that the minimum distance between any two points of the first m points is lower bounded by $1/n^2 h(n)$. Then

$$\Pr\left(D_{\min} \geq \frac{1}{n^2 h(n)}\right) = \prod_{m=2}^{n+1} \Pr\left(\mathcal{E}_m | \mathcal{E}_{m-1}\right).$$

First show that

$$\Pr\left(\mathcal{E}_m | \mathcal{E}_{m-1}\right) \geq 1 - (m-1) \times \frac{2}{n^2 h(n)}.$$

Exercise 11.8 (Minimum distance between neighbors in one dimension III) Consider dropping $(n+1)$ points uniformly at random in the interval $[0,1]$ on the real line. Let the distance between consecutive points be labeled $X_1, \ldots, X_n$, and define

$$D_{\min}(n) = \min\{X_1, \ldots, X_n\}.$$

Show that

$$\Pr\left(D_{\min} \leq \frac{g(n)}{n^2}\right) \to 1$$

as $n \to \infty$, where $g(n) \to \infty$ as $n \to \infty$. Hint: Divide the unit interval into subintervals of length $g(n)/n^2$, and consider the probability that no two points share the same subinterval.

Exercise 11.9 (Minimum distance between neighbors in two dimensions I) Suppose that n nodes are dropped in a unit square, with each node's location being picked uniformly (independent of other nodes) at random in the square. Let $D_{\min}(n)$ be the minimum distance between any two points. Show that

$$\Pr\left(\frac{1}{\sqrt{h(n)}\,n} \leq D_{\min}\right) \to 1$$

as $n \to \infty$, where $h(n) \to \infty$ as $n \to \infty$.

Exercise 11.10 (Minimum distance between neighbors in two dimensions II) Suppose that n nodes are dropped in a unit square, with each node's location being picked uniformly (independent of other nodes) at random in the square. Let $D_{\min}(n)$ be the minimum distance between any two points. Show that

$$\Pr\left(D_{\min} \leq \frac{\sqrt{g(n)}}{n}\right) \to 1$$

as $n \to \infty$, where $g(n) \to \infty$ as $n \to \infty$.

Exercise 11.11 (Two-dimensional Poisson process I) Instead of dropping nodes in a unit square, another model of wireless networks considers points dropped according to a two-dimensional Poisson process on the infinite plane. In this and the following problems, we will briefly introduce two basic results about two-dimensional Poisson processes. A two-dimensional Poisson process of rate λ is defined as follows: the number of points in a region R is Poisson distributed with mean $\lambda|R|$, where $|R|$ is the area of region R. If two regions R_1 and R_2 are disjoint, the number of points in the two regions are independent.

Consider a point generated by this process; for convenience, let its location be $(0,0)$. Let C be the largest disk centered at $(0,0)$ that does not contain any other points. Show that the area of C is exponentially distributed with mean $1/\lambda$. Remark: This result also holds if there is no point at the origin, i.e., the distance of the closest point from the origin has the same CDF given above, whether the origin contains a point of the Poisson process or not.

Exercise 11.12 (Two-dimensional Poisson process II) Consider the two-dimensional spatial process defined in Exercise 11.11. Let C_k be the disk around the origin such that the kth closest point to the origin is on its circumference. Show that the area of $C_k \setminus C_{k-1}$ is exponentially distributed with mean $1/\lambda$.

11.6   Notes

The geometric random graph model and the throughput scaling result for wireless networks was developed in [50]. The lower bound derived in this chapter is based on the proof presented in [164]. The model for wireless node distribution in a geographical area is an example of a spatial point process. A detailed introduction to spatial point processes is given in [28].

REFERENCES

[1] N. Abramson. The Aloha system: another alternative for computer communications. In *Proc. Fall Joint Computer Conf.*, pp. 281–285, Atlantic City, NJ, Nov. 17–19, 1970.

[2] T. Alpcan and T. Başar. A utility-based congestion control scheme for Internet-style networks with delay. In *Proc. IEEE INFOCOM*, pp. 2039–2048, San Francisco, CA, Mar.–Apr. 2003.

[3] E. Altman, K. Avrachenkov, and C. Barakat. A stochastic model of TCP/IP with stationary random losses. In *Proc. ACM SIGCOMM*, pp. 231–242, Stockholm, Aug.–Sept. 2000.

[4] T. Anderson, S. Owicki, J. Saxe, and C. Thacker. High-speed switch scheduling for local-area networks. *ACM Transactions on Computer Systems*, **11**:319–352, 1993.

[5] D. Anick, D. Mitra, and M. Sondhi. Stochastic theory of a data-handling system. *Bell System Technical Journal*, **61**:1871–1894, 1982.

[6] S. Asmussen. *Applied Probability and Queues*. New York, NY: Springer Verlag, 2003.

[7] F. Baccelli and D. Hong. AIMD, fairness and fractal scaling of TCP traffic. In *Proc. IEEE INFOCOM*, vol. 1, pp. 229–238, New York, NY, June 2002.

[8] J. C. R. Bennett and H. Zhang. WF^{2}q: worst-case fair weighted fair queueing. In *Proc. IEEE INFOCOM*, vol. 1, pp. 120–128, San Francisco, CA, Mar. 1996.

[9] D. Bertsekas. *Nonlinear Programming*. Belmont, MA: Athena Scientific, 1995.

[10] D. Bertsekas and R. Gallager. *Data Networks*. Englewood Cliffs, NJ: Prentice Hall, 1991.

[11] G. Bianchi. Performance analysis of the IEEE 802.11 distributed coordination function. *IEEE Journal on Selected Areas in Communications*, **18**(3): 535–547, 2000.

[12] T. Bonald and L. Massoulie. Impact of fairness on Internet performance. In *Proc. ACM SIGMETRICS*, pp. 82–91, Cambridge, MA, June 2001.

[13] T. Bonald, L. Massoulie, F. Mathieu, D. Perino, and A. Twigg. Epidemic live streaming: optimal performance trade-offs. In *Proc. ACM SIGMETRICS*, pp. 325–336, Annapolis, MD, June 2007.

[14] R. Boorstyn, A. Kershenbaum, B. Maglaris, and V. Sahin. Throughput analysis in multi-hop CSMA packet radio networks. *IEEE Transactions on Communications*, **35**: 267–274, 1987.

[15] D. D. Botvich and N. G. Duffield. Large deviations, economies of scale, and the shape of the loss curve in large multiplexers. *Queueing Systems*, **20**:293–320, 1995.

[16] J. Y. Le Boudec and P. Thiran. *Network Calculus: A Theory of Deterministic Queueing Systems for the Internet*. Berlin: Springer Verlag, 2001.

[17] S. Boyd and L. Vandenberghe. *Convex Optimization*. Cambridge: Cambridge University Press, 2004.

[18] L. S. Bramko and L. L. Peterson. TCP Vegas: end-to-end congestion avoidance on a global Internet. *IEEE Journal on Selected Areas in Communications*, **13**:1465–1480, Oct. 1995.

[19] H. Bruneel and B. Kim. *Discrete-Time Models For Communication Systems Including ATM*. Boston, MA: Kluwer Academic, 1993.

[20] J. A. Bucklew. *Large Deviation Techniques in Decision, Simulation and Estimation*. New York, NY: Wiley, 1990.

[21] C. S. Chang. *Performance Guarantees in Communication Networks*. London: Springer Verlag, 2000.

[22] C. S. Chang, D. S. Lee, and Y. S. Jou. Load balanced Birkhoff-von Neumann switches, part I: one-stage buffering. *Computer Communications*, **25**:611–622, 2002.

[23] H. Chen and D. Yao. *Fundamentals of Queueing Networks: Performance, Asymptotics, and Optimization*. New York, NY: Springer Verlag, 2001.

[24] M. Chiang, S. H. Low, A. R. Calderbank, and J. C. Doyle. Layering as optimization decomposition: a mathematical theory of network architectures. *Proceedings of the IEEE*, **95**:255–312, Jan. 2007.

[25] D. M. Chiu and R. Jain. Analysis of the increase and decrease algorithms for congestion avoidance in computer networks. *Computer Networks and ISDN Systems*, **17**:1–14, 1989.

[26] B. Cohen. "The BitTorrent protocol specification." Available at BitTorrent.org, Jan. 2008.

[27] C. Courcoubetis and R. R. Weber. Buffer overflow asymptotics for a switch handling many traffic sources. *Journal of Applied Probability*, **33**(3):886–903, Sept. 1996.

[28] D. R. Cox and V. Isham. *Point Processes*. Boca Raton, FL: Chapman & Hall/CRC, 1980.

[29] R. L. Cruz. A calculus for network delay. Part I: Network elements in isolation. *IEEE Transactions on Information Theory*, **37**:114–131, Jan. 1991.

[30] R. L. Cruz. A calculus for network delay. Part II: Network analysis. *IEEE Transactions on Information Theory*, **37**:132–141, Jan. 1991.

[31] J. G. Dai and B. Prabhakar. The throughput of data switches with and without speedup. In *Proc. IEEE INFOCOM*, vol. 2, pp. 556–564, Tel Aviv, Israel, 2000.

[32] G. de Veciana, G. Kesidis, and J. Walrand. Resource management in wide-area ATM networks using effective bandwidths. *IEEE Journal on Selected Areas in Communications*, **13**:1081–1090, 1995.

[33] G. dc Veciana, T.-J. Lee, and T. Konstantopoulos. Stability and performance analysis of networks supporting elastic services. *IEEE/ACM Transactions on Networking*, **9**(1): 2–14, Feb. 2001.

[34] G. de Veciana and J. Walrand. Effective bandwidths: call admission, traffic policing and filtering for ATM networks. *Queueing Systems*, **20**(1–2): 37–59, 1994.

[35] A. Demers, S. Keshav, and S. Shenker. Analysis and simulation of a fair queueing algorithm. In *ACM SIGCOMM*, pp. 1–12, Austin, TX, Sept. 1989.

[36] D. Dubhashi and A. Panconesi. *Concentration of Measure for the Analysis of Randomized Algorithms*. Cambridge: Cambridge University Press, 2012.

[37] A. Eryilmaz and R. Srikant. Fair resource allocation in wireless networks using queue-length-based scheduling and congestion control. In *Proc. IEEE INFOCOM*, pp. 1794–1803, Miami, FL, Mar. 2005.

[38] A. Eryilmaz and R. Srikant. Joint congestion control, routing and MAC for stability and fairness in wireless networks. *IEEE Journal on Selected Areas in Communications*, **24**:1514–1524, June 2006.

[39] A. Eryilmaz and R. Srikant. Asymptotically tight queue length bounds implied by drift conditions. *Queueing Systems*, **72**:311–359, 2012.

[40] A. Eryilmaz, R. Srikant, and J. Perkins. Stable scheduling policies for fading wireless channels. *IEEE/ACM Transactions on Networking*, **13**(2):411–424, 2005.

[41] G. Fayolle, E. Gelenbe, and J. Labetoulle. Stability and optimal control of the packet switching broadcast channel. *Journal of the ACM*, **24**(3):375–386, 1977.

[42] A. M. Frieze and G. R. Grimmett. The shortest-path problem for graphs with random arc-lengths. *Discrete Applied Mathematics*, **10**(1):57–77, Jan. 1985.

[43] A. J. Ganesh, N. O'Connell, and D. J. Wischik. *Big Queues*. Berlin: Springer Verlag, 2004.

[44] L. Georgiadis, M. Neely, and L. Tassiulas. Resource allocation and cross-layer control in wireless networks. *Foundations and Trends in Networking*, **1**(1):1–144, 2006.

[45] J. Ghaderi and R. Srikant. On the design of efficient CSMA algorithms for wireless networks. In *IEEE Conf. on Decision and Control*, pp. 954–959, Atlanta, GA, Dec. 2010.

[46] R. J. Gibbens and P. Hunt. Effective bandwidths for the multiple-type UAS channel. *Queueing Systems Theory and Applications*, **9**:17–28, 1991.

[47] P. W. Glynn and W. Whitt. Logarithmic asymptotics for steady-state tail probabilities in single-server queues. *Journal of Applied Probability*, **31A**:131–156, 1994.

[48] A. G. Greenberg, R. Srikant, and W. Whitt. Resource sharing for book-ahead and instantaneous-request calls. *IEEE/ACM Transactions on Networking*, **7**:10–22, Feb. 1999.

[49] R. Guérin, H. Ahmadi, and M. Nagshineh. Equivalent capacity and its application to bandwidth allocation in high-speed networks. *IEEE Journal on Selected Areas in Communications*, **4**:968–981, 1991.

[50] P. Gupta and P. R. Kumar. The capacity of wireless networks. *IEEE Transactions on Information Theory*, **IT-46**(2):388–404, Mar. 2000.

[51] B. Hajek. Communication network analysis. Available at http://www.ifp.illinois.edu/~hajek/Papers/networkanalysis.html, Dec. 2006.

[52] B. Hajek. Hitting-time and occupation-time bounds implied by drift analysis with applications. *Advances in Applied Probability*, **14**:502–525, 1982.

[53] J. Hale and S. M. Verduyn Lunel. *Introduction to Functional Differential Equations*, 2nd edn. New York, NY: Springer Verlag, 1991.

[54] H. Han, S. Shakkottai, C. Hollot, R. Srikant, and D. Towsley. Multi-path TCP: a joint congestion control and routing scheme to exploit path diversity in the Internet. *IEEE/ACM Transactions on Networking*, **14**:1260–1271, Dec. 2006.

[55] D. Heyman, T. V. Lakshman, and A. Niedhart. A new method for analyzing feedback-based protocols with applications to engineering web traffic over the Internet. In *SIGMETRICS 97*, pp. 24–38, Seattle, WA, 1997.

[56] I. Hou, V. Borkar, and P. R. Kumar. A theory of QoS for wireless. In *Proc. IEEE INFOCOM*, pp. 486–494 Rio de Janeiro, Brazil, Apr. 2009.

[57] I. Hou and P. R. Kumar. Admission control and scheduling for QoS guarantees for variable-bit-rate applications on wireless channels. In *Proc. ACM MobiHoc*, pp. 175–184, New Orleans, LA, May 2009.

[58] J. Y. Hui. Resource allocation for broadband networks. *IEEE Journal on Selected Areas in Communications*, **6**:1598–1608, 1988.

[59] J. Y. Hui. *Switching and Traffic Theory for Integrated Broadband Networks*. Boston, MA: Kluwer, 1990.

[60] V. Jacobson. Congestion avoidance and control. *ACM Computer Communication Review*, **18**:314–329, Aug. 1988.

[61] J. M. Jaffe. Bottleneck flow control. *IEEE Transactions on Communications*, **29**:954–962, July 1981.

[62] J. J. Jaramillo and R. Srikant. Optimal scheduling for fair resource allocation in ad hoc networks with elastic and inelastic traffic. *IEEE/ACM Transactions on Networking*, **19**:1125–1136, Aug. 2011.

[63] J. J. Jaramillo, R. Srikant, and L. Ying. Scheduling for optimal rate allocation in ad hoc networks with heterogeneous delay constraints. *IEEE Journal on Selected Areas in Communications*, **29**(5):979–987, 2011.

[64] L. Jiang and J. Walrand. A distributed CSMA algorithm for throughput and utility maximization in wireless networks. In *Proc. Allerton Conf. on Control, Communication and Computing*, pp. 1511–1519, Monticello, IL, 2008.

[65] L. Jiang and J. Walrand. Approaching throughput-optimality in a distributed CSMA algorithm: collisions and stability. In *Proc. 2009 MobiHoc's S3 Workshop*, pp. 5–8, New Orleans, LA, 2009.

[66] R. Johari and J. N. Tsitsiklis. Efficiency loss in a network resource allocation game. *Mathematics of Operations Research*, **29**(3):407–435, Aug. 2004.

[67] K. Kar, S. Sarkar, and L. Tassiulas. Achieving proportional fairness using local information in Aloha networks. *IEEE Transactions on Automatic Control*, **49**(10):1858–1863, 2004.

[68] F. P. Kelly. *Reversibility and Stochastic Networks*. New York, NY: John Wiley, 1979.

[69] F. P. Kelly. Stochastic models of computer communication systems. *Journal of the Royal Statistical Society, Series B (Methodological)*, **47**(3):379–395, 1985.

[70] F. P. Kelly. Effective bandwidths at multi-class queues. *Queueing Systems*, **9**:5–16, 1991.

[71] F. P. Kelly. Loss networks. *The Annals of Applied Probability*, **1**(3):319–378, Aug. 1991.

[72] F. P. Kelly. Notes on effective banwidths, in *Stochastic Networks: Theory and Applications*, F. P. Kelly, S. Zachary, and I. B. Ziedins, eds. Oxford: Oxford University Press, pp. 141–168, 1996.

[73] F. P. Kelly. Charging and rate control for elastic traffic. *European Transactions on Telecommunications*, **8**:33–37, 1997.

[74] F. P. Kelly, Mathmatical modelling of the Internet, in *Mathematics Unlimited – 2001 and Beyond*, B. Engquist and W. Schmid, eds. Berlin: Springer Verlag, pp. 685–702.

[75] F. P. Kelly. Fairness and stability of end-to-end congestion control. *European Journal of Control*, **9**:149–165, 2003.

[76] F. P. Kelly, P. B. Key, and S. Zachary. Distributed admission control. *IEEE Journal on Selected Areas in Communications*, **18**:2617–2628, 2000.

[77] F. P. Kelly, A. Maulloo, and D. Tan. Rate control in communication networks: shadow prices, proportional fairness and stability. *Journal of the Operational Research Society*, **49**:237–252, 1998.

[78] F. P. Kelly and R. J. Williams. Fluid model for a network operating under a fair bandwidth-sharing policy. *Annals of Applied Probability*, **14**:1055–1083, 2004.

[79] T. Kelly. Scalable TCP: improving performance in highspeed wide area networks, *Newsletter ACM SIGCOMM Computer Communication Review*, **33**(2):83–91, Apr. 2003.

[80] S. Keshav. *An Engineering Approach to Computer Networks*. Reading, MA: Addison-Wesley, 1997.

[81] G. Kesidis. *An Introduction to Communication Network Analysis*. Hoboken, NJ: Wiley, 2007.

[82] G. Kesidis, J. Walrand, and C.-S. Chang. Effective bandwidths for multiclass Markov fluids and other ATM sources. *IEEE/ACM Transactions on Networking*, **1**:424–428, 1993.

[83] H. Khalil. *Nonlinear Systems*, 2nd edn. Upper Saddle River, NJ: Prentice Hall, 1996.

[84] T. H. Kim, J. Ni, and N. H. Vaidya. A distributed throughput-optimal CSMA with data packet collisions. In *Proc. WIMESH 2010*, pp. 1–6, Boston, MA, 2010.

[85] T. H. Kim, J. Ni, R. Srikant, and N. H. Vaidya. On the achievable throughput of CSMA under imperfect carrier sensing. In *Proc. INFOCOM*, pp. 1674–1682, Shanghai, 2011.

[86] J. F. C. Kingman. On queues in heavy traffic. *Journal of the Royal Statistical Society, Series B*, **24**:383–392, 1962.

[87] J. F. C. Kingman. Some inequalities for the queue GI/G/1. *Biometrika*, **49**:315–324, 1962.

[88] L. Kleinrock. *Queueing Systems, Volume 2: Computer Applications*. New York, NY: Wiley–Interscience, 1975.

[89] A. Kumar, D. Manjunath, and J. Kuri. *Communication Networking: An Analytical Approach*. San Francisco, CA: Morgan Kaufmann, 2004.

[90] P. R. Kumar and S. P. Meyn. Stability of queueing networks and scheduling policies. *IEEE Transactions on Automatic Control*, **40**:251–260, Feb. 1995.

[91] R. Kumar, Y. Liu, and K. W. Ross. Stochastic fluid theory for P2P streaming systems. In *Proc. IEEE INFOCOM 2007*, pp. 919–927, Anchorage, AK, 2007.

[92] R. Kumar and K. W. Ross. Optimal peer-assisted file distribution: single and multi-class problems. In *IEEE Workshop on Hot Topics in Web Systems and Technologies (HOTWEB)*, pp. 1–11, Boston, MA, 2006.

[93] S. Kunniyur and R. Srikant. End-to-end congestion control: utility functions, random losses and ECN marks. In *Proc. IEEE INFOCOM*, vol. 3, pp. 1323–1332, Tel Aviv, Israel, Mar. 2000.

[94] S. Kunniyur and R. Srikant. A time-scale decomposition approach to adaptive ECN marking. In *Proc. IEEE INFOCOM*, vol. 3, pp. 1330–1339, Anchorage, AK, Apr. 2001.

[95] S. Kunniyur and R. Srikant. End-to-end congestion control: utility functions, random losses and ECN marks. *IEEE/ACM Transactions on Networking*, **7**(5):689–702, Oct. 2003.

[96] J. Kurose and K. Ross. *Computer Networking: A Top Down Approach*. Reading, MA: Addison-Wesley, 2000.

[97] T. V. Lakshman and U. Madhow. The performance of TCP/IP for networks with high bandwidth-delay products and random loss. *IEEE/ACM Transactions on Networking*, **5**(3):336–350, 1997.

[98] R. Leelahakriengkrai and R. Agrawal. Scheduling in multimedia wireless networks. In *Proc. ITC*, Salvador da Bahia, Brazil, 2001.

[99] D. Liben-Nowell, H. Balakrishnan, and D. Karger. Analysis of the evolution of peer-to-peer systems. In *Proc. 21st Annual Symp. on Principles of Distributed Computing*, p. 233–242, Monterey, CA, 2002.

[100] S. C. Liew, C. H. Kai, H. C. J. Leung, and P. B. Wong. Back-of-the-envelope computation of throughput distributions in CSMA wireless networks. *IEEE Transactions on Mobile Computing*, **9**(9):1319–1331, 2010.

[101] X. Lin, N. Shroff, and R. Srikant. On the connection-level stability of congestion-controlled communication networks. *IEEE Transactions on Information Theory*, **54**(5):2317–2338, 2007.

[102] X. Lin and N. B. Shroff. Joint rate control and scheduling in multihop wireless networks. In *43rd IEEE Conf. on Decision and Control*, vol. 2, pp. 1484–1489 Paradise Island, Bahamas, Dec. 2004.

[103] X. Lin, N. B. Shroff, and R. Srikant. A tutorial on cross-layer optimization in wireless networks. *IEEE Journal on Selected Areas in Communications*, **24**(8):1452–1463, 2006.

[104] J. D. C. Little. A proof of the queuing formula: $l = \lambda w$. *Operations Research*, **9**:383–387, 1961.

[105] J. Liu, Y. Yi, A. Proutiere, M. Chiang, and H. V. Poor. Maximizing utility via random access without message passing, Microsoft Research, Redmond, WA, Tech. Rep. MSR-TR-2008-128, 2008.

[106] X. Liu, E. K. P. Chong, and N. B. Shroff. A framework for opportunistic scheduling in wireless networks. *Computer Networks*, **41**(4):451–474, 2003.

[107] S. H. Low and D. E. Lapsley. Optimization flow control, I: Basic algorithm and convergence. *IEEE/ACM Transactions on Networking*, **7**(6):861–875, Dec. 1999.

[108] S. H. Low, F. Paganini, and J. C. Doyle. Internet congestion control. *IEEE Control Systems Magazine*, **22**:28–43, Feb. 2002.

[109] S. H. Low, L. Peterson, and L. Wang. Understanding Vegas: a duality model. *Journal of ACM*, **49**(2):207–235, Mar. 2002.

[110] D. G. Luenberger. *Linear and Nonlinear Programming*. Reading, MA: Addison-Wesley, 1989.

[111] N. McKeown. The iSLIP algorithm for input-queued switches. *IEEE/ACM Transactions on Networking*, **7**(2):188–201, 1999.

[112] N. McKeown, V. Anantharam, and J. Walrand. Achieving 100% throughput in an input-queued switch. In *Proc. IEEE INFOCOM*, pp. 296–302, San Francisco, CA, Mar. 1996.

[113] L. Massoulie. Structural properties of proportional fairness: stability and insensitivity. *Annals of Applied Probability*, **17**(3):809–839, 2007.

[114] L. Massoulie and J. Roberts. Bandwidth sharing: objectives and algorithms. In *Proc. INFOCOM*, vol. 3, 1395–1403, New York, NY, Mar. 1999.

[115] L. Massoulie and M. Vojnovic. Coupon replication systems. In *Proc. ACM SIGMETRICS*, pp. 2–13, Banff, Alberta, Canada, 2005.

[116] P. Maymounkov and D. Mazires. Kademlia: a peer-to-peer information system based on the XOR metric. In *Proc. 1st Int. Workshop on Peer-to-peer Systems (IPTPS'02)*, pp. 53–65, Cambridge, MA, 2002.

[117] R. Mazumdar. *Performance Modeling, Loss Networks, and Statistical Multiplexing*. San Rafael, CA: Morgan & Claypool, 2009.

[118] I. Menache and A. Ozdaglar. *Network Games: Theory, Models, and Dynamics*. San Rafael, CA: Morgan & Claypool, 2011.

[119] S. Meyn. *Control Techniques for Complex Networks*. Cambridge: Cambridge University Press, 2007.

[120] S. Meyn and R. Tweedie. *Markov Chains and Stochastic Stability*. Cambridge: Cambridge University Press, 2009.

[121] S. P. Meyn. Stability and asymptotic optimality of generalized MaxWeight policies. *SIAM Journal on Control and Optimization*, **47**(6):3259–3294, 2009.

[122] V. Misra, W. Gong, and D. Towsley. A fluid-based analysis of a network of AQM routers supporting TCP flows with an application to RED. In *Proc. ACM SIGCOMM*, pp. 151–160, Stockholm, Sweden, Sept. 2000.

[123] D. Mitra and J. A. Morrison. Multiple time scale regulation and worst case processes for ATM network control. In *Proc. IEEE Conf. on Decision and Control*, pp. 353–358, New Orleans, LA, 1995.

[124] J. Mo and J. Walrand. Fair end-to-end window-based congestion control. In *SPIE Int. Symp.*, vol. 3350, pp. 55–63, Boston, MA, 1998.

[125] J. Mo and J. Walrand. Fair end-to-end window-based congestion control. *IEEE/ACM Transactions on Networking*, **8**(5):556–567, Oct. 2000.

[126] J. Mundinger, R. Weber, and G. Weiss. Analysis of peer-to-peer file dissemination. *Performance Evaluation Review, MAMA 2006 Issue*, **34**(3):12–14, Dec. 2006.

[127] J. Mundinger, R. Weber, and G. Weiss. Analysis of peer-to-peer file dissemination amongst users of different upload capacities. *Performance Evaluation Review, Performance 2005 Issue*, **34**(2):5–6, Sept. 2006.

[128] B. Nardelli, J. Lee, K. Lee et al. Experiment evaluation of optimal CSMA. In *Proc. IEEE INFOCOM*, pp. 1188–1196, Orlando, FL, Apr. 2011.

[129] J. Nash. The bargaining problem. *Econometrica*, **18**(2):155–162, 1950.

[130] M. Neely, E. Modiano, and C. Rohrs. Dynamic power allocation and routing for time varying wireless networks. *IEEE Selected Areas in Communications*, **23**(1):89–103, 2005.

[131] M. J. Neely. Energy optimal control for time varying wireless networks. In *Proc. IEEE INFOCOM*, Mar. 2005.

[132] M. J. Neely, E. Modiano, and C. Li. Fairness and optimal stochastic control for heterogeneous networks. In *Proc. IEEE INFOCOM*, vol. 3, pp. 1723–1734, Miami, FL, 2005.

[133] J. Ni and R. Srikant. Distributed CSMA/CA algorithms for achieving maximum throughput in wireless networks. In *Information Theory and Applications Workshop (ITA)*, p. 250, La Jolla, CA, Feb. 2009.

[134] J. Ni, B. Tan, and R. Srikant. Q-CSMA: Queue-length based CSMA/CA algorithms for achieving maximum throughput and low delay in wireless networks. In *Proc. IEEE INFOCOM Mini-Conference*, pp. 1–5, San Diego, CA, Mar. 2010.

[135] J. R. Norris. *Markov Chains*. Cambridge: Cambridge University Press, 1998.

[136] J. Padhye, V. Firoiu, D. Towsley, and J. Kurose. Modeling TCP throughput: a simple model and its empirical validation. In *Proc. ACM SIGCOMM*, pp. 303–314, Vancouver, Canada, 1998.

[137] F. Paganini, K. Tang, and A. Farragut. Network stability under alpha fair bandwidth allocation with general file size distribution. *IEEE Transactions on Automatic Control*, **57**:579–591, 2012.

[138] A. Parekh and R. Gallager. A generalized processor sharing approach to flow control in integrated services networks: the single node case. *IEEE/ACM Transactions on Networking*, **1**:344–357, 1993.

[139] L. L. Peterson and B. S. Davie. *Computer Networks: A Systems Approach*. San Francisco, CA: Morgan-Kaufman, 1999.

[140] J. W. Pratt. Risk aversion in the small and in the large. *Econometrica*, **32**:122–136, Jan.–Apr. 1964.

[141] D. Qiu and R. Srikant. Modeling and performance analysis of BitTorrent-like peer-to-peer networks. In *Proc. ACM SIGCOMM*, pp. 367–378, Portland, OR, 2004.

[142] S. Rajagopalan, D. Shah, and J. Shin. Network adiabatic theorem: an efficient randomized protocol for contention resolution. In *Proc. Eleventh Int. Joint Conf. on Measurement and Modeling of Computer Systems, SIGMETRICS '09*, pp. 133–144, Seattle, WA, 2009.

[143] K. K. Ramakrishnan and R. Jain. A binary feedback scheme for congestion avoidance in computer networks with a connectionless network layer. In *Proc. ACM SIGCOMM*, pp. 303–313, Stanford, CA, 1988.

[144] J. Rawls. *Theory of Justice*. Cambridge, MA: Harvard University Press, 1971.

[145] J. Roberts and L. Massoulie. Bandwidth sharing and admission control for elastic traffic. In *Proc. ITC Specialists Seminar*, pp. 185–201, Yokohama, Japan, 1998.

[146] S. M. Ross. *Stochastic Processes*. New York, NY: John Wiley & Sons, 1995.

[147] S. Sanghavi, B. Hajek, and L. Massoulie. Gossiping with multiple messages. *IEEE Transactions on Information Theory*, **53**:4640–4654, 2007.

[148] A. Schwartz and A. Weiss. *Large Deviations for Performance Analysis*. London: Chapman and Hall, 1995.

[149] D. Shah and J. Shin. Randomized scheduling algorithm for queueing networks. *Annals of Applied Probability*, **22**:128–171, 2012.

[150] S. Shakkottai and R. Srikant. Network optimization and control. *Foundations and Trends in Networking*, **2**:271–379, 2007.

[151] S. Shakkottai, R. Srikant, and L. Ying. The asymptotic behavior of minimum buffer size requirements in large P2P streaming networks. *IEEE Journal on Selected Areas in Communications*, **29**:928–937, May 2011.

[152] R. N. Shorten, D. J. Leith, J. Foy, and R. Kilduff. Analysis and design of congestion control in synchronised communication networks. *Automatica*, **41**:725–730, 2005.

[153] M. Shreedhar and G. Varghese. Efficient fair queueing using deficit round robin. In *Proc. SIGCOMM*, pp. 231–242, Cambridge: MA, 1995.

[154] A. Sridharan, S. Moeller, and B. Krishnamachari. Implementing backpressure-based rate control in wireless networks. In *Information Theory and Applications Workshop*, pp. 341–345, San Diego, CA, 2009.

[155] R. Srikant. *Mathematics of Internet Congestion Control*. New York, NY: Birkhauser, 2004.

[156] R. Srikant. Models and methods of analyzing Internet congestion control algorithms, in *Advances in Communication Control Networks*, Lecture Notes in Control and Information Science **308**, C. T. Abdallah, J. Chiasson, and S. Tarbouriech, eds. Berlin: Springer, 2005, pp. 65–86.

[157] R. Srikant and W. Whitt. Resource sharing with book-ahead and instantaneous-request calls using a CLT approximation. *Telecommunication Systems*, **16**:233–253, 1999.

[158] W. R. Stevens. *TCP/IP Illustrated, Vol. 1: The Protocols*. Reading, MA: Addison-Wesley, 1994.

[159] I. Stoica, R. Morris, D. Karger, M. Kaashoek, and H. Balakrishman. Chord: a scalable peer-to-peer lookup protocol for internet applications. In *Proc. ACM SIGCOMM 2001*, pp. 149–160, San Diego, CA, 2001.

[160] A. L. Stolyar. Maximizing queueing network utility subject to stability: greedy primal-dual algorithm. *Queueing Systems*, **50**:401–457, 2005.

[161] A. L. Stolyar. Greedy primal-dual algorithm for dynamic resource allocation in complex networks. *Queueing Systems Theory and Applications*, **54**:203–220, 2006.

[162] L. Tassiulas. Linear complexity algorithms for maximum throughput in radio networks. In *Proc. IEEE INFOCOM*, vol. 2, pp. 533–539, San Francisco, CA, 1998.

[163] L. Tassiulas and A. Ephremides. Stability properties of constrained queueing systems and scheduling policies for maximum throughput in multihop radio networks. *IEEE Transactions on Automatic Control*, **37**:1936–1948, Dec. 1992.

[164] S. Toumpis and A. Goldsmith. Large wireless networks under fading, mobility, and delay constraints. In *Proc. IEEE INFOCOM*, pp. 609–619, Hong Kong, China, 2004.

[165] G. Vinnicombe. On the stability of networks operating TCP-like congestion control. In *Proc. IFAC World Congress*, Barcelona, Spain, 2002.

[166] P. Viswanath, D. Tse, and R. Laroia. Opportunistic beamforming using dumb antennas. *IEEE Transactions on Information Theory*, **48**(6):1277–1294, June 2002.

[167] J. Walrand. *An Introduction to Queueing Networks*. Englewood Cliffs, NJ: Prentice Hall, 1988.

[168] N. S. Walton. Proportional fairness and its relationship with multi-class queueing networks. *Annals of Applied Probability*, **19**:2301–2333, 2009.

[169] X. Wang and K. Kar. Throughput modelling and fairness issues in CSMA/CA based ad hoc networks. In *Proc. IEEE INFOCOM*, pp. 23–34, Miami, FL, 2005.

[170] A. Warrier, S. Janakiraman, and I. Rhee. DiffQ: practical differential backlog congestion control for wireless networks. In *Proc. IEEE INFOCOM*, pp. 262–270, Rio de Janeiro, Brazil, Apr. 2009.

[171] A. Warrier, L. Long, and I. Rhee. Cross-layer optimization made practical. In *BroadNets*, pp. 733–742, Raleigh, NC, Sept. 2007.

[172] T. Weller and B. Hajek. Scheduling non-uniform traffic in a packet-switching system with small propagation delay. *IEEE/ACM Transactions on Networking*, **5**:813–823, Dec. 1997.

[173] J. T. Wen and M. Arcak. A unifying passivity framework for network flow control. In *Proc. IEEE INFOCOM*, pp. 162–174, Apr. 2003.

[174] W. Whitt. Heavy traffic approximations for service systems with blocking. *AT&T Technical Journal*, **63**:689–707, May–June 1984.

[175] R. W. Wolff. Poisson arrivals see time averages. *Operations Research*, **30**:223–231, 1982.

[176] H. Yaiche, R. R. Mazumdar, and C. Rosenberg. A game-theoretic framework for bandwidth allocation and pricing in broadband networks. *IEEE/ACM Transactions on Networking*, **8**(5):667–678, Oct. 2000.

[177] X. Yang and G. de Veciana. Service capacity of peer to peer networks. In *Proc. IEEE INFOCOM*, pp. 2242–2252, Hong Kong, China, 2004.

[178] Y. Zhou, D. M. Chiu, and J. C. S. Lui. A simple model for analyzing P2P streaming protocols. In *Proc. ICNP*, pp. 226–235, Beijing, China, 2007.

[179] J. Zhu and B. Hajek. Stability of a peer-to-peer communication system. *IEEE Transactions on Information Theory*, **58**:4693–4713, 2012.

INDEX

For EU product safety concerns, contact us at Calle de José Abascal, 56–1°, 28003 Madrid, Spain or eugpsr@cambridge.org.

www.ingramcontent.com/pod-product-compliance
Ingram Content Group UK Ltd.
Pitfield, Milton Keynes, MK11 3LW, UK
UKHW050219090126
466816UK00009B/124